W9-CXS-439

FLYING WISDOM

FLYING WISDOM

by the Editors of *Flying* Magazine

VNR VAN NOSTRAND REINHOLD COMPANY
NEW YORK   CINCINNATI   TORONTO   LONDON   MELBOURNE

Copyright © 1979 by Litton Educational Publishing, Inc.

Library of Congress Catalog Card Number 78-18536

ISBN 0-442-22452-4

All rights reserved. No part of this work covered by the copyright hereon may be reproduced or used in any form or by any means—graphic, electronic, or mechanical, including photocopying, recording, taping, or information storage and retrieval systems—without written permission of the publisher.

Printed in the United States of America

Published in 1979 by Van Nostrand Reinhold Company
A division of Litton Education Publishing, Inc.
135 West 50th Street, New York, N.Y. 10020, U.S.A.

Van Nostrand Reinhold Limited
1410 Birchmount Road, Scarborough, Ontario M1P 2E7, Canada

Van Nostrand Reinhold Australia Pty. Limited
17 Queen Street, Mitcham, Victoria 3132, Australia

Van Nostrand Reinhold Company Limited
Molly Millars Lane, Wokingham, Berkshire, England

16 15 14 13 12 11 10 9 8 7 6 5 4 3 2 1

Library of Congress Cataloging in Publication Data

Main entry under title:

Flying wisdom.

    Includes index.
    1. Airplanes—Piloting.   2. Meteorology in aero-
nautics.   3. Instrument flying.   I. Flying.
TL710.F59   629.132′ 52′ 08   78-18536
ISBN 0-442,22452-4

CONTENTS

FOREWORD
by Norbert Slepyan

WHEN THE ENGINE STARTS, knowledge is golden. From taxi to shutdown, the principles and procedures of flight rule with an iron hand. Yet there are various ways of arriving at them. As a beginner, a pilot is taught *the way* to fly. Then, on his own, he discovers that there are *ways* to fly. From hour to hour, through his experiences and those of others, he gathers the multifaceted substance of flying wisdom.

When the crunch comes, that wisdom is his basic survival tool. The crunch can be as horrendous as a balking engine or a suddenly encountered thunderstorm; it can also be as innocuous as a complicated clearance delivered by a controller who talks like a tobacco auctioneer; in every case, the pilot's life or aeronautical self-respect is on the line, to be protected essentially by what he knows.

Ask any instructor worth his right seat what he values most in the people he teaches or checks out and he will answer that while finesse and coordination are very important, judgment is the key. And he will add that for every problem that a pilot may face, there may be more than one way at arriving at a safe solution, even though the decision-making process may be instantaneous.

This book brings together the combined experience of several of *Flying* Magazine's outstanding writers and editors as they have appeared in the pages of *Flying*. Like most pilots, they do not always agree in their solutions to problems or procedures—and we regard that as a virtue. Pilots often disagree about ways of doing things—landings, IFR flying, weather planning, et cetera, et cetera, and so forth. Often, from such disagreement, flying wisdom is derived.

Flying Wisdom addresses itself to four major areas that concern pilots intensely. First, there is weather—fronts, ice, utilizing information services. Then there is instrument flying, about which every pilot should know, even if he is still rated solely VFR. Finally, there are two areas that pertain to all pilots: making the most of one's equipment and perfecting one's airmanship, from short-field landings to getting the optimum efficiency from one's fuel.

We hope that this book will enable the reader to become a wiser and more secure pilot.

PREFACE /"WHY WE FLY"

by Peter Garrison

ESSENTIALLY, an airplane is a way of getting from here to there, like a car. It is simply a conveyance, a machine—more or less comfortable, rapid, and costly as the case may be, but nothing more, really, than a fast bus. You get aboard, fire it up, and sit it out. Nothing more to it.

Or is there? Indeed, there must be something special about airplanes—you can tell by the people who fly them. Not that there is anything special about pilots before they become pilots; like everybody else, they are for the most part quite ordinary people. Airplane manufacturers advertise that anyone can do it (while pilots try to convey the impression, tacitly or explicitly, that not everyone can). Again, it's like driving a car; everyone can do it, except a few who for one reason or another can't quite. And yet you never see someone at a cocktail party casually dropping mention of having *driven* here or there to a dazed and décolletéed blonde in the confident (and rarely disappointed) expectation of an apotheosis. Everyone drives; no one is impressed. Few fly, and lots of people are impressed—not least among them the fliers themselves.

Flying becomes the hallmark of the people who do it. Whatever else they are, they are pilots, and among other pilots they always find themselves at home, however little they may otherwise have in common. Some become incapable of talking about any subject other than flying; they long to bring the conversation around to airplanes, and when they get it there, they are loath to see it move on. They are in thrall to something—some element of the aviator's experience that acts upon them like a narcotic upon an addict. And addicts there are—genuine addicts, who have let marriages founder and careers pass them by because the spell of an airport was too hard to break, and the possibility of a chance to get behind the controls of an airplane again too precious to be passed up for something else. The aviator's experience, whatever it is, is a profound one—one that enters dim and out-of-touch recesses of the soul.

The explanations they usually give seem hollow. Gee, it's just great putt-putting around up there, where never lark nor even eagle flew, just swooping around, dancing with the clouds, great sense of freedom, touch the face of God, liberty, helluva feeling. Is that really it? For many people, the cockpit of an airplane is the antithesis of freedom: a tiny cell suspended in space, a very metaphor of all that is unsure and terrifying in the human condition; how can you bear it? Some are haunted by the thought of the fall. For others, the boredom of it all beggars imagining. How can it be, Joe Pilot, that you are so blissful when others quake or yawn? Or are you blissful? Are there not many pilots who, if they had it out with themselves, would have to admit that they don't really particularly *like* flying—that it isn't fun for them in the way playing ball or swimming or even watching television is fun? Who would have to admit that they have many a time forgiven the vacuity of one flight with the facile reflection that the next one will be different?

Yet they don't have it out with themselves, and despite the cost and the inconvenience and the noise and the crowding and the immobility, they fly and fly, and in the balance—when all their assets and liabilities are spun out in the centrifuge of some last judgment—they have not lost. But not because they wheeled with the eagles (an eagle passes a plane looking like a paper bag falling past a tenth-story window) and not because they danced with clouds (a cloud is a turbulent pudding as big as a mountain; one cannot dance with it; an airplane cannot dance, but only lunge in slow motion). Because of something else.

What is it?

Flying is a symbol. It is a symbol in a broad, historical sense—a Jungian archetype in good standing, a vision to which all men resonate in unison, whatever their age or place or time of birth. Remember the hawklike man, Daedalus? He was your interesting, romantic flier, a real cliché; cunning, egotistical, he had a way with women, and though he had a son who went about with him, he was not much of a father. He was too busy with his inventions and his contracts. The boy Icarus was headstrong and isolated; wouldn't take orders or even a piece of good advice. When in the course of a jailbreak Icarus fell to his death in the sea, Daedalus, cold soul with a head full of bitter ideas, must have breathed a silent curse, closed his mind, and flapped on.

Flight is jailbreak. It is also cunning and indifference—the virtues of wolves.

Angels have wings. They told us in Catholic school that angels' wings were a symbol of their mastery of space—angels could be where they wished and were always everywhere. (Or rather everywhere was in them, but the symbolist who dreamed up those fulsomely feathered limbs needed the simpler, more visual idea.) Birds, indeed, though they cannot be where they wish and everywhere at once, still are magical in their ability to make a straight line serve them for a highway. If gravity is original sin, then all dance, acrobatics, high-wire acts, and mountaineering are a seeking after salvation from that heavy sin; salvation to be a bird, have wings, be an angel. How helpless, what a sodden clay beast, feet planted forever on the ground, is man! (We admire birds so much, but what bird admires us?)

Flight is mastery, safety, success; also mystery and secession.

There is, too, a special perspective to it, both within and without. In the midst of games, talking, walking on the beach, one looks up and is spellbound by the silent arrowhead of a jet sliding by so far off, almost on the edge of space. It is hypnotic in its separateness, peaceful in its mysterious purpose, splendid in its simplicity and sureness, like a lone hawk hanging over a hill. And if the ideal of philosophic thought was once to see in the mind, in a crystal instant, all life and matter rolled up into a ball (as spacemen saw it rise over the horizon of the moon), then a pilot's hours are a string of such instants, looking down on the immobile earth, all come at once to the eye—towns and streams and highways and rails, all human life encompassed and mapped, ready to be understood as we understand what we can hold in our hand. Earth and the flier regard one another in a kind of mutual perception, in its best moments something like love.

The love colors all your memories. If you have traveled not by car or train but by plane, you have seen not tunnels, halls, and tables through and across which you crept or sped, but rather a tapestry rich in detail, myriad in its textures and colorings, vast as the horizons and vaster still, and heartless and peaceful as a sleeper's face. You have brooded over the vast abyss, swept across interminable fields that fanned past you like a spinning parasol, and hung in roaring stillness while the arcing line of night, like a monk's cowl, swung down overhead to quash before your eyes the last crescent of the day. You have flown through submarine caverns in which hanging gardens of mist and cloud turned blue-gray in the evening; over rusting deserts where an impeccable sun rose slowly over utter lifelessness; down valleys where shreds of cloud reached toward you from the backs of the windward hills; among thunderclouds whose miles-high cliffs broth-boiling and swirling on both sides of you let no sunbeams enter. In a Texas dawn you saw a grain elevator whose shadow stretched 80 miles to the horizon. If you miss anything when you die, it will be the glow of instrument dials in the darkness; the blissful sunrise; the long, hot, stormy afternoon; the calm and sadness and slowness of evening; the world as carpet, map, model railroad, standing at a remove from you, tinged and enveloped by the delight and melancholy of your own isolation.

Flight is knowledge and power. It is something other people can't do. That is, in fact, the vile part of it: that suffocating vanity, sometimes unconscious but often uncomfortably self-aware, that suffuses the pilot's motions as he starts the engines under the silent, mystified eyes of his passengers. He is a magician, a fabulous artificer—oh well, why not come right out and say it?—he's God Himself, taking you where you want to go, looking down on mundane concerns with a calm and detached eye, soaring philosopher, manipulating with nonchalance the instruments and controls that the layman dares not even brush with a careless cuff. It's a hard pride to give up, and many pilots find it difficult to teach their wives to fly. It is not to the advantage of heroes for the true nature of courage to become public knowledge.

The modern pilot capitalizes upon a myth that has its roots in religion and whose branches are the Wright brothers, the Red Baron, Charles Lindbergh, and all the aces of all the wars since a prop first turned in anger. If some pilots find low-wing airplanes more congenial than high-wing ones, it is doubtless because it is easier to daydream a low-wing plane into a P-51 as you climb aboard. One's vanity takes a long time to tire of indulging in small demonstrations of casual competence—supervising the takeoff surge with one hand and eye while the others see to some insignificant dial-twiddling, or conversing casually with the passengers during the landing roundout and touchdown. Never let them think you're worried (not that there would be anything to be worried about if one were as competent as one took pains to convince others one is).

The love of flying is not without its infantile aspects; but there is more to it than that. Flight is quintessentially a symbol of human ambition and daring, of the application of skill and intelligence to problems that seem to defy solution. While it has long since ceased to require either particular skill or particular daring, the art of piloting still glows with some remnants of the heat of its conception. Few pilots—few people, in fact—are insensible of this special aura, and it becomes in those who bask in its warmth a powerful agency, like money, of good or ill. There are many to whom it is only a sop to conceit, an empty snobbery, something to flaunt before strangers. But to the better it is like the great traditions of old craftsmen: a touchstone for inner growth, a blessing of which one must make oneself worthy, a kind of privilege that should inspire modesty. *Noblesse oblige.*

The remarkable thing about people who fly is not that they are so clever; it's that they are so lucky. The clever ones realize it, and even the least perspicacious catch some spark of a hint of their good fortune. It's what keeps them going despite poverty, divorce, and bad horoscopes. Sometimes, it keeps them going despite bad weather as well, and that is the dangerous side of the matter: the sirens' song to which few can stop their ears. The most difficult of the pilot's arts is that of knowing when not to fly. It is also the last test of the real pilot; for if flying is cunning and mastery, it is good sense and self-mastery as well. Until a pilot has learned to transcend the faint glory that shines through flight and to subordinate it to his own will, he has still not learned the most essential part of imitating birds: that birds fly because it is natural to do so, because they need to do so, and not because they admire themselves for doing so. Flying can be a door open to the transformation of a life from sheer plod to riding high; it can also be a *cul de sac* at the end of which a lot of bores sit talking at one another about their nonexperiences. Of all the risks that the joy of flight requires, that of falling prey to the misuse of it is the greatest of all.

PART I: WEATHER

1 TOOLS OF THE TRADE

by Richard L. Collins

PILOTS HAVE a lot of weather information to rely on—certainly much more than in the days when sailors depended on things like a mackerel sky to warn of an impending blow. The old weather signs are still as useful as ever, but things such as weather radar, teletyped reports from near and far, upper-wind soundings, and computer-generated forecasts add a lot to the picture we see out the window. Knowing *what* is available and *how* to put it together is important to weather wisdom.

Our hardware is the radar machinery, the observation equipment, and the computer; the information they produce hangs on clipboards in the National Weather Service office and flight service station, is broadcast, and is visible on the radarscope. We put it all together by using a transcribed weather broadcast, a telephone, a visit to an NWS office or flight service station.

There's a great collection of information in an NWS office; more, really, than a pilot can be expected to understand or digest. The familiar charts are hanging in the part of the office where the NWS man briefs you, but back in the other room, they have charts like you never saw before, with symbols and terminology that's not on any FAA written test. They're produced by a computer at the National Meteorological Center, at Suitland, Maryland, that digests a constant flood of weather data from all over—and far above—the world. These computer charts are relatively simple to a meteorologist, and the messages they display on vorticity and atmospheric stability are reflected on the even simpler charts used for preflight briefings.

The most basic of these is the surface-weather map. It shows the highs, lows, and fronts—the things that make and modify the weather. No pilot should fly without knowledge of how the systems are arranged on that map. Fresh charts come off the facsimile machine every three hours, starting at 0135 GMT, and are based upon data that is already about an hour and a half old. (It is always important to know the age of weather information and to check for changes when new information becomes available.)

Following the surface-weather map in basic importance is the weather depiction chart, issued every three hours starting at 0210 GMT. This depicts the actual weather, divided into three logical operational values. The basic "bad" weather is depicted by showing areas where the ceiling is below 1,000 feet and/or the visibility is below three miles. These shaded areas, surrounded by a solid line, can give the VFR pilot a clear, straight-forward message: IFR only. The next level of depiction—ceilings between 1,000 and 3,000 feet and/or visibilities between three and five miles—outlines where the VFR pilot *might* have a problem. Where the map is clear, the ceiling is 3,000 feet or better and the visibility five miles or greater.

During a personal visit to an NWS office or FSS, a pilot can study the surface-weather map and the depiction chart and get a good idea of how ceilings and visibilities are shaping up around the weather systems. It's not quite so easy to do this in a telephone briefing, but well-placed questions to the briefer should extract the information. The two prime things to seek would be the synopsis—the locations of highs, lows, and fronts—and the depiction of any adverse weather along or near the proposed route of flight. It's best to seek this information with your own blank chart at hand.

It should only take a moment to get that broad chart information, and with it in hand, the next thing you need is the actual weather along the route of flight. This comes in the form of sequence reports—actual surface-weather observations. The weather-wise pilot has a list of the desired sequences on hand and either looks at these in person or asks for them over the telephone. Sequences are updated hourly, on the hour. If the weather synopsis indicates excellent conditions and the depiction chart agrees, it might be okay to accept only the ceiling and visibility portion of the sequence. When there's doubt about anything, get the entire sequence; there's a lot of useful information in the temperature, dew point, wind, and the altimeter setting.

Weather-radar reports, shown on a radar summary chart, also give a clear message about what is going on in the real world. These summary charts must be used with caution, though, for if thunderstorms are the problem, they can change a lot faster than the radar summary chart. The chart is issued on an irregular schedule: out once an hour approximately on the half hour from the wee hours of the morning until 0830 EDT (1230 GMT); then it comes every three hours for two maps, drops back to a two-hour cycle for one at 2022Z, and then resumes hourly.

There is a good weather-radar service available at NWS offices and flight service stations that can help fill gaps in the radar-summary information. It is a digitized radar outline, printed over the teletypewriter and interpreted with the aid of a map overlay. It comes from the NWS computer and is derived from radar data from all over the country. The maps are produced by region, and cloud coverage and levels of precipitation intensity are shown. The information is updated hourly. Every NWS office and FSS should keep the information available, but some flight service stations don't seem to be calling for the hourly printout; they ask for it only on user demand.

Some flight service stations and most National Weather Service offices have their own radarscopes, and the best of these go out to a 250-mile range. What they show at the far reaches of coverage isn't complete information, because the radar might see only the tops of the meanest thunderstorms way out there and miss the lower but still disagreeable stuff because of the earth's curvature, but what's shown within 100 or 150 miles is the real scoop, and the equipment has the capability of measuring tops of the storms to get an even better idea of severity. (Only the NWS can measure tops; the radarscopes in flight service stations are simply repeaters of the NWS scopes and the FSS cannot stop the antenna sweep for top measurement.) Getting radar information over the phone is better than not getting it at all, but there's often a certain vagueness on the part of briefers who know full well that it will be awhile before you get the airplane cranked up, and that the weather they're looking at will move and could completely change in character before you get there.

Radar is one of the better items of current weather hardware, and the information it produces should be utilized in every briefing. The radar report covers the whole landscape between airports—something that surface-weather observers can't do—and there is an unavoidable message in a report of a solid line of precipitation echoes between here and there. Even general rain bears a message, for the IFR pilot as well as the VFR aviator; ceilings are likely to be too low for VFR whenever it is raining, and rain tells the IFR pilot that there is some vertical substance to the clouds and that tops are likely to be high—at least for normally aspirated piston airplanes. Pilots should get radar information for every flight unless it is severe clear all the way.

Pilot reports also are useful clues to actual conditions, though they are often few and far between. Pireps are available over the teletype circuit, and if you ask for them, a station can request and receive all current pilot reports in one summary.

Actual winds-aloft reports are available, too, though most pilots rely on the forecast winds. The actual winds are shown in chart form, with the charts produced at 0313 and 1513 GMT. The winds shown are three hours and 13 minutes old when the charts are produced, but if you think that's old, contemplate the way winds aloft are *forecast*: The forecast is based on information gleaned at 0000 GMT, it comes off the machine at 0900Z, and is for the 12-hour period beginning at 1200Z.

With the current information in hand—the surface map, the depiction chart, the sequences, the radar information, and the pilot reports—the pilot logically goes on to the terminal and route forecast. Some briefers and pilots put great faith in forecasts, which should be used as nothing more than an aid with which to wind up a briefing. The current stuff is fact, solid information; the forecast is nothing more than an educated guess.

The forecasters turn out three a day. The forecast times are varied by sections of the country, too, in order to make them more timely for periods of maximum use. Terminal forecasts are issued at 0940, 1440, and 2240 GMT in the Eastern and Central time zones and at 0940, 1540, and 2240 GMT in the Mountain and Pacific time zones. The Central- and Eastern-zone people have a fresh new forecast, issued at 4:40 or 5:40 A.M. (daylight time), to use in briefing the early birds. The Westerners don't have it quite so fresh unless they really get up early, for the first forecast of the day comes off at 2:40 or 3:40 in the morning there. This early-morning forecast is backed up with another one five hours later in the East and six hours later in the West. Late in the afternoon, there's one last forecast for the night fliers. Route forecasts, primarily for the transcribed weather broadcasts, are issued at 1040, 1740, and 2240 GMT in the East and 1040, 1840, and 2340 GMT in the West.

The forecaster uses a lot of information, much of which is developed in the National Weather Service computer, as a basis for his idea on what the flying weather is going to be like. The first step he takes is probably to look at the upper-air charts and how the computer is projecting changes in the upper-air flow. The 500-millibar chart (approximately the 18,000-foot level) is a favorite, and beside the current chart, the forecaster has computer projections for as much as 72 hours in the future at this level. The 500-mb flows have a lot to do with surface weather, and the forecaster is especially interested in the low-pressure centers up there, plus the ridges, troughs, and general flow of the westerlies. The vorticity of the atmosphere—the tendency toward rotation—is something we pilots don't hear or know much about, but a series of charts on this at the 500-mb level is especially important to a forecaster.

The surface-weather map is backed up by forecast charts showing what the National Meteorological Center thinks is going to happen on the surface and a discussion of why they think it will work that way. Upper-air soundings tell of the atmosphere's stability and moisture content, and satellite charts show cloud patterns. A severe-weather outlook chart gives an outline of areas where conditions might be conducive to the formation of severe activity, another forecast chart gives the mathematical probability of precipitation, and there's one on expected maximum and minimum temperatures. The computer even offers an opinion on expected surface winds. The effect of it all is to take the basic forecasting role away from the forecasters in NWS field offices; they do generate the ceiling and visibility forecast from the information but would seldom go out on a limb and issue a forecast that did not agree with the National Meteorological Center's guess.

The forecaster starts digesting the information about an hour and a half before issuing a forecast, reviews it immediately before issuance for any change, and compares the forecast and actual conditions after it's on the clipboard. If the forecast is grossly in error, it can be amended. (If you run afoul of an off-base forecast, remember that there's little individuality left in the business; you should curse the computer, not the forecaster.)

The only certain thing about weather is that it changes. Since the information that is available on the ground is equally available in flight (from the flight service station or by listening to a transcribed weather broadcast), an airman should always stay in touch no matter how much information he or she has amassed during a preflight briefing. Traffic-control radar sees airplanes and heavier precipitation, and there's nothing else (other than airborne weather radar) that actually shows the airplane's relationship to weather.

Traffic radar is designed to eliminate some of the precipitation when desired, however, for it is just too difficult to track little airplanes through a lot of precip. This does not erase all precipitation—rainfall rates in excess of one-tenth of an inch per hour will break through and begin showing on the scopes. Traffic radar does show differential in intensity, but it has no contour circuit to show storm cores. New digitized radar equipment that is becoming operational at air-route traffic-control centers gives the controller the option of displaying or deleting all weather return, but hopefully, controllers will display whenever there is doubt. The new equipment does present the weather return in a manner that allows easy tracking of aircraft in a precipitation area, so it shouldn't bother controllers to let it all show.

The controller is surely not the weatherman, but if a pilot seeks information on weather, he's likely to get it. Some controllers will even provide vectors through areas of bad weather, and my experience with these has been very good. The pilot must accept such vectors with an understanding of the radar's limitations, though, and not always expect a velvet-smooth ride. You can find a lot of turbulence around cumulus clouds from which the rain is light or not even falling. Traffic radar really does its best work in conjunction with a pilot who is more or less remaining clear of clouds and using the information in long-range planning of a route rather than in short-term avoidance of individual thunderstorm cells.

Combine that traffic radar with everything else, and the weather information service for pilots appears very complete. It is true that we sometimes feel frustrated because the person with the straight poop is hard to reach on the phone or because the data seems incomplete or inaccurate, but information is always there and patience is inevitably rewarded.

2 YOU CAN TELL IT BY ITS ACCENT

by Richard L. Collins

WEATHER ON A MAP is made up of highs, lows, and fronts. From the left seat of your airplane, it's more substantial and specific: solid cloud all around, rain pattering on the Plexiglas, static on the radio, turbulence, ice forming on everything, a cumulonimbus framed in the windshield, or a persistent fog that didn't lift at noon. Once we poke our nose into them, the meteorological phenomena are about the same whether you encounter them over Florida, Maine, Kansas, California, or Oregon, but the chances of encountering them change with geography. Mountains, plains, and large bodies of water (oceans, gulfs, and big lakes) have a tendency to make or modify weather.

Some years ago, when I moved my base of flying operations from the South to the Northeast, I had never heard of a "northeaster"—a wild and woolly storm that is a blizzard-producer in the winter and a basement-flooder at other times. (Most sources attribute the name to the strong northeast wind associated with the storm, but it both affects the northeastern United States and also moves to the northeast.) The first few northeasters that passed by as I was trying to fly into or out of the area left me perplexed. After a few batterings, though, I caught on and learned that many northeasters provide less than ideal IFR conditions.

The northeaster is a good example of a regional weather characteristic caused or modified by a body of water. The wildest northeasters tend to come from low-pressure systems that move eastward across the Gulf of Mexico, cross northern Florida and southern Georgia, and then start tracking up the Atlantic Coast. At Cape Hatteras, some mysterious force seems to pop either a mean or a sweet pill into the middle of the storm. If there is also a low-pressure area in the Ohio Valley around Cincinnati at the time the low reaches Hatteras, the resulting northeaster can develop exceptional characteristics. Forecasters usually do fairly well on a northeaster once they realize it is forming, but it does have to get to Cape Hatteras before the picture becomes reasonably clear. From Hatteras, it can pounce on the Northeast before the forecast makes it through the system.

Rainy northeasters seldom close airports. Ceiling and visibility usually stay well above IFR minimums, but the wind is strong and the lower-level turbulence (as seen from a light airplane) is moderate to occasionally severe. Snowy northeasters tend to do worse things to ceiling and visibility, and often bury airports beneath enough snow to close them. Predicting whether the storm will produce rain or snow is tough, too, because surface temperatures in a northeaster can often be close to freezing, and a miss of 25 miles in predicting the path of a storm center can mean the difference between 2 inches of rain or 20 inches of snow. That's a big difference. Thunderstorms can be embedded in a northeaster, and I've seen active ones in areas of heavy snow. This, though, is rare.

When lowering ceilings, precipitation, and freshening easterly winds are on the terminal forecasts or actual reports in the northeastern United States, especially in the fall and winter, buckle up tight. The storms almost always move rapidly, so patience is rewarded, but in the 12 years I watched northeasters closely, one did come to a dead stop. It was a late-winter northeaster, made up of a complex system of three low-pressure areas, and it remained stationary off the New Jersey coast for what seemed like days. Considerable beach erosion was caused, there was a good snowfall, and I spent the time in Indianapolis with absolutely no desire to fly into the storm. It clobbered the weather as far west as Pittsburgh.

The true northeaster is a coastal storm, but inland low-pressure areas can often move northeastward, or even in a northerly or northwesterly direction, along the coastal plain. These can carry plenty of turbulence, ice, and precipitation, but they often lack the widespread strength of a northeaster.

The Gulf Coast areas have their moments, too, with the most frequent inclement weather there connected with cold fronts that stop along or just south of the coast and either remain stationary or become warm fronts and move back north.

Forecasters usually do a fairly accurate job on Gulf Coast problems, though their job is made difficult by the lack of reporting stations out in the water. The front that becomes stationary right along the coast gives them their toughest job as far as the IFR pilot is concerned. Waves often form on these fronts and move eastward, with resulting widespread embedded thunderstorms several hundred miles inland. These waves can be ambitious enough to turn into the low that moves over to the Atlantic and becomes a northeaster, too. On the other hand, a front that becomes stationary along the Gulf might have little meanness in it and might just cause drizzle and low ceilings, or zero-zero conditions, along and to either side of the front.

The cold front that returns as a warm front (they call it a "return cold front") usually features low ceilings over a wide area, with drizzle or light rain 100 to 200 miles north of the front. Easterly moving waves can develop on these fronts, too, with the same results found in the waves that move along a stationary front on the Gulf Coast.

Tropical storms—deep lows—also move inland from the Gulf and bear watching. They are big rain-producers, and can contain a lot of thunderstorm activity.

The Gulf has a great bearing on the weather in the eastern half of the country, and the effect of Gulf moisture on precipitation can readily be seen by looking at rainfall charts. Little Rock and Kansas City, which are north of the center of the Gulf, average about 50 inches of precipitation a year; Oklahoma City and Omaha, slightly to the west, average only 30 inches of precipitation per year.

When there is a front along or south of the Gulf Coast, the wise IFR pilot recognizes that the forecaster might have a tough job deciding "what next" for the areas along and north and east of the coast, so he goes deeply into the radar-summary charts and the current weather before flying into, north, or east of the area. Also, when the winds aloft are blowing from the Gulf toward the area in which you are flying, look for deteriorating weather—especially if those winds are brisk.

Pacific Ocean storms usually don't move along the coast like the Gulf and Atlantic storms. The Pacific lows and fronts usually move onto the coast, tracking at right angles to the coastline. Some, especially along the Oregon and Washington coasts, can be quite severe, but forecasters do fairly good work on them, and there's a lot of traffic flying the coast to give pilot reports.

Most of the rain-producing weather systems come to the Pacific Coast in the wintertime, with fronts that approach from the southwest bringing more prolonged precipitation. The favored track for the low center curves from the central Pacific toward Vancouver Island, but the track can shift southward. As fronts move onshore and into the mountain areas, the surface frontal position becomes difficult to locate, and weather intensifies as the air mass ascends the windward slopes, while the descending air on the leeward side warms with a resulting decrease in activity. The actual frontal weather right along the coast can contain thunderstorms, but these are relatively rare; the West Coast has the lowest incidence of thunderstorms in the U.S.

Fog is a particular winter problem along the West Coast. A ridge of high pressure aloft behind a front can trap cold air in the valleys, and moisture supplied from the surface (still damp from rains of the front) can create a substantial fog situation. The Columbia River Basin and the Sacramento and

San Joaquin Valleys are typical areas for this phenomenon, and often another frontal system has to move into the area to destroy the temperature inversion and terminate the fog and low-stratus situation.

Stratus that forms over the ocean and moves onshore is a frequent problem, too, although coastal stratus doesn't necessarily mean a problem for the IFR pilot. In fact, the ceilings are usually acceptable and the tops low, so it can be an ideal IFR situation.

The western plateau is a great fog spot in the wintertime, too. Fog there is almost invariably preceded by precipitation, with ground wet from melting snow a frequent forerunner to a fog situation. This fog also often persists until a storm system with wind arrives to blow it out. Salt Lake City is the fog leader, and the visibility there was below one mile more than 12 percent of the time during a recent December-to-February period. Boise and Idaho Falls followed closely with about 8 percent. If you are headed for that area IFR and the forecast shows nothing but zero-zero for the next couple of days, don't pass it off as a stuck key on the teletype.

Fog along and well inland of the Gulf Coast is often widespread at night and in the morning, especially during the winter and spring. Forecasting the time this fog will lift to IFR minimums is really touchy, and it never pays to put many chips on the forecast. For some reason, zero-zero conditions plague the Southeast around Christmastime every year, and play havoc with everyone's travel plans.

East Coast fogs (except along the New England Coast) usually are neither as persistent nor as frequent as along the Gulf and Pacific Coasts. Forecasting the time they'll lift, though, can be just as difficult.

Fog can be a great problem even in the middle of the country—well away from the water—and there's one situation that can turn an instrument pilot's hair gray on one flight if he doesn't anticipate it. When a cold front moves across the plains and becomes almost stationary to the southeast of the Great Plains, a northeasterly flow often carries cool, humid air up the sloping plains. This classic upslope situation creates fog and stratus, and if a flow of humid air from the Gulf overruns the shallow layer of cooler air near the surface, multiple cloud layers can form. This creates very large areas of below-IFR-minimums conditions that persist for days, and a forecaster is hard-pressed to say when the ceiling will or won't make it up to a height that will allow IFR operations. The tops can be high, too. This is usually a wintertime phenomenon, so the surface temperature can drop below freezing and bring freezing drizzle and rain into the picture. The distances in the middle of the country can be astounding to an Easterner, and holding over Wichita for improvement can be unrewarding when the weather is also below minimums at every other station within 200 or 300 miles.

The Great Lakes have their very own effect on weather. When cold air blows in from the Northwest, over the relatively warmer Lakes, the warmth from the water quickly makes the cold air unstable, and a deck of stratocumulus develops and extends 50 to 100 miles downwind from the Lakes. With an extreme difference between air and water temperature (the latter around 32°F. in the winter) the cloud tops can build from 8,000 to 10,000 feet with moderate to heavy ice in the clouds, plus turbulence. Ceilings and visibilities drop in snow, and it can be a sticky IFR situation. Lake effect is not something that covers a large area, though, when compared with the range of our airplanes—provided you have the fuel in the tanks.

The Rocky Mountains probably have a larger effect on general-aviation IFR operations than any other geographical feature—simply because of their size (height, width, and length). The minimum en-route altitudes for IFR around the Rockies are often near the ceiling of our nonturbocharged airplanes, and when there is weather in the mountains, there is often ice in the clouds. When mountain weather is flyable, it is usually best flown VFR, unless you have a turbocharged and deiced airplane. There are exceptions, of course, but they are generally localized.

Wind is as much a weather consideration as ice is in the Rockies. If the wind aloft exceeds 30 knots at the ridge levels, turbulence is likely to be undesirable at all unturbocharged flight levels.

Possibly the two most notable things about Rocky Mountain weather center around upslope flow conditions and cold fronts. Any westerly push of air over the rising plains causes cloud formation as moist air is moved up the slope. This can cause low ceilings and visibilities along the eastern slopes of the Rockies and back into the mountains for some distance. It's often a condition that is slow to change.

Cold fronts coming across the Rockies from the West have to bring their moisture with them—the reverse of conditions farther eastward, where fronts start pushing moisture from the Gulf ahead of them. Thus, the weather is apt to *deteriorate* with a frontal passage in the Rockies. There's not a great deal of precipitation in the mountain areas, though: 12 inches per year in Helena, Boise and Lander, Wyoming; 16 inches in Salt Lake City; and just over 20 inches in Denver.

A phenomenon called a "cutoff low" is a prime cause of poor flying weather (precipitation, icing, low ceilings and visibilities, and strong turbulence) over the intermountain West. More of these lows occur in the spring and fall than at other times, and forecasting the formation of a cutoff low is difficult. These are also referred to as "cold lows aloft" and "Nevada lows," which explain a bit of their nature and usual location.

Logic tells us most of what we need to know about IFR flying in any mountain areas. Mountains disrupt airflow, and lifting action can develop. This means that the tops will be higher; there may be icing or lots of turbulence in the clouds; or, given the presence of moisture, thunderstorm formation becomes possible. The Appalachian Mountains are just about the right size to harass the nonturbocharged lightplane; they aren't so formidable as the Rockies in height and area, so they lead us on and they stir up the weather at our flight altitudes at the slightest provocation.

From an IFR standpoint, the best thing to do is to approach the Appalachians with the thought that things will surely change. If your flight has been smooth and on top of clouds at 9,000 while droning eastward across Ohio, watch for the tops to rise as the mountains are approached. (Moisture blowing off the Great Lakes can accentuate this elevation of cloud tops along the northern Appalachians in the wintertime.) If you have been in cloud but with little moisture and no turbulence at your level, watch for both to increase as you near the mountains; and always watch for thunderstorm activity along the mountains. It sometimes seems that there is a perpetual front that stretches from the Canadian border down to just north of Atlanta, with buildups aplenty much of the time. The saving feature of the Appalachians is that it doesn't take much time to cross them.

Thunderstorms are bad wherever you find them, but they do have regional characteristics. Our interest in their regional nature concerns how often they occur more than their strength and our ability to manage them.

The most severe thunderstorms are found in the Great Plains and the Midwest, although severe storms and tornadoes can occur anywhere. Thunderstorms are most often a problem in the Gulf area in the winter, and over the entire country east of the Rockies in the springtime. The prime "season" is May and June, carrying forward into July and August in the northeastern United States. Florida, being the most tropical state in the Continental 48, has the most thunderstorms, with about half of them occurring in the summertime. The area west of the Continental Divide has the fewest storms.

In that our operational procedure must be absolute avoidance of all thunderstorms, the severity of given storms isn't so important, but we do need to know if severe storms are predicted, so a wider (20-mile) berth can be given to any storm that does appear.

The forecasters usually do pretty well on predicting the general weather systems—the lows and fronts—that move across the country. They are not what would really be classed as regional weather characteristics, but some effects of the position and path of low centers on the weather might be noted.

Low centers that move out of the Southwest and across the southern tier of states tend to move with reasonable speed and don't usually develop into a storm that covers a large area. If the low takes a more northerly track, though, it's likely to cause bigger problems over a larger area. The low that moves from the Southwest into the Great Lakes area can turn into a major storm if the supply of warm and moist air from the Gulf is great and if there's a good supply of cold arctic air to feed in from the other side. The greater the contrast between the two air masses, the meaner the storm. The big winter disturbances usually have a lot of snow to the north of the low, and freezing rain and/or thunderstorms to the southeast where the cold air is shoving its way into the warm.

Today there's even bad weather in good weather—man-made bad weather. The smog of the West Coast is well known, and the air pollution of the East and some other areas reaches the crisis point with increasing frequency. This can make an IFR day out of a VFR day, and the problem is especially predominant along the East Coast when the Bermuda high (sometimes called a heat-pump high) settles in for a long stay well off the Atlantic Coast, with resulting light southerly winds aloft. This can foul up—literally—everything east of the Appalachians.

If thunderstorms develop while the air is heavily polluted, they pose a special problem. The tops of the haze and smoke frequently move on up to 12,000 to 15,000 feet along the East Coast, so there's no unturbocharged way to get up where you can really see the buildups; you're forced to just dodge them as they appear in the murk ahead. At least one air-traffic-control procedure makes the chore even more difficult: in one of their most hard-headed acts, and an admission of their occasional indifference to the needs of general-aviation IFR, the Air Traffic Service tends to route low-level IFR aircraft around the Washington and New York areas so far to the west that one ends up flying over the mountains, where the buildups are most likely to appear. There is a choice when flying around Washington—a good alternate IFR route to the east, around by Richmond—but the easterly IFR route around New York is V-139, which takes you way out to sea. When buildups over the mountains appear to be a problem and you are northeastbound past Washington and New York, the best procedure might be to file to an airport within the New York area for a fuel stop, and then go IFR from there on to Boston or wherever you are headed. It's a peculiar situation: the Air Traffic Service teaming up with a regional weather characteristic to make flying as miserable as possible for as many pilots as possible. Maybe Air Traffic Service will change someday. It's a cinch the weather won't.

③ CEILING AND VISIBILITY LIMITED

by Richard L. Collins

FRONTS AND AREAS of high and low pressure create and modify our flying weather, and it is important for us to be up on the big picture. The *exact* products of a weather system—the conditions that exist in the air mass we are using for the flight—are also important, and two things pilots constantly think about are ceiling and visibility. A layman's "nasty" day is a pilot's "500 and one." The VFR pilot hangs onto 1,000 and three as a salvation; the IFR pilot worries more over 2,400-foot runway visual range, or a few hundred feet of ceiling and half-mile visibility for landing minimums.

Pilots do a lot of wishing about low ceilings and poor visibilities, too. Countless millions of gallons of fuel have been used by airplanes circling and waiting for conditions to improve to IFR landing minimums. Untold square feet of floor tile have been worn thin by pilots pacing the FSS floor, devouring each new collection of sequence reports for an indication that the ceiling was lifting and the visibility improving. Many airplanes have been broken because the pilot read the tea leaves wrong, or read them through rose-colored glasses.

Ceiling and surface visibility are simple matters. They can be seen and measured from the ground, and are reported on the sequences we get from the flight service station. When flying VFR, we can look out the window and get a constant update on the ceiling and visibility in the area. It is easy to get those reports, and to see what is going on outside. Weather *wisdom* comes in learning what to expect next.

The sequence report reduces the actual weather to a collection of numerals and words for the convenience of the teletype machine or computer. After the station identification, the sequence gives the height of cloud bases, the visibility, and what is restricting the visibility if it is less than six miles or if precipitation is falling: "Birmingham, 300 scattered, measured ceiling 600 overcast, visibility three, light rain and fog,'' for example.

That part of a sequence can even tell us more than just the cloud heights and surface visibility. The nature of any obstruction to visibility can give a clue to other things. If drizzle or fog is making it fuzzy around the edges, the air is probably rather stable, and the instrument pilot might find low cloud tops and smooth flying conditions while the VFR pilot waits on the ground for improvement. If precipitation is falling, the air is less stable, and the potential for a bumpy ride increases. Showery or heavy precipitation is a signal to keep the belt extra snug, and the warning issued by a thunderstorm is obvious.

Precipitation can also be a warning signal of things to come in the way of lower ceilings and poorer visibilities. While rain seldom cuts visibility below a mile unless it is a real frog-strangler, rain can and often does promote the formation of a layer, or layers, of cloud below the deck from which the rain is falling. The lowest layer can often be perilously low. For example, if a station reports 5,000 overcast and rain has just started falling, don't be surprised to find the next report calling it 500 broken and 5,000 overcast with a few miles visibility. The report after that might be even simpler: 500 overcast. Warm rain falling into cool air can even produce dense fog when conditions are ripe. If you don't believe it, just run a hot shower in a cool bathroom for a few moments.

Rain-induced low clouds constitute a real trap for the VFR pilot. As he flies into the rain area, good ceilings might give him an excuse to press on. As rain increases, forward flight visibility decreases, though, and as he approaches those lower clouds, the pilot might not be able to see them coming.

The effect of rain (and of other forms of precipitation) on ceiling and visibility is a compelling reason for a pilot to always check the radar summary report. These reports show areas of precipitation and their movement (be sure to note the time the observations were recorded and make appropriate adjustments for that movement), and if precip is moving into an area, the pilot can learn this in advance and anticipate the possibility of lowering ceilings and visibilities. Sequences show only what is happening at selected locations; the radar summary reports help fill the gap.

Unfortunately, many pilots stop after getting the ceiling, visibility, and word of any obstruction to visibility on the sequence report. It's unfortunate, because these things tell us only what the weather was at a given time. The numbers on temperature, dew point, surface wind, and the altimeter setting give an insight into what the weather might be in the coming hours.

The temperature and dew point are fascinating numbers. The difference between the two is called the "spread," and is an indication of relative humidity. That has a lot to do with the ceiling and visibility, and a person with a good meteorological knowledge of an area can take the temperature and dew point, the wind, and whether or not any precipitation is falling and make a reasonable guess at the ceiling and visibility. He can probably make a pretty good short-term forecast, too.

Temperature and dew point are a key because clouds are formed when air containing water vapor is cooled below the dew point. (Cool air holds less moisture than warm air. Dew point is the temperature to which the air must be cooled, at a given pressure and water-vapor content, for it to reach saturation.) When the temperature and dew point meet, the resulting moisture condenses into droplets on microscopic dust particles in the atmosphere, and clouds form.

Cooling frequently occurs as a result of expansion as air rises, so there is often a logical relationship between temperature, dew point, and ceiling. The temperature and dew point converge at a relatively predictable rate of 4° to 4.5°F. per 1,000 feet in the kind of moist air that is conducive to cloud formation, and the height of cloud bases can often be determined by using the rule of thumb of 1,000 feet above the surface for each 4° spread in the temperature and dew point. If there's a 10° spread, there might well be clouds at 2,500 feet, for example. That rule can be highly modified by temperature inversions, where there is warmer air aloft, and by other factors, but it is still useful.

To get a picture of the relationship between temperature/dew point and ceiling/visibility, I recently collected 60 sequence reports that covered a large area of marginal weather. They showed IFR conditions at most stations with a spread of 2° or less; those with no spread tended to be at or below IFR minimums; a spread of 3° turned out to be marginal, and stations with 4° or better seemed to be over in the VFR column.

When the pilot on the ground who wishes he was aloft wants to look into the future, the temperature and dew point can often tell a more meaningful tale about ceiling and visibility than a terminal forecast. If they are together or within a degree or two, things aren't going to improve until they separate. And they aren't going to separate until something different happens. If it is early in the day, perhaps the sun will move the temperature up a bit as it rises. If it is morning fog, we commonly think of the sun "burning" the fog away. If it's midday and the weather is still cruddy, improvement will probably come only from a change of air. Poor conditions are likely to linger until a front of low center moves and causes a larger-scale modification in the air mass.

When systems are moving slowly, a forecaster has a difficult time pinpointing an exact time for weather improvement, and pilots must understand this and do a little work on their own. It's possible to study complete sequences, including those from neighboring stations in the direction from which improvement is likely to come, and make a decent short-term forecast.

In watching the slow progress of a low-pressure system away from one area, I recorded and compared reports from five A.M. until one P.M. to get an idea of what was happening as the storm moved on. At five A.M., one station reported 600 broken, 6,500 overcast, and seven miles visibility. That's not too bad, but there was only 1° between the temperature and dew point. By six A.M. the ceiling was down 200 feet, where it remained, and the spread stayed at 1° until one P.M. The spread increased to 2° as conditions improved slightly to 500 and three. The forecast said improvements by noon, but it didn't come true. With a 2° spread that late in the day (the date was December 21), and with conditions improving very slowly both west and north, the outlook for VFR was doubtful.

It is better to be on the ground wishing you were flying than to be flying wishing you were on the ground. The rapid deterioration of weather conditions noted at that station between five and six A.M. is a good demonstration of how dangerous a slim spread between temperature and dew point can be. This day, a number of reporting stations in the area reported good VFR conditions with a spread of 3° or less at one time or another, and all eventually went IFR—some with below-IFR-minimums conditions.

Regardless of how good the ceiling and visibility are at any given moment, a spread of 5° or less between temperature and dew point should alert the VFR pilot to the possibility of low ceilings and poor visibilities. A careful watch on the trend, as well as an alternate plan of action, should be standard operating procedure in such a situation. If IFR, and if the destination has a spread of 3° or less, better keep checking things both at the destination and at that solid-gold alternate, to make sure it isn't really brass.

Wind direction and velocity can also say things about the outlook. If the surface wind reflects a circulation from over the nearest large body of water, the ceiling and visibility are likely to do worse things than if the wind were blowing toward that water. Air that passes over water picks up moisture, and fog and low clouds can form when moist air is moved over increasingly higher ground. This is called upslope effect, and it can create huge areas of low ceilings and poor visibilities over the Great Plains and other areas of rising terrain. Upslope can also thwart the best plans of the VFR pilot in the mountains. For example, if the ceiling is 3,000 feet at the Helena, Montana Airport, simple arithmetic shows that the pass west of Helena should be clear; it's 2,000 feet above the Helena Airport. An easterly flow, however, can often bring about unexpected conditions.

The altimeter setting in a sequence can tell you a lot if you watch the trend. Is the barometer rising or falling? A rising barometer indicates better weather, and a falling barometer worse weather. Unless the report notes that the pressure is rising or falling rapidly, however, don't depend on the trend having any short-term effect.

Winds-aloft forecasts can also help you recognize potential weather problems while you're en route. Virtually all winds given by an FSS are forecasts, and since they are based on the expected location of lows and highs, it is reasonable to assume that if the winds forecast is terribly inaccurate, the other forecasts might also be wrong. If the winds are from a different direction or at a different velocity than you'd been led to expect, it's time to continually verify good VFR conditions ahead or, if you're IFR, double-check destination and alternate weather. (And don't forget to recompute the fuel requirement for the flight if the unforecast wind is slowing you down.)

Once airborne, the VFR pilot is continually challenged to judge ceiling and visibility, and this is not a particularly easy chore in marginal conditions. Even though these values are reduced to simple numbers for the sequence, we find that they aren't so cut and dried in reality.

To begin, problems often occur because scattered clouds do not legally constitute a ceiling—it takes broken to be classed as a ceiling—and many pilots don't follow the healthful practice of considering scattered clouds at least a *potential* ceiling. If conditions are such that some clouds form at 500 feet over a reporting station, a slight change in those conditions a few miles down the road could transform the scattered into broken clouds at 500 feet. The next step is an overcast at 500. The pilot who thought he'd make the trip VFR up on top of the scattered clouds would be in for a rude awakening.

Those scattered clouds can bedevil IFR pilots, too. A reported condition of 200 scattered and 800 overcast means that an approach with a minimum descent altitude of 300 feet agl should work out if the visibility is clear. But it also means conditions exist that are conducive to the formation of clouds at 200 feet, as well as at 800 feet, and a large hunk of scattered clouds at 200 feet might be situated at the critical point in an approach. There is no worse weather situation for a low approach than one with a fairly good ceiling, minimum visibility, and scattered clouds at or just below the minimum descent altitude or decision height.

Cloud bases aren't textbook-perfect, either, and the lower the clouds and visibility, the more ragged and indefinite the bases tend to be. This bothers the VFR pilot who tries to fly as near the ceiling as possible. The cloud doesn't just start abruptly, and by flying near the point at which the cloud becomes absolutely dense, he's actually scooting along in the cloud base. This cuts visibility drastically, and if he happened to be passing near an airport where instrument approaches were in progress, he'd hardly have time for a proper introduction before meeting an aircraft descending out of the soup. The best visibility is found as far below the ceiling as you can safely fly and avoid terrain and obstructions.

The indefinite nature of cloud bases causes what is becoming a very common IFR problem, too. A pilot conducting an instrument approach is in a high state of readiness to see the ground as he descends toward minimums, and he probably has the reported ceiling value in mind. When he moves into the thinner bottoms of the clouds, and houses or trees start becoming peripherally visible beneath the airplane, the natural reaction is to look up to see if ground contact has been established.

There also seems to be a tendency to descend more at that point, on the basis of being able to see the ground straight down. Many an instrument approach has terminated early because a pilot who thought he had ground contact disregarded the all-important rule: don't leave the MDA or DH unless the runway or associated light system is in sight. Any time the ceiling is reported to be below 500 feet and the visibility a mile or less, the ceiling is probably indefinite and is not something you can fly beneath. Rather, you just reach a point where the runway appears through the murk ahead. Or it never appears, and you go somewhere else and log some more time in the process.

Most important, the pilot must be realistic about the relationship between ceiling/visibility and uneventful flying. Wishing has no place in an airplane, and the weather-wise pilot learns to be aware of weather signs and to take them at face value. There is a clear-cut line between hazardous and safe weather for any pilot, and he must learn to draw the line and stay on the right side of it without regard for pressures to get somewhere or do something. A pilot knows that he is tempting the fates when his thoughts or words are: "I *think* the ceiling and visibility will be good enough. . . ."

4 CLOUDS
by Richard L. Collins

WEATHER CAN BE fickle and mysterious when we try to decide what it is going to do in a few hours or tomorrow or next week. The weather always remains in view, though, ready for our inspection, and what can be seen of it can tell us a lot about flying conditions. Clouds are especially important, and the pilot should learn to read them on a minute-to-minute, hour-to-hour basis.

Clouds are formed when air containing water vapor is cooled below the dew point. Cool air holds less moisture than does warm air. The dew point is the temperature to which the air must be cooled, at a given pressure and water-vapor content, for it to reach saturation. When the temperature and dew point meet, the resulting moisture condenses into droplets on microscopic dust particles, and clouds form. The altitude at which clouds will form and the number of clouds that will form depend primarily on the amount of moisture that is available and on the stability of the atmosphere. Stability is the atmosphere's ability to resist a narrowing of the spread between temperature and dew point.

The water vapor for clouds is provided by evaporation from water surfaces (oceans, for example), moist ground, sublimation from ice and snow surfaces, and solid precipitation. The necessary lifting comes from convection (heating of the surface); large circulations, such as those associated with fronts and low-pressure areas; and air being forced up the slopes of mountains.

There is a direct and basic relationship between atmospheric stability and cloud types. A cloud is made as moisture-laden air is cooled. If the air is unstable so that its temperature drops rapidly with increase of altitude, clouds with vertical development are likely to form. These will be cumulus types; the term "cumulus" comes from the Latin word meaning "pile" or "heap." (All cloud names are derived from Latin.) In contemporary Americanese, a big pile of cumulus can mean a heap of trouble for a pilot. If the air cools slowly with increasing altitude, it is stable, and any clouds that develop in it will lack vertical development and be spread out in a layer called stratus.

Also, warmer air flowing over a colder surface will tend to be stable because it is cooled from below. Low stratus clouds are thus likely to form. On the other hand, colder air flowing over a warmer surface is more likely to be unstable because of heating from below, causing cumulus types to develop.

Another important factor can be temperature inversion, meaning that the air aloft is warmer than the air below it. Warm air atop cool makes for total stability, that is, total resistance to lifting. The point of inversion will mark the tops of stratus clouds that form beneath it. If the inversion happens to be fairly low, and if there is a good supply of moisture beneath it, the resulting clouds could be right on the deck: it will be foggy. Fog is often thought of as a separate phenomenon from clouds, but it is really ground-based stratus cloud.

Add another basic form of cloud to stratus and cumulus: cirrus, or streak cloud. Cirrus clouds might be considered the opposite of fog, for they are thin clouds at high altitudes, are made up of ice crystals and do not harass pilots to any extent.

Cumulus and stratus clouds are low-altitude types. Stratus bases will be found less than 6,500 feet above the surface and often at 1,000 feet or less; cumulus bases are usually found below 6,500 feet, but the clouds develop vertically. Cirrus clouds stay above 16,500 feet and are often found as high as the 30,000s and 40,000s. All other cloud types tend to be variations on these three basic themes. Cirrostratus and cirrocumulus are stratus- or cumulus-type clouds found at the high levels typical of cirrus. Altostratus and altocumulus are layers or heaps of cloud found in the middle (6,500 to 23,000) feet levels. Stratocumulus are low-level clouds, a combination of stratus and cumulus that forms in what might loosely be called rather unstable air. Nimbostratus are ragged, low clouds that develop as a result of rain. The granddaddy of them all, the cumulonimbus, is the thunderstorm that can seemingly start at the tip of your toe and extend upward beyond 50,000 feet. There are other variations, some with names that make you think the weather observer is out to snow you with his knowledge of Latin. There are altocumulus castellanus, mammatocumulus and cirrus intortus, to name a few, but the basic types are what pilots need to know about.

Clouds tend to defy understanding, so they are isolated and studied separately. During a flight, a pilot seldom sees or flies through only one type of cloud. We think, instead, of weather in terms of fronts, lows or highs, so we must understand clouds as part of weather systems.

What clouds tell us about warm fronts is very helpful. If we start north of the front, in clear skies, we would come upon cirrus first, which would be at high altitudes. Scattered cirrus arranged in a random fashion don't tell much of a tale, but cirrus that align in bands and tend to thicken with time are a definite indication of warm, overrunning air. The indications are that there is moisture to the south—a lot of moisture—and that a warm front is probably on the way. Further study of the cirrus can reveal a little about the nature of the front. If the cirrus begin as streaks and then thicken into a solid layer, creating a high, thin, cirrostratus overcast, the indication is one of relatively stable, overruning warm air. Formations of cirrocumulus—high, puffy, white clouds—are an indication of instability in the overrunning warm air: an active warm front might well be anticipated.

As we progress toward the warm front, next could come altocumulus clouds in the 6,500- to 23,000-foot range, also an indication of instability in the approaching front. These clouds take on the appearance of closely aligned, parallel, rounded masses. The next clouds might be altostratus, in the same altitude range. These have a uniform appearance, like a rather dense sheet, although the clouds are thin, with light patches between dark patches. They are an indication of stability in the overrunning air. In either case, light rain is likely to begin as the altocumulus or altostratus thicken and become solid. Heavier rain would follow as showers and thundershowers in an unstable warm front and as steady precipitation in a stable one.

The clouds we have looked at so far—cirrus, altocumulus, altostratus, and any cumulonimbus that might pop up in an unstable-warm-front situation—are all formed when warm, moist air was forced aloft, up the slope of the warm front. Next would come low stratus clouds in the cool air beneath, in the area where rain is falling from the altostratus, altocumulus, or cumulonimbus. This stratus forms when warm rain falls into cold air below. The standard example of how the low clouds form is that of the warm shower in a cold bathroom. The total meteorological explanation is a bit complex, but basically, the warm rain falling through cool, stable air adds moisture by evaporation, thus raising the dew point and forming cloud. If the temperature contrast is great, the stratus cloud formation will probably be on the ground as fog. If the temperature contrast is not large, meaning that the rain falls from warm air into cooler air and then back into somewhat warmer air near the surface, stratus will probably form, based at 1,000 feet or below. The base of the stratus will be found where the temperature of the falling rain equals the air temperature.

Think about that, because it can be an important thing to watch for when waiting for very low ceilings to lift. For purposes of illustration, let's assume that the temperature at 6,000 feet is 50° F. and the surface temperature is 33° F., an extreme but not totally uncommon situation in the south-central and southeastern United States in winter. Any drop of drizzle or rain that falls from that warm air above will be in air colder than the drop until it reaches the surface, so it'll be making cloud all the way to the ground; the day might well be a foggy one. Heating from the sun can sometimes cause ceilings to lift a little by midday, and forecasts in such areas usually reflect a hope that this will happen. However, real improvement is not likely to come until a change in circulation, brought on by a frontal passage or substantial movement of a low-pressure area, changes the temperature pattern or moisture content of the air.

A somewhat more common situation, in which stratus instead of fog forms in the warm-frontal rain area, might find a 40° F. raindrop falling into a layer of cooler air, cooling some as it falls, evaporating and making cloud. It may then enter air near the surface that is warmer than the drop. When it reaches the warmer air, its role as creator of stratus cloud stops; the cloud base is there, and the drop has to be content to end the day as a plain old raindrop falling on someone's head.

When stratus-type clouds form, there is always some sort of temperature inversion above them. When there is an inversion, the possibility is strong that low clouds will form.

Fog that forms outside areas of precipitation provides another good example of stratus-type development. The ground cools at night and cools the air just above it, but the temperatures aloft don't make a corresponding change in the dark. When the surface wind is calm, the cold air at the surface remains undisturbed. So an inversion forms, warm air atop cool. If air near the surface cools to its dew point and moisture is present, you have fog, a cloud on the ground. When the sun comes up the next morning, its heat begins to reverse the process. Fog seldom exceeds 1,000 feet in thickness and is usually much thinner than that. The fog first lifts from the surface, becoming a thin layer of cloud, and then dissipates quickly as the temperatures rise. The key to morning fog or stratus is the temperature inversion. The strength of the inversion determines how long the condition might last.

The IFR pilot's dream situation is often one of widespread stratus, with relatively low ceilings below and low tops, allowing flight in the clear. This means that the IFR pilot can go where the VFR pilot can't, with the only actual instrument flying required being during the climb to on top and the descent at the destination. Unfortunately, these conditions are the exception rather than the rule and are found only where there is a large area of cool, moist air overrun by warmer, drier air. If the overrunning air is moist and unstable, other clouds will be found above the stratus, rain will fall, and the whole messy bunch of clouds might merge with tops to 20,000 feet. There's one operational item worth adding here, too: a pilot might be flying serenely along on top of stratus in relatively warm air without much thought about icing, while plenty of it awaits him below. The basic inversion requirement for this situation means that the temperature will drop when the clouds are entered and that if it drops below freezing, ice will be encountered.

Stratocumulus clouds can pose quite an ice problem for the general-aviation IFR pilot, too. These are a combination of stratus and cumulus, meaning that they are layered but with some vertical development. I've most often found problems with them behind slow-moving or stalled cold fronts. They can be fairly thick—if you have a report of bases from 1,000 to 2,000 feet and tops around 6,000 feet or higher, chances are the clouds are strato-cumulus. These clouds are exceptionally moist toward the top; the tops are often just above the cruising levels most frequently used by singles and light twins (6,000 to 9,000 feet); and the fact that they are in cold air completes the setup for lots of ice. When stratocumulus appears to be the cloud of the day and the freezing level is below the cloud tops, the flight must be either on-top or one below the freezing level and low in the cloud system. A pilot flying from the warm to the cold side of a cold front might have a particularly icy time with stratocumulus. If cruising at, say, 8,000 feet, the pilot might make it through the front with no ice, only to find himself thoroughly frosted after passing the front. The probability would be that he had flown within the top few thousand feet of stratocumulus behind the front.

There's another bit of basic weather wisdom to add when planning a climb through to the top of any cloud, whether ice is a problem or not. Except in the tropics, precipitation heavier than drizzle isn't likely to fall unless rain-producing cloud is above the freezing level. (Ice crystals, it seems, have something to do with the precipitation-triggering process.) Knowing the freezing level can at least tell you that the cloud tops are above a certain altitude. If the freezing level is 14,000 feet and it is raining, the tops of all clouds are likely to be above 14,000 feet.

If we see a low deck of stratus-type clouds and no precipitation is falling, that is an indication that either on-top or between-layer operation would be possible at non-oxygen-demanding altitudes. The tops of the low clouds probably would not be high, either, because precipitation of other than light intensity requires clouds to be more than 4,000 feet thick, and stratus layers are generally thinner than that. The air is usually smooth in stratus clouds, but some bumps might be encountered on climb and descent. Wind-shear turbulence is not unusual in the inversion situation necessary for stratus formation. When the view of stratus is from the ground, before takeoff, the pilot must also consider that the airplane will be in the observed clouds for only a few minutes and that the situation could change quickly. That's why the wise ones know where the lows and fronts are located and check the radar reports for echoes. If precipitation is falling anywhere nearby, the cloud-forming process is again active in the air atop the stratus, and the pilot should find out why *before* launch.

Nimbostratus clouds are a good example of turbulent stratus types. These form at low levels, in precipitation, and appear as dark clouds with a ragged lower surface. Fractostratus, usually called scud, form below the nimbo-stratus, and the effect is one of a dark, chaotic, and forbidding sky. The conditions in which they form (heavy rain and often a strong surface wind) can stir up a good measure of turbulence. The extreme example would be in the nimbostratus that form beneath thunderstorms.

Cumulus clouds are self-revealing, too. If, during the summer, little cumulus clouds grow into big ones of cauliflower shape by midmorning, chances are the afternoon has a thunderous potential. The rate at which they grow is an indication of the instability of the air; the fact that they are big and juicy means there's moisture present. Lots of fat cumulus and some cumulonimbus in an area south of a warm front suggest that the weather in the frontal zone will be turbulent, too, for that moist and unstable air south of the front is the same air that is being forced over the cool air that lies on the surface, ahead of the warm front. Plenty of cumulus ahead of a cold front portend an active frontal passage.

An example of how one can learn about the weather from clouds occurred recently while I was studying a situation of widespread bad weather. There were several frontal lines and lows on the weather map, but the picture was far from clear; an assessment of the conditions I might encounter was difficult. A pilot report stated that the tops between two points were at 7,000 feet, with a line of swelling cumulus reaching to 10,000 feet along the way. You could just see the vast carpet of stratocumulus at the pilot's feet with a pouch developing in its middle. If the cumulus kept swelling, they would soon be above the no-oxygen ceiling. If they built on through the freezing level, which was at about 18,000 feet, they would probably turn into thundershowers or thunderstorms. The fact that the cumulus were in a line indicated some sort of front or low-pressure trough in the area. The pilot who reported the condition actually had a clearer picture of the situation than did the weatherman gazing at all his charts.

Those building cumulus clouds can even give the pilot directions about where to fly. I once heard a controller complain about pilots continually wanting to deviate from their assigned route, to go around "buildups," when he wasn't showing any weather on his scope and there probably wasn't a real thunderstorm within miles. The pilots weren't going around thunderstorms that didn't exist; they were going around building cumulus. The turbulence in a 12,000- to 15,000-foot-tall cumulus can be tooth-jarring, and the controller who is peeved because pilots want to avoid such shaking ought to have *his* chair pumped up and down vigorously every few minutes as a reminder that the environment in which the pilot works is not as serene as his. Any cumulus that has built vertically more than a few thousand feet is worth circumnavigating.

Cumulonimbus are easily visible from lower operating altitudes when they occur within an air mass or in connection with the passage of an active cold front, but in a warm front, they are generally embedded in other clouds. The message from any thunderstorm is to stay a safe distance away. It is possible to assess how mean a cumulonimbus is by estimating its height—severity is proportional to height—but even a runt thunderstorm can be more than a light airplane can manage. Other cloud types form around thunderstorms, too; the whole stack isn't just cumulonimbus. Plain cumulus might be clustered around the big one, with nimbostratus lying below, in the rain area, and altostratus gathered around the storm in the middle levels. When an air-mass type of storm dissipates, it leaves cirrus-type clouds behind.

Some cloud formations downwind of mountains also warn of turbulence. These are lenticular clouds, small, lens-shaped clouds that form at the peak of a mountain wave. High winds flowing over the mountains create the wavelike effect. The lenticular cloud is a sure sign of mountain wave and its accompanying turbulence; a rotor cloud beneath the lenticular guarantees memorable turbulence. A lack of these cloud types, however, does not suggest even a reasonably smooth ride if the wind is strong as it pours over the mountains. If there's no moisture in the air, the clouds won't form, but the turbulence will.

A pilot can also get some clues from the effects of clouds on sunlight. If the cloudy sky is very bright, the clouds are thin. If the cloudy sky is dark, the clouds are thicker and there may be multiple cloud layers. If the view in one direction is very dark, rain is probably falling there. Rainfall from some cumulonimbus clouds can become so heavy that there is a blackish-green hue in the area beneath the storm, and in a squall line, this can form an impenetrable wall in the sky. That is probably the best example of all on which to end, for while there are clouds to fly over, under, around, and through, the line of cumulonimbus is one place where none of those flight plans would work.

5 COLD FRONT
by Richard L. Collins

THE COLD FRONT is a tough one—a huge bulldozer trundling across the countryside, pushing clouds up and then sweeping them away as it passes and grinding up airplanes that dare fly before it has finished with its work. Even a well-equipped instrument pilot might view the blue cold-front line on the chart much as an ant regards the Great Wall of China.

A cold front is actually nothing more than the dividing line between two dissimilar air masses when cold air is advancing on warmer air. (When the reverse is true—when warm air is advancing on cold—the dividing line is called a warm front.) The cold front is not something of set proportion that moves on a timetable. It can be mean as hell and spawn storms that wipe out whole towns, or it can move by with hardly a ripple. And while it is true that most weather comes in conjunction with a front (cold, warm, stationary, or occluded), a front does not always mean that the weather will be bad. Nor is frontal movement always predictable; a front can move quickly or slowly, and change speed anytime during its life span. Mountains can have an especially dramatic slowing effect on frontal movement.

We tend to think of fronts as individual things, but there is one primary polar front that undulates across the Northern Hemisphere, separating air coming down from the polar region and up from the tropics. It tends to stay north, and often becomes hardly identifiable in the summer; it starts flexing its muscles in the fall, usually becomes aggressively southbound in the winter, and often has some dramatic last flings in the spring. When circulation causes it to move south, it's a cold front. When it moves north, it's a warm front. When it stops, it's a stationary front. It is *the* front, although other fronts can and do exist ahead of and behind it.

Low-pressure areas—cyclones, as the textbook calls them —often form on fronts and affect the movement of fronts. (All low-pressure areas do not form on fronts, however. Hurricanes, for example, are not associated with fronts.) Fronts themselves are essentially areas of low pressure with high pressure on each side. The usual pattern of a falling barometer until frontal passage, followed by a rising barometer, illustrates this clearly.

Low-pressure formation on a front is usually described in terms of the phenomenon of a wave in the water, and the lows are initially identified as "waves" on a front. Just as wind blowing across water can make waves, wind blowing in opposite directions on each side of a front can make low-pressure waves. Such low-pressure waves can curl over, just like whitecaps, thus establishing a complete counterclockwise circulation. Given strong development, the wave turns into a full-scale low-pressure system.

Let's follow a wave formation. Our front has a good supply of cold air behind it. The air mass ahead is warm and moist. The circulation pattern isn't moving the cold air southward, though, and we end up with northeasterly winds at the surface behind the cold front and southwesterly winds ahead. The opposing winds lead to the formation of a wave or a series of waves on the front. If the waves are weak and don't curl over into a full low-pressure circulation, the front might remain stationary with dollops of stormy weather moving eastward along the front in connection with the waves. If a wave develops into a full low-pressure center, the circulation around it could cause frontal movement. The disturbance could next turn into a full low-pressure storm center, with the cold front trailing in a southerly to southwesterly direction and a warm front off in an easterly direction. It is this pattern that we most often see and study, and it is always worth a review.

First, a little geography. Storms that form on the mid-Pacific polar front hit the West Coast but seldom make it across the mountains as organized storms. Instead, they redevelop east of the mountain ranges. Strong lows often develop over Colorado, to the east of the mountains, and move across the central and eastern United States as major storms. Storms frequently redevelop east of the Canadian Rockies, too, with cold waves and blizzards blowing down into the U.S. as the storm moves eastward. The Great Lakes are also in an area that favors the formation of disturbances, with the relatively warm water of the Lakes contributing to atmospheric instability. Lows form over the Gulf of Mexico, too, and in the Atlantic, the favorite point for storm generation is Cape Hatteras. The fact that these storms usually form on an identifiable slow-moving cold or stationary front gives the pilot something to watch for.

Next, take a typical newspaper weather map and sketch in the classic patterns: place a low-pressure center that developed in Colorado and moved northeastward over Lake Michigan, with a cold front trailing down through Illinois, Missouri, western Arkansas, and eastern Texas. Visualize the counterclockwise circulation around the low and you'll see how it brings large quantities of cold air down from Canada; the southerly circulation ahead of the cold front brings moist air up from the Gulf of Mexico.

For flying weather, the pilot ahead of that cold front in the southeastern

United States will probably find southern winds, scattered to broken clouds, and scattered showers. That weather might tell the pilot something about the cold front and what to expect there if he or she happens to be flying west.

The strength of the southerly circulation is one clue. If it is strong, the cold front is more likely to be strong. Prefrontal cloud cover also carries a message. If there are few clouds, the moisture supply is not plentiful. If there are a lot of clouds and they are building into showers, there is moisture and instability, with both likely to increase closer to the cold front. Add some knowledge of the weather behind the front to what can be seen ahead of it, and the picture becomes even more complete. A big temperature drop on the cold side and strong winds perpendicular to the front suggest that the frontal zone will be tough. The winds behind the front also tell something about frontal movement. If they are perpendicular to the front and strong, the front is moving rapidly—in rough proportion to wind speed. If the wind behind the front is light or is beginning to parallel the front, movement would be slow and the frontal zone probably wouldn't be as turbulent.

Several things "make" the weather we find in connection with a cold front. Convergence—the simple coming together of the northwesterly and southwesterly winds—for example, leads to "lifting" and is one cause of inclement weather in the frontal zone. Atmospheric instability (more rapid cooling with altitude than normal) would probably be present in winter when abnormally warm air is drawn up in the lower levels ahead of the cold front and would also contribute to the problem. A moisture supply ahead of the front would provide another ingredient, and it is usually present in a southerly flow—especially over the eastern half or two-thirds of the United States.

When a cold front is moving rapidly, the bad weather created by it usually covers a relatively small area. The fast-moving cold front has a steep slope; that is, the frontal lifting is done in a short distance, with the warm air tending to retreat rather than sliding smoothly up over the cold air. This limits the area of convergence and spends the weather's energy quickly, if violently—often in squall-line thunderstorms. Push hard on that warm air, as the rapidly converging cold front does, and there's going to be a big bulge ahead of the front. Put it all together—moisture, instability, and convergence—and you have the makings of a spectacular frontal passage.

Squall lines can form in the unstable air from 50 to 250 miles ahead of a cold front as well as in the frontal zone. Squall lines ahead of a cold front are more likely to form in the afternoon, when heating is at a maximum. My own observation is that a front is usually good for only one fully developed squall line, so if there's one ahead of the front, there's often not one with the front. I've also noted that squall lines ahead of fronts appear to be of a somewhat more broken nature than squall lines accompanying a front, which are often solid, tall, and mean for hundreds of miles. Upper winds play a major role in thunderstorm, squall-line, and tornado formation, too, and a old and rapid flow at high altitude above warm and moist air at low altitude means instability and tells the forecaster that conditions are ripe for storm formation.

Take some of the troublemaking ingredients away from a cold-frontal situation and the tiger can turn into a relative lamb—at least for the IFR pilot. A shallow frontal slope is characteristic of a slow-moving front, with the inclement weather covering a larger area but with less violent activity. You might say that the same action is spread over more territory. Take out the moisture and the front might go through dry. If the temperature is close on both sides of the front and the winds are light, the frontal zone might be tame. If it reached a point where there was no temperature differential, the front would disappear.

On the other hand, a slow-moving cold front can also be as much of a problem to a pilot as a fast-moving one. The weather covers a larger area and moves slowly, so the pilot might not be inclined to afford the luxury of landing, chaining the airplane to the ground and waiting for the storm to blow through if he's trying to go west through it. The slow-moving front makes more of a demand on the pilot: either you fly through or you stay on the ground for a day (or for days if you're VFR). The price for penetration can be high unless the front is tame.

From a pilot's point of view, the problems found in a very slow-moving cold front are probably more directly related to instability, moisture supply, and low-pressure-system development along the front. The front might just become a handy thing for the weatherman to use in giving a geographical picture of the problem. It might have light winds on each side and little temperature differential. But there could still be action. A very slow-moving cold front that took almost a full week to move across Texas in late September 1973 provides an interesting illustration. One that moves slowly is tame, right? Wrong, in this case. It spawned tornadoes to the north, in Kansas. Torrential rains fell there, too, and thunderstorms were forecast all week ahead of the front in Texas and Arkansas. The activity remained more to the north, though, in Oklahoma and Kansas. The weather map kept showing a low there; perhaps it was a series of waves developing on the nearly stalled front that caused the problem.

Things changed on Thursday of the week, and the southern portion of the front moved and developed some action. Perhaps a low formed south and moved northward along the front. Thunderstorm activity built up Thursday afternoon east of Dallas and formed a line as it moved eastward, harassing all the pilots who were going in that direction on their way home from the National Business Aircraft Association annual meeting in Dallas.

The forecasters were a bit slow in spreading word of this development, thus leaving pilots to depend on their own weather wisdom. One pilot who was northeastbound from Dallas told me the forecast he got called for rain, but that the briefer had added that the weather was horizontal (as opposed to vertical, which best describes thunderstorms), and that pilot reports had indicated good rides through the area during the afternoon. My friend got to the rain area about three hours after his briefing, and the elements made it plain to him that it was far from stable. His 310 climbed 2,000 feet with power off at one point, and when he emerged on the northeast side after a few very turbulent minutes, the storm had left him a reminder: the rain had peeled the paint off the wing leading edges. A couple of hours later, a Texas International Convair was lost in western Arkansas as it attempted a VFR diversion around this line of thunderstorms. The same front kept moving at a snail's pace, and by the following Monday had not even progressed to the western slopes of the Appalachians.

When we start flying, nobody promises that we will be given a smooth, rain-free path to fly. It is up to the pilot to provide that. In the case of this front, which threatened for days and finally made good the threat, perhaps there was a key that would quickly have told of its mean streak. The front was definitely there, with a light northwesterly flow behind it and a somewhat stronger southerly flow ahead of it. There *was* convergence. Heavy rainfall made the presence of moisture obvious, and it would have been reasonable to anticipate turbulence in the frontal zone. Perhaps the only thing not anticipated was the severity, which might have been caused by a low-pressure disturbance forming on the front.

A low-pressure center's location in relation to the flight path is important when making a judgment on a front. If there's a strong low to the north—over Lake Michigan, say, as it is on the weather map you drew—and you are in Texas flying through the front that stretches to the south of it, weather will probably be good except in the actual frontal zone. Move on north, closer to the low, though, and things change. The low itself becomes more dominant in determining what weather will be found, and conditions are likely to be bad (or worse) over a larger area. Lows tend to be meanest in the direction of their movement; lows generally move in an easterly direction, so that would suggest potential problems ahead of the cold front and to the southeast of the low, as well as to the east and northeast of it. On your map, the problem areas would be Michigan, Indiana, Ohio, and the southern half of Illinois.

Knowing about the low center also tells you something about weather behind the cold front. If the low is far to the north and the front moves through briskly, the weather is likely to be good after frontal passage. Plenty of moisture can circulate around the top of a low, though, so when the low center passes close to the reporting station, a lot of wind, cold air, and moisture can be mixed up in what is usually called a blizzard. The best blizzard-producer is probably the low that moves by just to the south of an area.

The strength of a low has a very direct bearing on fronts, too. When a low is deepening, which means it is becoming stronger, the movement of the low across the surface slows but the circulation around it increases. Even though the low slows, the portion of the front a distance away from the low would tend to accelerate because of the increase in circulation. When a low is filling, or becoming weaker, its movement will accelerate but the circulation around it will tend to decrease, slowing frontal movement a distance away from the low center.

If a cold front moves very rapidly, as would be the case when the low is far to the north and very strong, there's a chance it can be followed by a little brother. The cold air moving quickly over the warm terrain is warmed some, and another temperature demarcation line might form a few hundred miles behind the primary front. This is called a secondary cold front, and it isn't as pronounced as the first one. It can still be squally enough to halt VFR flying and to give the low-altitude IFR pilot a rough and perhaps icy ride.

People living in the eastern part of the country tend to think of weather usually clearing behind a front, but the script can change somewhat in the West. The moisture ahead of and in a cold-frontal zone in the East is generally brought up from the Gulf of Mexico. This source isn't available to feed moisture up ahead of the front in the West. There, the front is more likely to bring moisture along with it, which means inclement weather *behind* the front.

One other item germane to a discussion of the cold front is the low-pressure trough. While a trough is not a cold front, it often presents the same picture and problems to a pilot, and many identify the passage of a trough as a cold-frontal passage. A trough, though, appears in an area where air is coming together from two directions without a marked temperature difference in the converging air. The convergence causes lifting, and troughs are often ideal locations for thunderstorm activity. The storms can line up and give the appearance of the kind of squall line that might precede a cold front. Troughs are not depicted on weather maps, though they can be readily identified on the chart: the isobars—lines of equal pressure—kink away from the primary low-pressure center to identify the trough, much in the same manner that the isobars kink away from the low along frontal zones.

The cold front is but one of the basics we deal with in developing the weather wisdom we need for flying. Passage of a cold front does bring about a marked weather change, but the cold front is still just a part of the big picture. A pilot should seek answers to a lot of questions when a cold front is approaching or is across the path of flight. Low-pressure center location, temperatures and winds on each side of the front, and cloud conditions on each side are all things to explore. Current information can combine with the forecast to give the pilot a good idea of what the front will be like. Some cold fronts fulfill the promise of clear skies after passage; others might stop over or just south of a location and make things miserable for days. And the pilot finds that some fronts can be penetrated VFR or IFR with hardly a bump—while others do, for a fact, present an obstacle as formidable as that Great Wall of China.

⑥ WARM FRONT
by Richard L. Collins

COLD FRONTS are usually sharply enough defined that a pilot can easily identify and work his way through, around, or over them. If he should choose to, the pilot can deal with the cold front by waiting until it moves away. It usually isn't a long wait. By contrast, the warm front is a huge blob of slow-moving, messy weather, with widespread low ceilings and the chance of embedded thunderstorms. Indeed, the warm front has something for everyone: its low ceilings take care of the VFR pilot, and embedded thunderstorms can give pause to the IFR pilot who lacks airborne radar. Warm fronts can bedevil forecasters, too. The National Weather Service admits that it can forecast the time of passage of a warm front with only half the accuracy with which it can forecast passage of a fast-moving cold front.

On the weather map, put a low-pressure center in eastern Oklahoma, extend a cold front down through east central Texas, and drape a warm front through central Arkansas, northern Mississippi, central Alabama, and southern Georgia. Counterclockwise circulation around the low gives us a southwesterly flow, and there's warm air in the pie-shaped wedge that centers on Louisiana. Cold air squats to the west of the cold front, and there's also surface cold air to the north of the warm front and the low center. The surface circulation north of the warm front is southeasterly to easterly, circling around the top of the low and becoming northerly to northwesterly behind it.

The cold-front picture is one of cold air aggressively pushing its way into warm air. Friction tends to retard the cold air at the surface, contributing to a steep frontal slope, or maximum lifting in minimum distance. That spawns strong storms. (Try that for a tongue twister next time you are waiting for a cold front to pass.) The warm front's surface position can be envisioned as the trailing edge of the cold air being overtaken, with the action coming from the warm air sliding up over the cold air as it advances. The surface friction that makes the cold front so abrupt also gives the warm front a more gradual incline because it tends to retard the retreating cold air at the surface. Also, if the prevailing westerly wind velocities increase with altitude, they tend to make for a shallower warm-front slope

35

(and a steeper-sloped cold front).

The slower movement and shallower slope of the warm front help explain why a warm front usually means a larger but less violent area of bad weather than does a cold front. We might experience pretty good weather as a cold front approaches, a relatively short period of crash-and-bang and then a return to pretty good weather. Just ahead of the warm front, rainy weather has prevailed for quite some time, IFR or nearly so all the while, and the chance for better weather will have to wait until the warm front moves north.

Cirrus clouds are the vanguard of a typical warm front, appearing where the warm air has been forced up the slope of the cold air to an altitude of 20,000 feet or more. The slope of an average warm front is between 1 percent and .3 percent, so the first cirrus might appear from 400 to 1,200 miles ahead of the front itself. Approaching the front from the north, the warm air is found at progressively lower levels; seen from the surface, clouds run the gamut from high cirrus down to low stratus before the front passes. The pilot flying on a course perpendicular to the front at 8,000 to 10,000 feet might fly into cloud 200 miles from the front and remain IFR until flying out on the southwest side. Cumulonimbus can be embedded in there, too, representing a definite hazard we'll discuss later. Viewed from the side, a cross section of the warm front would show that cloud tops might remain stable from the first sign of cirrus to nearly the rear of the front, with more cloud filling in below the tops to make a roughly triangular hunk of cloud with the long point of the triangle aimed in the direction the front is moving. Tops of an average warm front can be found at 20,000 feet above the ground, with cumulonimbus tops poking out above that altitude. Piston airplanes without turbocharged engines are not likely to do much on-top flying in a warm front. When they reach that area of the front where embedded thunderstorms lurk, it can be impossible to spot the storms visually.

Behind the warm front, in what's called the warm sector, flying weather is likely to improve, but it's not absolutely guaranteed. Passage of the warm front should bring a wind shift from east or southeast around to south or southwest, and temperatures should rise somewhat, although the air may have been relatively warm before frontal passage. The pressure, which had fallen with the approach of the front, would become steady or even rise slightly. Weather in the warm sector can be showery if the air is moist and unstable. You may also find low ceilings and drizzle (and low cloud tops) after frontal passage if the air near the surface is very stable.

As the VFR pilot contemplates a warm front, there isn't much temptation to try to penetrate the front from behind the flow, from the southwest traveling in a northeasterly direction. Looking at weather a couple of hundred miles northeast of the front is equally discouraging. The weather reports generally signal a clear no-go for VFR a couple of hundred miles northeast of the active warm front and along it. An unwary VFR pilot flying toward a warm front from an area of good weather to the north, though, may be lured into real trouble because conditions, at least initially, don't look that bad. Rain starts falling from altostratus or altocumulus clouds that are usually found a couple of hundred miles ahead of the front. Cloud bases usually sit at 6,000 feet or more when the rainfall starts, and the ceilings and visibilities remain pretty good for a short while. Then low stratus clouds begin forming as the warm rain falls into the cold, stable air below. These might be scattered at first, but they soon solidify into a low, solid overcast. (To demonstrate the cloud-forming ability of warm rain falling into cold air, observe your hot shower in a 60° bathroom.) Once rain begins, it's likely to continue until the front passes, and the VFR ceiling and visibility are probably shot for the duration. When flying toward a warm front, only a naive pilot would press on in hope of finding better conditions.

A VFR pilot waiting on the ground for weather to improve can watch the sequence reports to the south for a sign of wind shift from an easterly direction to a southwesterly direction for a clue to coming improvement, but remember that it might be slow and that improvement after warm-front passage isn't guaranteed. Also, once the warm front passes, the cold front is close behind.

For IFR pilots, it is business as usual in a warm front, except for the question of embedded thunderstorms. It is true that the storms found in warm fronts are the least severe of all frontal thunderstorms, but that is a relative judgment. Flying non-radar-equipped airplanes with light wing loading in *any* thunderstorm area, regardless of severity, carries some risk. You can almost bet that the forecaster will include at least the chance of embedded thunderstorms whenever a warm front is working its way into an area. So nagging doubt about the risk will almost always be present.

There are ways the weather-wise pilot can develop some feel for a warm front. Look at station reports from behind the warm front: is the air there warm with lots of clouds or unstable and moist? It's basically the same air that is found in the wedge over the underlying cold air in the frontal zone, and if it's wet and unstable before it slides up over the cool air, action is likely north of the warm front. Another key is the appearance of the first high clouds ahead of the warm front. Are they cirrus or cirrocumulus? The latter is an indication of instability in the warm air. Sailors dubbed cirrocumulus sky a "mackerel sky" and reefed for a blow. Sailors weren't as lucky then as aviators are today; they had to sail on and accept what came. We can either detour or land and go hide until the storm goes away. One thing worth noting is that warm-front thunderstorms do not appear so turbulent when viewed from the surface as cold-front storms do. A warm-front storm will bellow and belch and pour out a lot of rain, but it lacks the surface turbulence of the others. This is because the cooler air has a damping effect beneath the warm air. The storm base is on the slope of the warm front, and the downdraft isn't likely to penetrate with its full fury the cooler and stabler air below. It'll be plenty bumpy aloft, though.

Thunderstorms are more likely and more intense when the warm front's slope is steep. The slope will tell you where to look for the likely bases of the actual cumulonimbus clouds that are embedded in other clouds. Thunderstorms 100 miles ahead of a surface front with a one-percent slope would probably have bases about 5,000 feet above the ground. Fly at the lowest possible altitude in a warm front. ATC procedures often prohibit use of the lower altitudes en route; this is unfortunate because back when much airspace was uncontrolled and you could legally fly instruments there without a clearance and with 1,000 feet of terrain clearance, I recall a young aviator managing some reasonably smooth trips out of warm fronts at low altitude. Help in avoiding the worst of the embedded storms can be had from the ground-based traffic radar, which might partially make up for the controller's unwillingness to assign the lowest possible altitude, but traffic radar doesn't do too well at picking out the real thunderstorms in widespread rain areas; in other words, don't count on it.

There's convergence and lifting along the slope of a warm front whether or not thunderstorms are present, so the IFR pilot shouldn't anticipate a velvet-smooth ride through a front. It'll probably be smooth in the cool, stable air down low, but once the front slope is penetrated and the airplane enters the warm air mass, some bumps will probably be encountered.

No two fronts are alike. So far we've dealt with the classic pattern of the warm front: first, the cirrus, then the developing lower clouds and rain, and finally, the frontal passage and an idyllic warm-sector picture of fat cumulus in a pretty blue sky. It does happen this way sometimes, but variations on the theme are more likely, and these variations can have quite an effect.

The location of the low-pressure center is a key point. We know that weather is apt to be meanest in the direction the low is moving in—generally eastward or northeastward. A pilot penetrating a warm front near the low center may find a lifetime supply of turbulence. In the wintertime, the low could slip by just south of you, thus depriving you of a look at the balmy air at the surface south of the warm front.

A low is also a blizzard producer when it moves just south of you. The warm air is pushed up over the cold and around the top of the low, and the precipitation from the warm, moist air aloft falls into cold air below, causing snow. I watched another variation recently as it developed on TV weather broadcasts and the TWEB. Some tornadoes had torn up some towns a few hundred miles to the west. The low was moving to the north, according to the forecaster, with a trailing cold front moving to the east fairly rapidly. As is so often the case, the forecaster didn't mention a warm front, but pointed out a southerly flow ahead of the cold front. The warm front was apparently there but ill-defined in the large area of warm air ahead of the cold front and east of the low. Strong southeasterly winds were reported at stations ahead of this vague warm front and light southwesterly winds were behind it. Showers were the norm, although ceilings weren't as low as expected in the warm-front zone.

Even though a tornado watch and severe thunderstorm warnings were in effect, there was little static on the low-frequency TWEB, and the radar reports progressively downgraded the area of precipitation from a line of heavy thunderstorms to a broken area of thunderstorms to some thunderstorms mixed with rain showers during the course of the morning. The cold front seemed to have passed, the wind shifted to more of a southwesterly direction than a westerly, temperatures dropped a little, and the pressure rise was slight. Improvement in the weather was hardly dramatic, with ceilings increasing from around 1,000 to 3,000 or 4,000 feet with persistent showers. Temperature and dew point behind the "front" remained close; they should have been spreading out after a cold- or warm-front passage.

Had it been a warm-front passage instead of a cold-front passage? Probably not. The cold front had overtaken the warm front and had formed an occluded front. Occlusions are important to the pilot seeking weather wisdom because they are frequently the cause of busted forecasts.

In this case, the hardly identifiable warm front was moving northeast, and the cold front was moving east. The low was strong and deepening, which meant it wasn't moving across the surface rapidly, although circulation was increasing, thus accelerating the fronts—especially the cold one. It simply caught up with the warm front. As it closed with the surface position of the warm front, the clearly definable surface position of both fronts went fuzzy. The cold air coming around the top of the low had, in effect, merged with the cold air to the east of the low, resulting in a low center that was completely surrounded by cooler air at the surface. There was still warm air aloft, and the warm-front weather still existed above the colder air at the surface. The thunderstorms were tame because of lack of convergence caused by the strong cold front at the surface. The front passed with a little wind, a lot of rain, no thunder, and the wind shift from southeast to southwest.

A few days prior to that piece of weather, we had a classic "mackerel sky" late one afternoon. The situation went from clear to a thorough rash of cirrocumulus, and even though the forecasts made no mention of an approaching warm front, something was obviously on the way. By midnight, it was thundering. That persisted until late morning of the next day, when the storms moved off to the northeast. The afternoon was warm with low clouds and drizzle. Cloud tops were low—4,000 feet—after the storms passed. The forecaster stuck to his guns about a southerly flow over the area, but made no mention of a front. The organized storm activity indicated that something had provided the convergence and lifting to build some thunderstorms, however. The pilot's key to the true picture was widespread thunderstorm activity in an area where only a chance of storms was forecast. That indicated a very basic miss on the part of the forecaster and was a sufficient signal to distrust his product until he had time to adjust to the situation.

The warm front is at times perplexing, hard to define precisely, and difficult to deal with, either VFR or IFR. It is like any other piece of weather, though. The pilot who studies it and makes sound decisions based on knowledge will find it less cruel than the pilot who makes a flip decision based more on the need to go somewhere than on his knowledge of the elements.

7 FRIDAY'S DOUBLE FEATURE

by Richard L. Collins

THE BEST CLASSROOM for studying weather is the airplane. Theories, lines on the map, and the classic patterns blend into the view out the window: sometimes a flat expanse of stratus beneath, even though the forecast had been for soggy and bumpy; at other times soggy and bumpy when the forecast called for a flat expanse of stratus. Some flights, such as a recent trip I took, combine many of the weather lessons into a day's activity in a way that fixes them in your mind.

To make a flight on Friday, I begin to contemplate the weather on Thursday. This particular time, I had a head start on my weather study because I had flown on Thursday and had sampled the situation firsthand. The synopsis on Thursday showed a stationary front in the area with a cold northeasterly flow at the surface and a southwesterly flow aloft. Thunderstorms had been forecast in the frontal zone Thursday, but they were scattered and never did match the expectations of the severe storm watch that was issued. The instrument approach to home base was slightly turbulent, with a ceiling of 500 feet and a visibility of two miles. There was a basic stratus deck in the area, with some cumulus poking up to 10,000 or 12,000 feet. The Thursday evening forecasts indicated that the front would begin to move southeastward that night as a cold front, leaving a smooth path for a southwestbound flight the next morning. I doubted that. When fronts become stationary, they tend to stay that way until something happens that will move them. In my experience, that something is usually a combination of a low-pressure formation on the front and a stormy period before the improvement. Also, the northeasterly flow at the surface—and clear, better weather to the south and west—made it look more like my home airport was north of what was becoming a warm front rather than south of a cold front.

At five on Friday morning, the forecast on the transcribed weather broad-cast gave the synopsis of the evening before: a nearly stationary cold front but now with a low expected to develop on the front to the southwest. If that materialized, the front in our area would indeed become a warm front as the circulation around the low developed. I drew the map as I listened and copied some sequence reports. The TWEB synopsis had been recorded at 10 the night before, so it was old and subject to error, but the sequences didn't give good reason to doubt it. There was no static on the radio—no thunderstorm was nearby—but the mention of widespread embedded thunderstorms on the TWEB dictated a call to the National Weather Service. Most of the activity was to the southeast, east, and north. The weather should be okay for IFR, with a lot of stratus forecast but not much zero-zero around. A low was now showing on their map 300 miles southwest, just south of my destination, and it appeared that it would develop and move on.

When I called the flight service station to file, the man there wanted to brief me even though I told him I had talked with the NWS. In ominous tones, he reported several airmets and sigmets calling for embedded storms with high tops and widespread low ceilings. I had made the go decision while talking with the NWS briefer, and the lack of precipitation along my proposed route of flight was an overwhelming plus sign. There also were plenty of places to go if I didn't like what I saw aloft. I filed IFR two plus 30 en route, six hours of fuel. I'd have lots of range to play with if the early-morning forecasts proved inaccurate.

After takeoff and a solid IFR climb, I broke out on top at 4,700 feet. There was some higher broken above, but the top of the stratus deck below was flat and the visibility was excellent. There was a strong temperature inversion below the top of the stratus. The reading had been 38° F. on the ground and was up to 50° F. at 3,000 feet. The layer of cool air at the surface was very shallow, and I filed this in my mind without attaching any significance to it, although it suggested that the surface position of the front wasn't far to the south.

Down the line, things began changing. Another layer formed, based at my cruising level of 6,000, so 10,000 was requested and approved. This altitude was good for eyeballing some distant, rapidly swelling cumulus that were trying to form a line. A weather check verified a fresh northwesterly surface wind west of that line and a southeasterly wind at a station just east of it. A cold front must have been there, with enough convergence to provide lifting and the resulting cumulus formation. I spoke with the traffic controller, and he reported some weather return in the area, scattered along the line. I flew through a low spot, clear of cloud at 10,000 with only some light turbulence when passing a couple of cumulus. Tops of the highest cumulus were esti-mated to be at 18,000. It was downhill the rest of the way; the destination was

1,000 overcast with breaks and an improving trend.

As I rode to my business appointment, I briefly reviewed the situation and wondered about the flight back home in the afternoon. It was now my opinion that the cold front would move eastward rather slowly—slowly, because the circulation was weak—and that the main problem would be with home base and a warm-frontal situation there. I thought the low had developed and was actually north of the route instead of south as forecast. At least the circulation suggested this. The trip home would probably have to penetrate both the cold and the warm fronts, making the afternoon an interesting one. The morning air had seemed to be very stable—the temperature dropped slowly with altitude up to about 10,000 feet.

I briefed myself at the FSS at about 3:30 P.M. First, I studied the radar summary chart, which was an hour and a half old. It showed only scattered to broken activity with maximum tops of 18,000 feet close to home. The surface-weather map positioned the low-pressure center south of course; nonetheless, the circulation patterns said it would be north of course. It was a weak low, with only mild circulation, so the cold front had moved barely east of its morning position, and the surface position of the warm front was now about 50 nautical miles before home base on the northeastbound flight. North of the warm front, ceilings remained in the 300- to 500-foot range, but there were plenty of places within range with excellent weather. Forecasts called for ceilings of 1,200 feet at my destination on arrival, but the forecasts were old and could be older by the time I was ready to shoot that approach. To me, the terminal forecast is one of the least important items during a briefing just prior to takeoff. The actual weather, the trends, the radar reports, and the synopsis tell the true tale. In this case, at 3:30 in the afternoon, the terminal forecast was nine and a half hours old and nothing to bet on—especially when the destination weather was running below what was forecast. I filed two hours en route, less than the trip down because of some tailwind component. A close alternate, legal according to the forecast, was indicated on my flight plan, but I'd have fuel aplenty to return to an area with much better weather and alternates.

I picked 7,000 feet for my cruising level, and the cumulonimbus along the cold-frontal line were soon visibile. The distance from which you can see the real cauliflower has always been interesting to me; these were clearly visible from a position about 90 nm away, and they had tops of only about 18,000 feet.

As I neared the cold front, I studied two possible paths through gaps in the line. Both were tolerably wide, and I finally picked the southerly deviation from course and asked the controller if that looked good on his radar. He described the precipitation return he saw in the area, concurred that a southerly deviation would be best, and okayed a turn off the airway.

The cumulonimbus in the area were interesting to study as I moved between them. They were some of the meanest, most turbulent-looking clouds I've ever seen, but the tops were low, probably not over 18,000—the front wasn't a strong one, so there wasn't much lifting. There appeared to be a rather strong southerly flow aloft, above the tops of the storms—a possible indication that warmer and more stable air aloft was stunting the vertical growth of the storms. Combine these factors with a rather low freezing level of about 1,000 feet and the fact that the clouds probably started precipitating as the tops moved through the freezing level, and you begin to see why the storm clouds couldn't grow any taller than they had.

I remembered a conversation I had had about thunderstorm tops with a pilot who operated on the theory that whenever thunderstorm tops were below 20,000 feet, penetration of the storm in a light twin would create no problems; a little moderate turbulence maybe, but nothing you couldn't handle. I wondered if he would penetrate something that had the appearance of these clouds I faced now. If he ever chose to try, it would probably be a one-shot effort.

The controller broke my train of thought with a call to change frequencies. He added that the next man could probably tell me about a bunch of weather on up the line, so I switched to the new controller.

The warm front was apparently creating the problem, and I'd have a good bit of weather for the last 50 miles of my flight. I was still 150 miles out when I learned this, but the sky had been hinting at it all along. Once I was past the cold front, the air was noticeably less stable than it had been before reaching the cold front. There was some light turbulence at 7,000 feet. Cumulus was building at two levels—plain old cumulus from below, and altocumulus based at about 10,000 feet. The sky was dark in the direction I was headed, and the wind at 7,000 feet moved from light northwesterly to strong southwesterly, as if it wanted my airplane in that dark piece of forbidding sky.

My theory always has been to take a look, and this afternoon I had enough fuel to fly closer to the worst weather, study it, and retreat if necessary. There were several things in favor of continuing, too: I would be near home base when I reached the weather, and the center and terminal radar facilities there generally do excellent work defining areas of heavy precipitation. Also, the system did not appear to have a lot of depth or true thunderstorm activity. I often fly along listening to the TWEB just to stay constantly abreast of weather all over the area; today, there was little organized static, and no station reported a thunderstorm. Finally, the radar reports I had studied in the FSS indicated only scattered to broken activity in the system.

As I moved on up the line, a plane flying toward me came on the frequency. The controller asked him about the weather; he reported a good ride at 6,000 feet but added that his airborne weather radar had helped him to pick a path around some showers.

For some unknown reason, pilots believe that a change in altitude will reinforce one's store of weather wisdom. My rationale for requesting a climb to 13,000 feet when I was only 100 nm from my destination was the fact that 13,000 looked like it would top a layer of clouds and allow me to eyeball the path along which I would be letting down for an approach. The controller cheerfully approved the climb, and I plugged in the oxygen and started climbing. I didn't intend to stay up there long; I just wanted to look.

The temperature did not drop rapidly, going from plus 7°C. to minus 5°C. in 6,000 feet—an average drop of two degrees per thousand feet, indicating some measure of stability. The view from 13,000 wasn't very rewarding, although I could see that this level would allow flight through a saddle between a couple of swelling cumulus that would have been uncomfortable at a lower altitude. Staying clear of these clouds also ensured an ice-free passage. The sky on the other side of those cumulus was a rather dark and shapeless mass; but the afternoon would soon be fading into evening, so sky and cloud color were not reliable indicators.

Upon my questioning him, the controller reported a heavy precipitation echo both to the left and right of my flight path, but added that a direct course to the vortac looked good. I have found that controllers are usually both willing and able to help with weather information as long as the pilot does not give the impression that he is trying to transfer the decision-making responsibility from the cockpit to the ground. An acceptable question is: "Do you show any weather on your scope?" If the answer to that is "Yes," an acceptable follow-up is: "Does my present heading look okay to avoid it?" If it doesn't, he'll say so, and you can work from there.

The pilot should always remember that there are definite limitations to using traffic-control radar systems for weather information. These systems are usually operated with a feature known as circular polarization, which maximizes signal returns from airplanes and minimizes weather returns. With CP, only heavy precipitation paints on the controller's scope, and there is no way to identify the cores of storms. Despite these shortcomings, this information on precipitation is better than no information at all.

I requested a descent from 13,000 because it appeared that I'd be moving into the area where smoother air would be low rather than high; this was approved, and I started down. There were higher clouds above with cumulus types building from below. Tops were about 9,000, which was below the freezing level.

When I switched to the approach controller, I rechecked the weather. He confirmed what the center controller had said, but he added that there was a large area of precipitation about 10 miles west of the airport, and it was moving eastward.

My plan was to shoot a VOR approach to my home base, which is five miles northwest of the terminal airport. The ceiling at the terminal was 700 feet; at home, where the elevation is 200 feet higher, the ceiling would be 500—right at minimums. The weather that was moving in gave me an incentive to make the first approach a good one. As I descended and maneuvered through some rain showers and light turbulence, the controller advised that the visibility had dropped from two miles to one. Then he called again with word that it was down to three-fourths. With word that the ceiling was 500 feet and the visibility one-half, the decision was easy to make: I would shoot an ILS at the big airport. Just a few days before, I had shot the VOR approach to home late on a misty afternoon and missed it because of weather. The big airport had almost folded to below IFR minimums before I could complete the ILS approach there. When weather begins deteriorating rapidly, a pilot has to think fast. I put it straight down the chute without delay, and the approach lights were soon in view.

The last part of the ride had been smooth—only a couple of light showers and bumpy moments. When the heavy rain did come, about 20 minutes after I landed, there was no lightning or thunder with it.

The airplane had been a perfect weather classroom as it moved twice through a cold front and twice through a warm front. The weather cooperated, too, displaying a good picture of a wide range of weather characteristics without producing anything forbidding enough to make the flight uncomfortable or to cause a delay or diversion.

It had been an educational Friday, one that brought into play many of the elements of weather wisdom that a pilot must be able to call forth.

⑧ HOW TO HANDLE WINTER FLYING

by Thomas H. Block

WHEN WINTER'S HERE and Mother Nature's cold shoulder can't be shrugged off, even the clearest of winter skies need some special flight-planning.

When the thermometer has sunk to basement levels, even the simple act of preflighting isn't what it was a few months ago. A heated hangar is the most effective way of forgetting the hostile environment outside, and a warm place becomes nearly a necessity if any unusual poking, prodding, or troubleshooting is called for. (Beware, however, of dragging a snow-covered airplane indoors just long enough to melt the snow, then launching into a subzero sky that will refreeze everything that has puddled—some of it in awkward places.) If it's cold but calm and sunny, the ramp may provide more comfort than you'd find in a cold-soaked T hangar, but wherever you find it, the preflight goal is reasonable comfort. Without it, the exterior inspection becomes more a test of the airman's mettle than the airplane's.

Wearing proper clothing during walk-arounds will extend a pilot's tolerances. A snowmobiler's insulated one-piece suit is easy to get into and out of, even when pulled over your favorite three-piece pinstripe, and it's the perfect protection not only from arctic blasts but from the oil, gas, and soot that a clean white sleeve inevitably finds quicker than a dowser's stick after an artesian well.

Getting an airplane ready for flight takes longer in January than it does in June, so give yourself lots of time. The ideal solution is to arrive in your thermal outfit, mittens, and insulated boots 30 minutes or so before the passengers, and to have the aircraft warmed and ready. The next best arrangement is to leave everyone inside operations while you do the preflight.

The worst situation is to attempt to get a cold airplane fit for human habitation while the cargo is dancing around the ramp complaining about cold feet and frostbitten ears. The rush to get going can all too easily cause errors and oversights.

Conspiring with the need to hurry is the fact that cold airplanes are in no rush to go anywhere. Metal surfaces are sometimes difficult to move, plastic and rubber fittings are brittle, and lubricants flow like lava. Gas drains are particularly prone to staying shut when you first try to open them and then sticking open when you want them to close. Cockpit-actuated gas drains should be tugged on carefully and then inspected from the outside to see that they've shut. If the drain is frozen and you have no way to thaw it, a decision to go becomes a calculated risk. You'll obviously be flying with water in the fuel tank, but is it an amount that could be crucial, should it choose to thaw at altitude? Going without draining means you've decided to depend on your guess as to how much water is in there.

A few drops of water can immobilize all sorts of aircraft systems. Throttle, mixture, and carburetor-heat cables; door latches; flap tracks; and retractable landing-gear mechanisms are the sort of components upon which ice loves to get a grip. Sometimes the frozen equipment can be broken loose by brute force, but you stand an even chance of breaking it both loose and into pieces. Deicing fluid can be sprayed on areas such as the landing gear, flap tracks, and door hinges, but heat is often the only answer. Deicing fluid is available in quantity at most cold-weather FBOs, and small cans of it can be bought at automotive-supply stores.

It will take 45 or more minutes inside a warm hangar to thaw an ice-laden throttle or mixture cable, and even when it's finally working again, your problem has yet to be solved. It's easy to forget, while standing in a warm hangar, that water refreezes, that ice adheres like superglue, and that the most effective way to distribute it to every nook and cranny on an airplane is to first melt it and then take it back outdoors. Unless everything you thawed is also *dry*, there is a potential for real problems. Wings, windshields and props can be wiped clean, but many other areas—tail cones and control cables, for example—must rely solely on evaporation. If refreezing has been a problem, have the airplane spend a few days in a heated hangar or treat it to a Florida vacation.

Leaving ice or snow on a wing or control surface is unhealthy because it alters the shape from NACA so-and-so to God knows what. Even a light dusting is dangerous because it will affect the wing's boundary-layer airflow more than you might imagine.

Removing the ice and snow is a matter of time, patience, money, and local conditions. Dry snow is easily swept off with a broom, while an overall coating of freezing rain calls for a heated hangar and lots of time, or a thorough soaking in deicing fluid and lots of money. If you decide to scrape or beat the ice off the airplane, remember that you are also doing the same number on the paint.

Starting procedures for cold-soaked engines vary, but most north-country pilots seem to agree on using lots of prime and being quick enough to catch the engine on the first cough. Rather than draining the socks off the battery, it's best to admit that evil spirits have intervened to keep the powerplant from coming alive, and if you don't succeed after a reasonable time, give it up before you've also ruined a starter and scheduled yourself for a battery recharge. Hike back to operations and borrow or rent an engine preheater or clear out a spot inside the heated hangar.

While heated hangars and portable preheaters get the job done quickest, anything that *keeps* the engine oil from coagulating is worth the effort. Dipstick heaters, lightbulbs, and heating pads will work as remedies for cold-soaking. By planning ahead and plugging in the night before, even a few watts will go a long way toward keeping the oil limber. Some people recommend pulling a cold engine through by hand before you start it, but others point out that coagulated oil won't work its way around the engine anyway. Preheating is the best answer; other techniques seem mostly wishful thinking.

Jet pilots have a handy solution to their winter engine problems. While turbine powerplants aren't especially prone to the cold-weather blues, the springs, slides, and genies in the fuel-metering and miscellaneous controls will sometimes act sluggish when air temperatures are very low. This often results in a hung start: the engine will ignite but won't accelerate beyond some low percentage. The solution is to drink a cup of coffee before trying again to give the heat from the first light-off a chance to soak through the engine's parts. Chances are good that the second or third attempt will be successful. On a piston engine, unfortunately, a botched first start will sometimes cause the moisture in the air sucked into the cylinders to ice over the sparkplug gaps, rendering them useless until they're thawed.

Winter is very rough on engines. Turbines don't suffer much—unless they inhale chunks of ice tossed up by the wheels or from the intake cowl—but if reciprocating engines could talk, they would begin to cry in late November. Temperature change is the real enemy of piston powerplants, and careful warmups and cool-downs will provide them with maximum life and minimum problems.

The first minute of a cold engine's operation is critical. Too many frostbitten pilots will try to help an engine through its warmup by running it too fast—which seems like a good idea while you're praying for more cabin heat—but running at anything more than a vigorous idle until all the oil has been thawed and the metal parts have felt the first flush of heat is akin to jumping out of bed and running the Boston Marathon. The cold-weather kits supplied by the manufacturers are important additions. The baffles and plates are designed to solve known problems, and if you haven't installed them, you stand a chance of discovering precisely why the kits were recommended.

Cycling a controllable-pitch propeller is not intended to build the prop's muscles but rather to circulate cold oil. On a big radial-engine powerplant, with its large oil reservoir and more accurate gauging, the engine oil temperature will fall off the shelf when you first squeeze what's in the cold prop dome back into the crankcase. Run the props through a few times to be sure that circulation is up to snuff. Much has been said about the evils of supercooling an engine with a prolonged idle-power descent, and all of it still applies: always come down with enough power to keep cylinder-head temps in the green, even if it means flattening your descent or doing a big 360, and make sure the oil and head-temperature gauges are well off the pegs *before* you begin the takeoff.

Gyros and other moving parts in the cockpit were not designed for the deep cold, but more often than not, things on the panel that aren't working properly when first turned on will come up to snuff by the time the warmup is completed. Blasting off into a gray unknown with a half-tumbled horizon is a complete case of the stupids, however: sometimes things don't work properly because they're broken.

If there's ever a time to check the controls for full travel, it's just before a winter takeoff. Rudder, aileron, and elevator must be moved fully to their stops to be certain that ice or cold haven't inhibited them. It is not uncommon to be able to move an elevator or aileron only a few inches.

Retractable landing gear can be particularly troublesome. Slush on the runway is the biggest criminal, since it packs nicely in the wells and freezes solid after you've climbed to an even colder altitude. Leaving the gear down awhile after liftoff, or recycling after climbing above the airport's obstructions, will allow the slipstream to blow away your potential problems. If the feet have been frozen into the well, spend some time at a low and warmer altitude to loosen the ice's grip.

Searching for and lingering in warmer air takes fuel, and unless you've added some winter reserve to your fuel planning, you may be in trouble. Everything drags on, in fact, when it's wintertime. Darkness lingers forever, and so will snowplowing crews, or so it will seem as you watch your fuel gauges sink toward E. Sudden snowshowers, broken ground equipment, and even incapacitated airplanes on the runway are more likely at this time of the year. The possibilities for delays are nearly limitless, so extra gas must be trucked along to cover all bets.

To say that winter weather is changeable is to say the obvious, yet many of the pertinent changes are not easy to see. Winds aloft can be double or triple what has been predicted, and for almost anyone flying at slower than jet speeds, having an extra half a hundred against the nose wll effectively throw fuel rules of thumb out the window. Not only is careful planning a must, but so is the need for en-route monitoring of actual groundspeed and fuel consumption.

Low-level flying can be a touchy business during the winter. Except for those dry and blue-skied days, forecasts will often be trumpeting the chances for local snowshowers. Nothing can turn VFR into IFR quicker than whirling snowflakes, and the urge to go lower to maintain ground contact has been the last urge of many a pilot. A snowshower is a deceiving devil, and if you aren't instrument-rated, stay clear of them and be prepared to do an immediate 180 if you find yourself engulfed by one.

Wintertime is instrument time, but though clouds are often present, the tops are usually low; and airframe icing—the nemesis of cold-weather piloting— can generally be dealt with by planning and good judgment. Snow seldom causes airframe icing, and most other sources of wing, tail, and prop accumulations can be handled by a timely change in altitude. Prefrontal icing conditions may sometimes be severe enough to create real problems, but day-to-day instrument flying will usually be into conditions containing enough low tops, nonicing altitudes, and multiple layers to make the mission workable.

Knowing how to dock a sailboat is good practice for parking on icy ramps or negotiating narrow, snow-lined, crosswind taxiways. Aerodynamic control means more during winter taxiing than it does during summer, and a properly placed aileron or rudder will sometimes make all the difference. The best overall precaution when things are slick is to go slow. An icy spot, a brisk wind, and a dinged prop tip are a common winter trilogy.

Takeoffs and landings suffer most, of course, when the airport is slick. Contact with the runway guarantees nothing when braking action is nil, since the tires will act as if they've never been put on the pavement. The pilot should act in a similar fashion by continuing to work the flight controls.

Instrument approaches through blowing snow, crosswind landings on icy runways, never-ending decisions about how much fuel and what alternates to pick are the meat and vegetables thrown into a pilot's winter stew. Once we can digest these ingredients, we'll appreciate how much truth there is in one particular bromide: flying in the summer takes guts and seat belts, while flying in the winter calls for patience and skill.

⑨ WHERE THE ICE IS

by Richard L. Collins

ICE IS A BASIC item that we use to advantage in cooling drinks. Many a pilot has spent reflective moments watching those simple ice cubes swim around in a sea of Jack Daniel's and wondered how something that can be so beneficial to a little drink can be so hostile to an airplane. Ice has always been one of our most difficult aviation weather problems and one for which time doesn't seem to have produced many solutions. The general-aviation pilot flying a piston-powered airplane today faces the same problems, and probably thinks the same thoughts, as did the airline pilot of the late 1930s—back when airliners flew at altitudes most of us use today.

Why is the relationship between airplanes and ice so complicated? One reason might be that ice is difficult to forecast accurately. The weatherman may know that there are both below-freezing temperatures and clouds aloft, so he forecasts icing in the clouds and in the precipitation above the freezing level. We pilots want to know, with a fair amount of precision, not only what types of clouds exist over the route but the temperatures inside them. The forecaster may make an educated guess that misses by only a few miles—and we may fly down the route, collecting ice and cursing the forecaster, even though he was off only slightly. Or we might let the forecaster scare us into staying home, only to run into a buddy the next day who flew and had no problem. We might even decide to go, then get up there and find little or no ice, concluding thereby that forecasters are prone to cry wolf.

Ice behaves illogically, anyway. If the air is cold, the water in it should already be frozen, so that the airplane could brush past it without bother. That's not the case, though, and to explain airframe ice in ground school, they teach us about supercooled water droplets. The air temperature may be below freezing, but the undisturbed droplets are still liquid, and they stay that way until some unfriendly airfoil knifes through the cloud and splatters them. In retaliation for the splattering, the droplets freeze.

Once the supercooled-water-droplets theory is accepted, ice almost seems logical. It all begins with clouds. Stratus clouds are stable—that is, they have no vertical development—and they are made up of tiny water droplets. Big droplets would fall out of the cloud as rain or snow. Temperature has a direct effect on the droplets. You find the greatest number of supercooled water droplets in a stratus cloud when the temperature is close to freezing. As the temperature goes lower, there are fewer droplets and more ice crystals. When a wing barges through a stratus cloud, some of the droplets impact on the leading edge. If it's an icy day, these droplets tend to freeze quickly on contact and become a rather rough ice coating on leading-edge surfaces. It looks like frozen milk and is called rime ice.

No ice is nice, but the ice found in pure stratus cloud is usually easy to escape. Ice accumulation in stratus is not likely to be rapid, and if a pilot gets out quickly, he's likely to have little trouble. Stratus-cloud layers are usually not thick, and an altitude change of a few thousand feet may move the airplane to ice-free air. The fact that there is a direct relationship between temperature and icing in stratus cloud is an aid in combating ice, too. Climbing to cooler temperatures aloft will cut the ice accumulation. At $-15°$ C., pure stratus cloud is almost wholly ice; at that temperature, there's no airframe icing.

Cumulus clouds are a different story. The droplets in them are larger, and the vertical development of the cloud propels the droplets upward. This nullifies the direct relationship between the temperature and the amount of ice that might be encountered in a cloud. In fact, you may find more ice in the middle or top of a cumulus cloud, where the temperatures are very low, than near the base, where the temperature is just around freezing. The icing in cumulus probably won't coat the airplane with the slow but steady buildup that is typical of a stratus cloud. Instead, it'll come on you in blasts as you fly through the most saturated areas of the cloud. The bigger droplets splatter on the leading edges of the wings and other surfaces, with few of them being deflected by the advancing wing. Freezing isn't as instantaneous as with the smaller droplets, and the result is a smoother and clearer coating of what is appropriately called clear ice. The only virtue of clear ice lies in its being smoother than rime ice, meaning that it creates somewhat less drag.

The options are limited when you face an ice problem in cumulus clouds: either stay clear of cloud or fly where the temperature is above freezing.

Trying to distinguish between stratus and cumulus in flight can be difficult because you often see a mixture of types, appropriately called stratocumulus, at the lower levels. The icing found in such clouds is called mixed icing because it is a mixture of rime ice and clear ice.

The degree of turbulence is often a clue to the types of clouds and ice you can expect. If the air is smooth, the cloud is more stratus than cumulus, and climbing will probably either get the airplane to cold air or on top of the clouds. If the air is a bit turbulent, then cumulus content prevails, and more severe ice problems can be expected. The tops will be higher, the droplets will be bigger, and it may be hard to find air that's cold enough to help with the problem. If the air is very turbulent, the problem may become severe. One thing to note: when you are climbing in clouds with any vertical development, icing can be most severe in the tops just before you break out. Don't tarry there.

Mountainous terrain can also cause ice problems. Ice accumulation reduces the service ceiling of the airplane, and in the high mountains, it might quickly drop you to a level below the ridges. The mountains themselves tend to create more severe icing than would be found in similar air over flat country. The vertical currents found atop and to the windward side of ridges help produce those big, juicy, supercooled water droplets and to propel them upward. Many a pilot has watched what had been a perfectly good, ice-free altitude over flat country turn into an icy nightmare once he trespassed into mountainous terrain.

Probably the most feared of all icing conditions and the creator of those legendary collections of an inch of ice a minute is freezing rain. Freezing rain falls when warm air overspreads cold. The raindrops are supercooled as they fall into and through the cold air, but they do not freeze until they hit something—be it a tree or an airplane in flight.

After encountering freezing rain, the solution—says the textbook—is to climb to the warmer air from which the rain is falling or to descend if weather reports indicate that warm air can be found near the surface. Both these measures work, but when you are VFR, clouds may preclude climbing; when you are IFR, the controller may not have instant relief to offer. A better solution is to use your knowledge of weather basics to stay out of freezing-rain situations in the first place. Freezing rain can be found immediately ahead of warm fronts and where occlusions present warm air aloft and cold air at the surface. Be wary of any precipitation when flying in such an area, and flee at the first sign of a raindrop.

A good knowedge of a front's position and alignment is essential to a pilot trying to find the solution to any icing problem. While ice can be found far from a front, it can be found over a wider altitude range and greater area in the immediate frontal zone. Also, when you are penetrating a front, conditions will change as the flight progresses—and not always for the better. For example, a pilot flying perpendicular to a warm front, going from the warm side to the cold, may find himself droning comfortably along with an outside air temperature of 4° C.—but he shouldn't count on staying comfortable. The temperature is going to drop as he moves toward the cold air, and it may drop drastically and abruptly. A pilot who is flying parallel to a front can look for conditions to change more gradually if they change at all.

The weather-wise pilot has a lot of information other than the forecast to use when looking for a possible ice hazard along a route. Temperatures are a mainstay. Surface temperatures are important because they tell you whether or not you can find relief from a problem by moving lower. If it is below freezing at the surface, none of the ice will slide off the airplane before the landing. The pilot will have to make it to the hangar with the ice aboard. To look at the bright side, if it is very cold at the surface, it can also be very cold aloft.

The winds-aloft forecasts include temperatures aloft. These are subject to the same errors as any forecast but are still helpful. It's not uncommon in the wintertime to have cold air near the surface, with accompanying icing conditions, and warmer air aloft that will melt any ice accumulated during the climb. The temperature forecast should tell you this; it will also tell you what altitude will put you above icing conditions in stratus-type clouds. Actual temperatures aloft are available from some stations; these are valuable, but time changes all things, including the temperature aloft, so the age of the observation is important, as is the proximity of the station to any front that might have altered the temperature considerably.

Remember, frontal positions are important because they identify the potential ice boxes. Terrain is also important. If mountains—even relatively small ones such as the Appalachians or Ozarks—underlie the route of flight and there is wind of any consequence forecast at the ridge level, beware.

The effect ice has on an airplane is an integral part of the ice-airplane relationship. Any form of ice effectively changes the shape of the airfoil. Rime ice, which collects mostly at the leading edge, is the worst offender; drag increases greatly as it accumulates. Clear ice tends to conform to the shape of the airfoil, creating less drag than rime ice but adding more weight to the airplane as it collects. One inch of clear airframe ice on the leading edges of the wings and other forward surfaces of a single-engine airplane might weigh 300 pounds or more. Even with deicer boots, ice will collect on unprotected areas, increase drag, and add weight.

Ice on the propeller can be a spectacular problem. It affects propeller efficiency, but worse, it creates an imbalance when the vibrating prop sheds it unevenly. Just a few ounces more on one blade than on the other makes the engine and prop shake. Fortunately, that is easy to avoid. If the leading edges of the prop are filed smooth and coated with some of the available spray-can ice repellents, the prop will generally stay clear of ice. Unfortunately, the spray-can magic doesn't work on the wings; vibration is what enables it to work on the propeller, and wings don't vibrate.

Ice can also do bad things to antennas. If the airplane is bedecked with the wire type, ice can silence the radios as it collects, induce vibration, and eventually break the antennas off the airplane. Improved antennas are available at avionics shops and are better bought before an icing encounter than after.

Engine induction-system icing is always an area of concern. Snow, especially the wet kind, can plug a pretty sizable hole in a hurry. There are many cases on record of total power failures in singles and twins in icing conditions, and the wise pilot should be familiar with the induction system on his airplane and the proper method of handling induction-system icing.

Ice changes the aerodynamics of the airplane, and certain ice-related accidents suggest that flaps not be used on an iced landing. Given just the right (or wrong) set of circumstances, an extension of flaps can cause a loss of elevator effectiveness. This means that the approach speed will have to be considerably higher—add some for the ice and some for the lack of flaps—and you should choose a long runway. Stall speeds also rise with ice accumulation. Low-altitude maneuvering is therefore particularly hazardous, and long straight-in approaches are the best.

Ice can quickly and totally overpower the defroster and blank out the windshield on an airplane. That's why fully deiced airplanes have windshield alcohol or heat. Some pilots report that putting a Jeppesen book atop the panel to deflect as much hot air as possible to the windshield helps. Doing this probably helps melt a little ice, but it could do the same to a plastic windshield if you have a defroster with very hot breath.

Ice can also create strange noises as it collects on the airplane—noises you never heard before. Some airplanes shriek; others buzz. It's almost as if there were a spook out there telling the pilot that the time has come for him to exhibit some weather wisdom and get the airplane out of the icing conditions. "Getting out of there" was the only solution to the ice problem in the days when the airplane was young; it is the only solution now, and it will probably continue to be the best solution when the airplane reaches its hundredth birthday in the year 2003.

10 ICE
by Don Jonz

THE THOUGHT of inflight structural icing inspires the crazies in a lot of airmen. In my opinion, most of it is a crock.

During the great Alaskan oil boom of 1968, I had the job of hustling a Beech Queen Air back and forth to the North Slope of Alaska. Late that December, an unusual siege of "icing conditions" clobbered northern Alaska. Surface temperatures at Fairbanks, albeit cold by civilized standards, warmed to a fiery 20° F. Snow fell and stuck as quick as you could shovel it off a wing. Nonetheless, a little reflection produced an interesting suspicion: this wet, tacky crud, born in the cold aloft, was probably dropping through a warmer level. On its trip down, perhaps near the surface, it recooled—but not enough to completely resolidify.

Any flight would therefore probably encounter a temperature rise during climb, which would aggravate ice adhesion. A pilot *might*, however, climb like a tiger through the worst of it, find a warmish altitude, and while away his time munching apples and peanuts. The situation begged a look-see.

The Queen Air was *very* lightly loaded, doused with deicing syrup, its radios set up for a retreat ILS back to Fairbanks, and the whole shebang launched into the night at maximum rate of climb, with my eye glued to the thermometer. Sure enough, temperature warmed—about 1° F. per 1,000 feet. Ice light on. Yes, as the experts suspected, ice started to build up on the wings and windshield. But—an important point—temperature rise was steady. Rate of climb stayed solid, and a good ILS was close at hand. Under these circumstances, the rate of ice accretion was acceptable.

At 9,000 feet, the thermometer peaked at 35° F. Icing stopped. I climbed to 11,000 feet to experiment; temperature cooled. Icing began again. Center gave the okay to slide back down and maintain 9,000, and there I sat for a couple of hundred miles. The ice disappeared. Fifty miles north of the Arctic Circle, clearance was received to climb to 11,000 feet for the shunt over the lunarlike Brooks Range. Again no ice. I unloaded and came back home flying the same profile.

During the next week, I flew two trips to the North Slope every day—about 60 hours of flying—never encountering more than moderate icing during climb and descent. Once, I heard a C-46 captain declare an emergency due to severe icing. He was at 12,000 feet, and, I presume, in the thick of things. The next day, an F-27 at 20,000 feet turned around north of Fairbanks because of severe icing. The Queen Air, the only light twin flying, zipped up and down like a yoyo. Folks still remember it as "The Week of Ice." I'm not so sure. For my money, the escapade was numerically safer than shooting rollers at the local patch on Sunday afternoon. Soup isn't crowded.

So much for war stories. The question is: "How do you *safely* learn ice flying?" The same way you learn a lot of things. Read. Reason. Take little jump-ins. Playing with ice is like playing with the devil: fun, but don't play unless you can cheat. If you are sneaky, smart, and careful, you can fly 350 days a year and disregard 99 percent of the nonsense you hear about icing.

As you explore the techniques of ice flying, remember these points.

There isn't an airplane alive that can handle prolonged heavy icing, 707s included. Boots, hot wings, spray cans, alcohol, and dudes that paint their wings black notwithstanding. Gadgetry occasionally buys a little more time, but a sophisticated wing degrades quickly under a load of ice. A thick chubby wing—the kind we don't have anymore—packs ice better. The slower your aircraft, though, the more aggravating icing will become because of the length of exposure. Aerodynamic heating and slipperiness at jet speeds don't give ice much of a chance to hang on.

Always keep your exposure to a minimum. If you want to ford a torrent, make your dash at the narrows.

In order of usefulness, deicing/anti-icing accouterments are: 1/Thermometer and timepiece. 2/Heated pitot. 3/An ice light (a flashlight will do); I like to switch on my landing lights occasionally to analyze precipitation texture. 4/Heated props; if the props are kept deiced, the rest of the machine will handle a respectable load of ice. 5/Heated windshield. 6/Boots; personally, I find them a pain and ineffective, but some pilots seem to like them.

Most bad icing in Alaska happens at temperatures between 0° F. and 32° F. I've had it stick at − 73° F., but the rate of accretion below 0° F. rarely causes concern. Snow and ice crystals are usually harmless, except when the temperature is in the slushy numbers close to 32° F. If snow does stick, it means there is a cooler region somewhere above. If practical, climb into it.

If the temperature at your altitude is between 0° F. and 32° F. and you're in the weather, you're apt to get icing whether it's January or July. I have always considered spring and fall most conducive to icing—too cold for rain, too warm for dry ice. But it depends on the part of the country you're in. The same goes for rime ice and clear ice. Most ice is a mixture.

Sublimation is the process by which ice evaporates from the solid state to the vapor state without passing through the water state. The food industry calls it freeze-drying. Experienced pilots with more class say ice "wears" or "burns" off. The faster your airspeed, the faster ice sublimates. The colder the outside temperature, the slower sublimation works. At 0° F. and 200 mph, it takes a two-hour flight to burn off a quarter inch of ice. At − 60° F., don't count on any help at all.

Stay away from the top 1,000 feet of clouds. There is more ice just under the roof than in the rest of the whole cloud. It blows my mind to see a grownup pilot level off at 12,000 feet, all his electrical gadgets hanging on the line, when 13,000 or 14,000 feet would put him in the clear—fast, comfortable, bright-eyed, and bushy-tailed at the end of his trip. It's hard to convince some people that the sun is always shining on top.

Keep an eye on the thermometer. It is your most valuable icing instrument. For example, make a mental note of the temperature at takeoff. Figure the approximate lapse rate during climb. Note the altitude when the temperature crosses 32° F. (to get an idea of the altitude to which you must retreat if you get into trouble). Notice those little temperature plateaus where life is in a wonderfully steady state. Catch the point where the bottom falls out of the mercury. Wonder at the squiggly temperature vacillations that announce wind shifts. Up here in American Siberia, we frequently see a household thermometer taped to a wing strut in view of the pilot. It works well, and the price is right.

My company, Pan Alaska Airways, used to operate a pair of Short Skyvans. It was interesting to dispatch both ships to the same destination at the same time. One pilot would return behind schedule, haggard, grim, primed with a hairy hangar story, dripping ice, his anti-icing fluid exhausted. "Instruments all the way," he'd say. The other pilot, whose alcohol tank rarely needed filling, was already home and showered. Feel the moral?

Atmospheric cooling, called temperature lapse rate, averages about 3½° F. cooler for every 1,000 feet during climb. The common temperature inversion befuddles this neat generalization. Temperature inversions are the primary cause of low-altitude haze layers. The top of the haze layer usually marks an abrupt temperature change. Temperature inversions in the Arctic can be extreme; I have seen it go from − 60° F. at the surface to 0° F. at 500 feet. These extreme inversions produce an interesting but harmless flight phenomenon known as instant allover frost. When climbing a brittle-cold airplane into a sharply defined layer of warm air, the machine—windows and all—frosts over. It dissipates as the airframe temperature equalizes with the outside air temperature.

The weight of ice is not what debilitates. It is the shape. Twenty strategically placed pounds of ice will sink a good light twin. Ice reduces rate of climb dramatically, pulls down cruising speed, and raises stall speed. Worse yet, *it*

changes stall characteristics. A rule of thumb says that an eighth of an inch of
rough ice or frost will reduce rate of climb by 50 percent and cruise by 10
percent.

I was jamming along in a Cessna 180 once when approach control gave me
a descent into freezing rain. Not being checked out on thermometers, I
complied. A good Cessna 180 ski-plane indicates around 130 mph and stalls
at 50. My ship was soon dragging its tail. Ice shrouded the windshield. It
glistened from the struts and skis and glomped over the wing. Since the tail
was low, ice formed curious layers on the wing undersides. "Wow, how
interesting!" I turned up the wick. Again. And again. Little pussyfoot affairs.
Soon I was descending at 100 mph at full throttle. Luckily, the runway was
straight ahead. I rounded out—guessing the altitude, as I couldn't see
through half an inch of windshield ice—when bang, at 90 mph she stalled.
That "landing" bared a lot of shorts in the left front seat. The stall had been
instantaneous and indescribably violent.

Don't accept holding patterns in icing levels. Basically, if the temperature is
between 0° F. and 32° F. and the assigned altitude will put you on the gauges,
give the controller an honest "No!" And hang onto your safe altitudes
unaware of the pilot's in-flight conditions. If ice is evident, insist on another
slot. It is your neck. Rarely does icing span more than 6,000 vertical feet in
thickness; but like all generalizations, this one will waste you if you lean on it
too much.

Flying between cloud layers is chancy. If you're not traveling too far, okay.
Typically, layers fuse together. When that happens, guess who gets crunched?
Unless the layers are distinct and conditions stable, you are better off going
under or over.

The rate of ice accretion versus time to go is a great index. Know when the
icing began. If ice hits you hard and heavy, such as freezing rain or airborne
slush, a 180-degree turn is your best ace. If rate of ice accretion is slow,
distance short, and there is a comfortable ceiling below, hang in there; you'll
be okay. It sounds like I'm hedging, but therein lies safety. A pocket full of aces
makes any game fun. Always have a mental picture of the weather ahead.
Keep reserve altitude; happiness is knowing you can go lower. Have gobs of
gas. Keep lots of approach plates up your sleeve. Don't fly into obviously bad
conditions. I've seen dudes with apparent death wishes barge right into a real
mess when all they had to do was look out the windshield. If you see dark
precipitation areas ahead at critical temperatures, don't pole your nose in.

Before takeoff, determine the height of the tops. Look out the window. You
are your best forecaster. The lack of reliable top reports is the single biggest
void in aviation weather, a chain without a master link, probably because
high-flying airlines couldn't care less, which means the FAA couldn't care at
all. But here is what I do. First, I call the weather bureau for their guess. Then, I

call center or approach control for a pilot report. If that doesn't work, I call one of the airline dispatch offices. They're usually glad to query one of their pilots. Top reports indicating a uniform layer makes ice flying a piece of cake.

When picking your way through icing, use commonsense radar. Climb through holes. You'll be surprised how often this can be done if you give it serious study. There is nothing wrong with flying in VFR conditions on an IFR flight plan. Simply frame your ATC requests for slots that will put you in sunshine. It is a concept not widely practiced. Too bad. Weather accidents are easily preventable. Stratus and cumulus clouds are analogous to plains and mountains. Sometimes they even follow the contours of the land, like a papier-mâché mask. Cloud formations have plateaus, bluffs, ranges, rifts. Practice flying this high topography. Usually, you can spot a nearby fissure, or at least a canyon. Get a clearance from the center and zip through. Some years I've flown 365 days, a lot of it IFR, in so-called weather, yet 90 percent in the clear. Ergo, no ice. If holes can't be had, keep in mind that lighter areas contain less moisture. Less turbulence, too. Next to the thermometer, commonsense radar is your best instrument. Be wary of the pilots that say it can't be done. *They* can't.

Floatplanes, ski-planes, and generally dirty airframes make bum ice haulers; the ice collects on spreader bars, struts, wires, protruberances. Clean lines, uncluttered profiles, smooth paint, and recent waxing make ice work harder to hang on.

If you don't like ice, stay the hell out of IFR conditions. You *have* to penetrate fog, clouds, wet snow, freezing rain, or general mugginess before you can get structural ice. I'm forever surprised to find VFR pilots and new IFR pilots who haven't been taught this simple rule. Ice doesn't jump out of the clear. You must be IFR-rated, have an IFR clearance, and be IFR-current when you start playing with IFR weather. If you receive your instrument rating in sunshine, better get a few hours of wet dual in the hands of an experienced instructor. After that, you're probably ready for baptism.

Here's how to begin: 1/Review your manual on how and when to apply carb heat or alternate air. 2/Remove or smooth all frost or ice forms from props and wings (*top and bottom*); be especially attentive to the forward half of the wing. 3/Load well below gross weight; rate-of-climb reserve is one of your better aces. 4/Get on top through a hole. 5/Be certain there are a couple of thousand feet between the ceiling and the ground. 6/Wait until you near your approach fix, get an IFR clearance, and turn the pitot heat on; if you have them, activate prop and windshield anti-icing. 7/Jump in. 8/Preface all changes in altitude and heading with a clearance from ATC. They are usually glad to oblige except in large terminal areas, which is reason enough to practice off the beaten path. 9/After maneuvering a bit to get the hang of things, request an approach clearance. Carry reserve airspeed on final. Put your windshield defroster on hot to melt a peek hole. If it gets dicey, you may have to look out the side window. 10/After landing, inspect the ice buildup. Analyze the nooks and crannies where ice stuck. Pay particular attention to the props, sharp radii on the windshield, and the antennas. Note the length of time it took to build up. Critique is half the fun and three-fourths the education. I think you'll agree a little ice trip gives a fine sense of achievement.

The crux of flying "icing conditions" is that seldom does one need to hassle *actual* ice. Except for really far-out occasions (in which case you'll have a warning from the weatherman), ice isn't found very low or very high. It is an effort to sympathize with pilots who insist on flying oceans between 2,000 feet and 12,000 feet when the highest obstacle is a mast. Yet they do. Ice up. Lose airspeed. Burn more fuel. Is there security in being able to glide if something goes wrong? Glide to what? As a nonsked pilot, I used to shuttle ice over the Atlantic—most of it unnecessarily. Ignorance was and still is no excuse.

If you violated all the rules of common sense, punched into icing too long, and collected a severe load, what do you do? Well, your whole day stands a fair chance of getting scrambled. After getting a clearance, your escape routes are: Hasten a 180-degree turn. Descend. (Sitting fat and happy at freezing level waiting for something to happen is unfair to insurance adjusters.) Or climb. This means forward on the knobs. Smartly. I've seen guys drag their feet with too little power and come tumbling out of the sky. If you're going to lay the whip to her, do a good job.

I prefer climbing out of trouble, but it is a critical decision. First, it is a decision that must be made early. It is fraught with the probability that conditions will degrade near the top, not to mention that you will have a hard time getting up there. Supercharger and oxygen are almost a must for this type of remedial flying. If your bid for blue sky is unsuccessful, you have a monkey on your back. The trip down won't be pleasant.

I remember an ill-fated Aero Commander in Southern California several years ago that picked up 12 inches of ice in 12 minutes. Needless to say, he landed short. Ice can be, and is, consistently deadly. An aircraft under severe ice loads suffers on both ends of its speed range. Stall goes up. Cruise goes down. When the two meet, the world comes to an end.

If you live west of Denver, use extreme caution. A lot of weather pushes in from the West Coast. It continually works up the Continental Divide. Tops are usually wet and high. East of Denver, heaven has dumped its load and is starting downhill. Tops are easier to handle.

Don't charge off into ice when your destination is doing poorly and going downhill. An approach with a load of ice to ILS minimums is dumb, a go-around problematical.

In short, be brave. Defensively.

11 IN THE DARK
by Richard L. Collins

A COUPLE OF acquaintances who oversee groups of people traveling around the country in single-engine airplanes made the observations that have prompted this essay. With some determination, one of them said that he lets his people fly at night and fly IFR but draws the line on night IFR. Forbidden. Period. The other, who is active in aviation but is not a pilot, remarked that he likes to ride along on night flights and on IFR flights, but that night IFR makes him quite uncomfortable.

How did night IFR get such a bad reputation? Instead of representing something ominous, it is to me a matter of self-preservation. I fly IFR at night because it is the best way to minimize the risks of flying in the dark.

Not only after this subject was raised in my mind, a day's activity underscored my thinking. A round-trip flight, 280 nm each way, was on tap. I'd be departing late in the afternoon, attending a dinner meeting, and then flying home that night.

The weather for the daylight leg of the trip posed a dilemma that is frequently confronted in a Skyhawk. There was some IFR weather around the destination and it seemed likely to persist past the estimated time of arrival. The normal call would be for IFR all the way. But there was a 30-knot cross-headwind at 3,000 feet and a 45-knot headwind at 6,000 feet, and it is not possible to get 3,000 feet as an en-route cruising level in the area. That prompted a VFR flight. There were some areas of rain, but the ceiling remained comfortably high to a point within 40 miles of the destination. I obtained an IFR clearance there and concluded the flight in a routine manner. I contemplated the weather while driving to the meeting—I certainly had no intention of flying back through it VFR in darkness.

Later, when I checked the weather for the flight home, the briefer indicated that it would be okay for VFR. The reported ceiling was 3,000 feet or better and the visibilities were good. A chance of thunderstorms was forecast along the route, but no storms were mentioned on the weather radar summary charts. I still wasn't interested in VFR. The stationary front that had produced the scattered rain late in the afternoon was still showing on the map, and fronts can be notorious between-station weather producers. Also, it has been my experience that weather observations taken at night aren't always accurate. The darkness problem is reflected in the oft-stated meteorological notion that clouds tend either to form or to dissipate at dawn, when light is put back on the subject.

For the return trip, the winds aloft had, naturally, shifted and there would be more cross- than tailwind. The IFR flight plan I filed reflected a much smaller tailwind component than the headwind component I had experienced during the afternoon.

The departure point was an uncontrolled field, under a fold of a TCA's fat. There was no way to get a clearance on the ground by radio, so I would either have to get it by phone or take off VFR. The reported weather suggested the latter course, but as I looked at the black sky, the difference between day and night operational considerations became obvious. By day, I can stand beside the airplane, estimate the visibility and cloud height, and say: "Yes, Richard, you can take off VFR and then get that IFR clearance." That night, I could only look up at the dark sky and say: "Richard, the Feds said the ceiling is 3,000 feet at an airport 20 miles away from this one." As I started the airplane, I wished that I had taken the trouble to get the clearance over the phone. If a layer had moved in at 500 feet, things could have become rather restless in the cockpit as I worked into the combination act of maintaining VFR and getting a clearance.

The weather report was close enough. My IFR clearance was in hand well before I reached the cloud bases. I was initially cleared to 4,000 feet, an altitude that would be in cloud. The air was slightly turbulent—a result of the strong winds. Static on the ADF indicated thunderstorm activity some-where—in the other direction, I hoped. Reported storms well to the west were the probable cause of the static.

The next clearance, to 7,000 feet, resulted in a real test. A full moon came up on top at about the same time I did. Flying above the clouds at night with a full moon is one of life's more beautiful gifts.

The frequencies become less congested as the hour grows late. Things seem more informal, and as I requested approval to contact a flight service station for new weather, the center controller volunteered to procure the information for me. He also checked with the next center to see if they were showing any precipitation on up the line.

The frontal zone was still there, characterized by an increase in the height of the cloud tops. I had been well on top at 7,000. Then I was skimming the tops for a while. Finally, the clouds enveloped the airplane.

It is always interesting to leave the strobes and the beacon on for a moment when flying into clouds to see how much spatial disorientation they can produce. I've never felt disoriented by them, and this night was no exception. Flashing lights are not the type of things to leave on intentionally, though, and

I doused them after a moment or two. They are quite a distraction.

The clouds were a little bumpy, and there was some water in them. Light rain could be heard gently splattering on the windshield.

My thoughts turned to VFR. I wondered if the people who frown on night IFR would feel better flying VFR on such an evening as this. Would they have gone up on top? It would have been possible to top the weather at or above 7,000 for a while, but the clouds eventually moved higher. How high? No pilot reports were available, but I'd occasionally get a glimpse of the moonlit side of a cloud when I was passing through a break, and I guessed the tops were well above 10,000 feet. I had no intention of going higher than 7,000 anyway, as there was enough oxygen in my bottle only for part-time use, and 7,000 is my limit at night without oxygen.

What about VFR beneath the clouds? It would be awfully dark for visual flying down there. There wouldn't be any way to see clouds or rain. What happens if you are flying along VFR at night and rain starts beating on the windshield? Is that VFR? What about the surprise that comes when the strobes start illuminating the innards of a cloud? No, I was quite pleased with being IFR at 7,000 feet on this evening. It was the best possible deal.

I have instrument post lights installed in my Hawk, and I feel the lighting is good. I have read that turning the pane lights to full bright helps your instrument flying when distractions arise at night, and some turbulence that developed on my flight home prompted a test of this theory. With the lights turned up bright, the panel did attract my undivided attention. My scan seemed better, and the turbulence was less bothersome.

Closer to home, the descent and approach provided another good illustration of why night IFR is so far ahead of night VFR. The ceiling at the nearest reporting station was advertised as 8,000 feet, but as I flew level at 3,000 feet, waiting to intercept the final approach course, the airplane was still in cloud. The ceiling in the area turned out to be about 2,500 feet. Some of the hilly country out to the west would be obscured at that value, and it would have been a bad night for VFR. It was a snap IFR. I won't mention the thumpy landing, which was complete with wind shear and crosswind. It would have been equally bad after a VFR flight.

There are a number of considerations that must be recognized if night IFR is to work smoothly. Weather is a big one. Where the night VFR pilot is bedeviled by clouds and terrain that can't be seen in the dark, the night IFR pilot has to deal with departure and arrival ceiling-and-visibility conditions that are less predictable, and that are reported with less accuracy. In this flight, that was no problem—an anticipated 8,000-foot overcast that turns out to be an actual 2,500-foot overcast is just no big deal to an IFR pilot. But if it had been a 250-foot overcast instead of an 800-foot overcast, the story would have been different.

I had a problem with the weather as reported in my old Cherokee Six one evening. The reported ceiling only five miles from the airport I was using was good, and I started into the nonprecision instrument approach to home base with some confidence that I would soon be on the ground. However, when I was level at the MDA, 500 feet above the ground, there were still clouds beneath my Cherokee. An occasional light would slide by beneath but nothing else. When I was one mile from the airport, I called a missed approach and asked for an ILS to the nearby large airport.

The controller thought it strange that I would give up before flying over the field and said I could have the ILS or another shot at that VOR approach to the small airport. I've read too many accident reports, though. There are a lot of IFR problems at night, and a very high percentage of them involves a descent into the ground short of an airport. If one wishes to be led astray, those lights moving by below can be awfully tempting. Down another 100 feet, mister? It's the devil himself talking. Descend a little more? Trouble is, that descent might continue until something hard intervenes. The key to success is not to leave the minimum descent altitude on a nonprecision approach until and unless the lights of the runway are clearly visible and things look ripe for a normal landing.

Even when you're flying clear of clouds at night, the game is harder to play and the stakes are higher than in the daytime. This is especially true on a circling approach. By day, avoiding the ground is simple—peripheral vision will handle the task as we look at the runway. At night, the altimeter is the only thing that will keep us out of the trees when flying over unlighted areas.

One other point about weather versus night flying. I don't think I have ever seen anything documenting the fact that weather is generally worse at night, but I know it is true. A relatively high percentage of the times an airport is below IFR minimums comes during the hours of darkness. Scud seems to form more often at night, too, and a piece of scud in the wrong place is as effective as a ceiling.

The vagaries of nighttime weather leave no question that IFR during the hours of darkness is more difficult than during the day. But it sure beats VFR. If you are going to fly cross-country at night in anything other than clear weather, it's a cinch that IFR is the way to go.

12 WHERE'S THE SILVER LINING?

by Richard L. Collins

THUNDERSTORMS are fascinating to pilots all year long. They may be scarce in the winter, but spring and fall provide a generous supply embedded in warm fronts and in majestic squall lines ahead of cold fronts. These frontal areas and squall lines are usually well forecast and clearly defined. Summer thunderstorms are different, though. They often occur *within* an air mass, not along or to one side of a known line between two dissimilar air masses, so accurate forecasting of these storms is difficult. We see more of them, too, and "chance of thunderstorms" is frequently included in summer forecasts. They are difficult to predict, though; some days they form, other days they don't, and it is sheer guesswork to try to pinpoint exact locations that might be affected by them.

Summer thunderstorms can be an operational consideration for VFR as well as IFR flights. When conditions are ripe, they form in clusters and short lines. Hail can fall, very strong surface winds often develop—sometimes even tornadoes—and the turbulence found in and around these storms will kink your wings. For practical purposes, these summer thunderstorms demand as much respect as those found in a squall line.

Only a couple of simple ingredients—conditionally unstable warm air and moisture—are necessary for the formation of summer storms. What is conditionally unstable warm air? Air is said to be unstable when it cools with altitude more rapidly than is normal. Summer mornings usually dawn clear and cool at the surface, where the temperature is probably around 70° F. The air is stable, and there is no turbulence. As the sun heats the surface, the temperature increases rapidly, though the temperature aloft changes little. Warm air is less dense than cold, so the air at the surface begins rising as it is heated. As we fly on summer mornings, we note that the characteristic turbulence of such a day starts when surface warming begins in earnest. Nice little bubbles of warm air are making their way upward. Heating was the condition that would make the air unstable, and heating has occurred.

Moisture is also necessary for the production of a storm. The elementary rules of meteorology tell us that clouds are formed when air containing water is cooled below the dew point. As moist air is lifted by solar heating, it begins cooling. When air is cooled to its dew point, the moisture condenses into droplets on dust particles in the atmosphere, and clouds form. On the summer day we are considering, these clouds are the cumulus that begin forming in the morning.

The air has now gone from a stable condition in the cool hours around dawn to an increasingly unstable condition. The moisture that was visible as light morning fog and haze is collecting up in those cumulus clouds by 9 or 10 o'clock, and the visibility away from clouds is much improved. This good visibility (outside precipitation areas) is a characteristic of unstable air because the moisture in the air that might otherwise restrict visibility is collected, lifted, and converted to clouds.

As the summer morning moves on, is there any way we can tell whether or not forecasts of thunderstorm activity are likely to come true? There's no sure way, but pilots can get some clues from the air-temperature gauge while climbing to cruise altitude, and from the appearance of the clouds.

If the temperature decreases uniformly and rather rapidly as you climb, this is an indication of instability. If the temperature remains relatively stable or decreases only slightly during the climb to cruising altitude, storm formation is less of a possibility. If a temperature increase—an inversion—is noted, that is an indication of extremely stable air, and thunderstorm activity is unlikely.

The appearance of the clouds also tells us a bit about what to expect. If cumulus begin forming soon after the sun comes up, and if they develop vertically and take on the appearance of healthy cauliflower by midmorning, the chances of storm development are better than if clouds are late-forming and appear to lack vertical ambition.

If conditions are ripe for storm formation, a process starts that accentuates the vertical development of the cumulus clouds that form. The rising particles within the cloud become warmer than the surrounding air and move onward and upward with increasing vigor. If you have flown by a building cumulus, you know what this process looks like. The rapid upward push is easily visible and can even fool a pilot. He might look from a distance and decide that his altitude was adequate to clear the top of the cloud comfortably, but it might grow enough in height to change this in a matter of minutes.

If you have flown through building cumulus, you appreciate the activity within them. You go through with a bang, and the turbulence comes in sharp jabs. A budding cumulus cloud has not developed the intensely strong and well-defined updrafts of a mature storm, and there are no downdrafts, but there is plenty of action. Vertical velocities might reach 50 feet per second in the top part of the cloud. That's why it is best to avoid all cumulus buildups, and especially the upper part of them.

There's one big hurdle for the cloud to clear if it is to grow up and be a thunderstorm: the top must build up through the freezing level. If that doesn't happen, there won't be a thunderstorm. If the top of the cumulus does go through the freezing level with a high degree of ambition, it is very likely to grow into a mature thunderstorm within 10 or 15 minutes.

Remember now, all the action in the cloud—especially toward the top—has been upward, and the cloud has been collecting moisture all the while. This means that moisture carried up by the updrafts must eventually come back down as precipitation. The beginning of rain marks the maturing of the storm. The cloud has held the moisture aloft as long as it can, and the time has come to let it fall. The domino theory prevails, because as the cold raindrops from aloft fall down through the cloud, they bring cool air along with them. This descending air is cooler than the surrounding air, and as a result, its downward movement is accelerated—often to as much as 40 or 50 feet per second. This downdraft comes out the bottom of the cloud, and its downward velocity is transformed into horizontal velocity as it fans out over the surface. The downdraft becomes that blast of gusty, cold air you feel as a storm approaches. The strength of that first surface gust, and the following breeze, is directly related to the height of the storm as well as to the speed of its movement across the ground.

Our storm went to a lot of trouble to achieve its prestigious position in the world, and it isn't about to give up the minute the downdraft starts. In fact, the updrafts around the edge of the cloud actually increase as the downdraft starts, as if they think they could afford to feed it moisture indefinitely. This is probably the meanest moment in the life of a thunderstorm, with updraft velocities spurting on up to 100 or more feet per second. Lightning begins. Hail is most likely to spew out the top at this time, and to be carried by the upper-level winds so that it falls ahead of the storm. Turbulence is at its greatest. A lot of pilots profess to have flown through thunderstorms, but few have successfully penetrated a big one at maturity. As the cold air cascades down through the middle of the cloud and the updrafts strengthen, the level of turbulence is simply beyond the structural and controllability limits of most airplanes as well as beyond the comprehension of most pilots.

In the lower levels, there can also be a tremendous wind shear—even in clear air—that can be as dangerous as anything the storm can throw at you. Warm air is rushing up to feed the storm, which is spending its fury by letting the precipitation fall and the cold air rush downward and outward; the area where the up and down rub together is an area of maximum turbulence. Obviously, this happens inside the cloud, and in the roll cloud that often forms ahead of a powerful storm on the advancing side. But there is another area of shear, in clear air between the roll cloud and the point on the surface that the first gust has reached. The wind-shear turbulence in this area can be severe. Newspaper accounts of thunderstorm accidents often tell of a witness having seen the airplane come down *before* heavy rain started falling. In other words, the airplane was in trouble before it actually reached the storm, and it was probably flying in the area of severe shear turbulence near the storm, usually on the advancing side. (Severe storms often create severe turbulence in all quadrants.) Storm-avoidance guidelines suggest that a 5-mile berth be given normal storms. Severe storms should be given 20 miles.

There are no soft spots in a thunderstorm, but we do learn from the structure of one that both the updraft and downdraft action are likely to be stronger in the upper levels. The updrafts accelerate as they climb, and the downdraft begins decreasing at about 5,000 feet above the surface and finally fans out and becomes horizontal to the surface. Before the days of airborne weather radar, some pilots would actually penetrate thunderstorms in larger airplanes. When they did so, the softest penetration spot was found to be from 4,000 to 6,000 feet above the surface. If a roll cloud was present, this would generally take the airplane over the top of the low-level shear turbulence that it indicated and into the side of the storm where the action was vertical and at its lowest level. I'd rather turn around and go back to a comfortable bar somewhere, regardless of the size of my airplane, because I can think of several airline airplanes, including four-engine jets, that have found destructive levels of turbulence in thunderstorms.

Let's go back to our mature storm now and see what happens to it next. We left it at a peak, which only lasts for a few minutes. A summer storm is usually a one-shot affair—a slow-motion explosion, in effect—since it does not have the continuous lifting service of a front to perpetuate its development and activity.

After rain begins falling, the downdrafts soon become the predominant feature, and the storm begins to die. The moisture is all falling out, the updrafts are weakening, and no more moisture is being carried aloft. The storm is spent.

The life of a summer thunderstorm is short—from 20 minutes to about an hour and a half—but when conditions are ripe for the development of storms, they can form in clusters, with new storms forming as old storms die. This can lead to the impression that storms live longer than they actually do. Finally, though, as the solar heating diminishes with the last light of day, any summer thunderstorms left will also begin fading away, and no more will form because the air will usually return to its stable nighttime state.

If you're flying along wondering how to avoid the thunderstorms that have appeared ahead, it's good to know some characteristics other than just how the storms were manufactured. Thunderstorms tend to move with the middle-level winds, which are most often from the southwesterly quadrant in the summertime. Strong winds discourage the formation of air-mass storms, so middle-level winds are generally light when such storms are around, and the movement of air-mass thunderstorms across the surface will be at a relatively lazy pace. The reason it is good to know which way storms are going is because you should give them the widest berth in the direction of movement, which is where the clear-air shear turbulence at lower levels will more likely be found.

If storms are developing in clusters, new storm development generally takes place on the side from which moisture is being fed to the cells. If there is an appreciable low-level wind flow, it is probably from the south, and new development will be on the south or southeast side of a cluster. In such a case, the south is often still the best side of a cluster for a detour, because the feeding effect moves moisture into the collection of storms, and there might be fewer obscuring clouds on that side of the cluster.

There's another form of summer storm that has some of the same ingredients as the common air-mass storm triggered by solar heating. It is the orographic thunderstorm, formed as wind moves moist air up mountain slopes, with solar heating also contributing to the lifting. This effect is the reason you often find the buildups highest and the thunderstorms most numerous along and over mountains and mountain ranges.

Summer thunderstorms: they are beautiful, violent and necessary, for their rain constitutes *the* summer rainfall in many parts of the country. While they are a real operational consideration for pilots on many afternoons, they should not be allowed to pose any hazard. The summer storm is easy to see and avoid, and its nature is such that change will come soon. All a pilot need do is remain well clear and be patient.

13 CAUTION: THUNDER- STORMS

by Richard L. Collins

ONCE UPON A TIME, pilots avoided thunderstorms with luck, intuition, and liberal use of the eyeball. They would sometimes stumble into one of the beasts, though, and when that happened, luck began to play a much larger role, intuition became an ex-virtue, and the eyeball got very busy trying to focus on the gyrating instruments and send the proper messages to the hands and feet. It's amazing the system worked as well as it did.

Fortunately, the good old days are gone, and now even pilots without airborne weather radar have something that can do a good job of helping them avoid thunderstorms: traffic-control radar shows areas of precipitation, and at times, it seems almost as good as airborne radar. However, the use of traffic-control radar for weather avoidance does have a lot of drawbacks.

Before getting advice from any kind of radar in thunderstorm areas, we must first understand—and never forget—that there is no way to actually penetrate storms without being dashed against the ground occasionally. Several years ago, after a rash of storm-connected airline accidents, guidelines were issued on avoidance: stay 5 miles away from garden-variety storms, and stay 20 miles away if severe thunderstorms are predicted. Those distances were developed after it had been painfully proven that anything closer was unsafe. Obviously, these guidelines preclude penetrating a solid squall line. (When jets do manage squall lines, it's usually by flying over the top of them.)

Storms are visible on a radarscope because rain is dense enough to reflect the radar waves and show as a target. (So is hail.) The heavier the rain, the denser the target. Radar especially designed for weather detection and avoidance shows most rain, from light to heavy. To positively identify cores of thunderstorms, weather radar has what's called a contour function. This blacks out areas where the rainfall rate exceeds a certain value. The most severe turbulence is usually found where the rain is heaviest, and where a rapid change in rainfall rate is occurring over a short distance. If there's a big blob of rain on a scope with a black hole close to one edge and a distinct white line between that hole and an area of no precipitation, that's a bad place. Stay five miles away unless severe storms are predicted. Then stay 20 miles away.

Traffic radar has no contour function. It does show harder rain as a brighter area, but it lacks the ability to pinpoint the core of a cell. Where rain covers a wide area, as in a warm-front situation, and a few thunderstorms are embedded in the clouds, a traffic radarscope might or might not identify the areas of most intense precipitation.

Traffic radar does show precipitation as well as weather radar does, but it has a circular-polarization feature (CP) that can be used to eliminate most of the weather return so that rain won't hide the traffic they're trying to control. Center controllers usually can't turn off the CP for a moment to get a better picture of the weather for inquiring pilots, for the CP affects the radar transmitter as well as the receiver, and thus affects all the radarscopes using one antenna. Several sectors in a center use the same antennas, so turning the CP off momentarily might foul up controllers all over the big center-control room.

(We will be getting much better weather service from centers after new automated equipment is installed. This will permit display of a computer-generated outline of weather areas on the scope, so the controller can see both his traffic and an outline of the weather.)

Approach-control radar has the same limitations as center radar, but a couple of things make it more useful for weather avoidance. For one thing, it doesn't have the range of center radar, and there is more detail to what an approach controller sees on his scope. This might give him a slightly better picture even with the CP on. In a smaller approach-control facility, they can often coordinate between all the controllers and turn the CP off for a moment to have a look at the weather. If you're flying within 30 or 40 miles of an airport with a terminal radar facility, you might call them and get better weather advice than center will be able to provide.

Despite these shortcomings, the present radar system can be very useful to the radarless pilot. You just have to know and heed the rules of the game.

The eye remains important. In most storm areas (except warm-frontal areas), cumulonimbus clouds are visible from a reasonable distance at many normal operating altitudes. As air pollution increases, thunderstorms do often become embedded in the smaze, but you can still see them to some extent. The high rainfall rates associated with a storm give the sky an unmistakably menacing look. Even pilots with radar in the nose use the eyes in their head a lot, for if you stay clear of clouds and precip in thunderstorm areas, the chances of a smooth ride are better. Try to stay at an altitude where you can see what's ahead, and then get as much information as possible from the man on the ground.

As we ask controllers about the weather ahead, we find a considerable variation in the service they provide. Some air-route traffic-control centers seem to give more help than others with their radar, but each time you think you've found an uncooperative facility, it turns friendly. Until recently, I would have named Washington Center as one of the less helpful ones—not because they aren't nice guys but because they are usually busy controlling a lot of traffic and have little time for other things. Yet Washington Center was especially helpful to me just awhile ago. A big storm was perched atop the Elkins, West Virginia, VOR as I approached it from the southwest. I was fairly low, and while Center could see my transponder return, they couldn't read my calls as I tried to check in with them level at 6,000 feet. Finally they relayed, through another aircraft, the message that I was in radar contact and should fly a heading of 110 degrees for vectors around weather. It was a storm that could have been avoided by eyeball, as most are, but to have done it by eye would have meant flying up closer and then making a considerable detour with a 50-50 chance of picking the best way around. The controller, however, could look down on me, the storm, and his other traffic, giving a way around that would not foul up my position in relation to his other traffic, yet making the detour as short as possible.

The same general degree of helpfulness is true of all centers, and the variation from hand-holding to cold-shouldering rather applies to individual controllers. Some will even volunteer information on the location of weather return on their scopes and suggest headings to avoid it. Others will give information if you ask for it, and still others will just say that theirs is not weather radar.

If a controller does not volunteer information about weather, but you suspect some might be ahead, the way you make a request is important. A reluctant controller is more willing to give you *information* than he is to *help* you do what you are supposed to do yourself. Many controllers feel that pilots calling for help in getting around weather have taken off with little knowledge of what's out there and are trusting the man on the ground to lead them down a safe path. They can't accept that responsibility, and it's easy to understand why: if they should agree to vector an airplane through a storm area, they have implicitly agreed to provide safe passage, and sometimes it doesn't work. Even if a controller keeps the airplane clear of the really tough areas, there might be enough turbulence elsewhere to upset a not-too-talented instrument pilot. While the controller certainly wouldn't be at fault in such a case, he could be heavily implicated any time an airplane he was vectoring didn't make it through a storm area.

The key is in just asking for information. Ask the controller if he's painting weather ahead. If his answer is yes but he doesn't add any advice, then ask him if it looks better to the left or to the right. Guess at a heading or figure a detour route along another airway and ask if that is clear. Even the most reluctant controller will answer specific questions, and you do everyone a favor when you don't try to put him on a spot.

It's a good idea to ask for information on weather as far in advance as possible, too. A very high percentage of the thunderstorm activity we encounter is visible from afar, and even if we can't see it, we can hear other pilots on the frequency talking about it. If the stuff is pinpointed in your mind well before you reach it, a more efficient detour can be planned. If we wait until we can count the wrinkles in the cumulonimbus and then start asking questions, about the best a controller can do is suggest we turn around and go home.

Remembering some characteristics of thunderstorms can help when planning a way around them. Individual storm cells tend to move with the middle-level winds (more often than not out of the south, southwest, or west in this part of the world), and the part of the storm needing the widest berth tends to be in the direction of movement. If storms are developing in clusters, new storms will usually develop on the side from which moisture is being fed into the building cells. This is often the southeast side of the cluster, because of the frequently found southwesterly middle-level winds and southerly low-level moisture-bearing winds when thunderstorms are around. The south is still often the best side of a cluster for a detour, for the formation of new cells takes up much of the available moisture, and there might not be as many obscuring clouds there as around on the north side.

Severe thunderstorms sometimes don't follow these rules, for tornadoes are often found on the right *rear* flank of a severe storm, which is contrary to the rule that the meanest area is to the front. Also, severe storms often move to the right of the steering middle-level winds, because their demand for moisture is so great the storm actually works its way toward the supply.

The severe-thunderstorm situation, though, is usually easy to identify. The possibility of these is well advertised, they can generally be seen from a distance and are spectacular in appearance, and they will bust through the circular polarization on anybody's radarscope. When they are forecast, stay well away and don't be tempted to cut a corner. Hail can fall from that beautiful anvil that spreads off the top ahead of the storm, and can punch holes in your airplane when you are miles away from the area of heavy rainfall.

When your eyes and the controller's advice or information agree on a way to go around storm areas, don't expect a velvet-smooth ride every time. Remember that if he has his circular polarization on, he's looking only at heavier rain; it takes a rainfall rate of between a quarter- and a half-inch per hour to break through the CP. If something ahead looks unhealthy, don't fly through it; ask for another deviation, and don't wait till the eleventh hour to do so. A building cumulus that will bust loose any minute, or even a pretty good little thundershower, might not show on his scope at all—yet it could provide a memorable few minutes for you.

If things do deteriorate, don't start chewing on your controller and asking for new advice after the bumps begin. He has done his best, and straight ahead is usually as good a way out as any. Just relax and concentrate on flying the airplane.

Remember, too, that the time often comes when it is best to switch to VFR. If the clouds look menacing at all available IFR flight altitudes, if there is no way to remain clear of cloud most of the time and pick a path through the cauliflower, and if the controller says he's not painting any weather in the area, the time has come to be an IFR dropout. If there is a good ceiling underneath, the best ride can often be found below the cloud bases, where you can see the rainful areas and steer around them.

The most important thing to remember about thunderstorms is that they are supremely powerful and can be the master of *any* airplane. While they appear to move rapidly when you are standing on the ground, their speed is very slow when compared to that of an airplane. There's no way for a thunderstorm to get you—you have to fly into it, so don't be bashful about delaying a flight while storms move on to harass another route. While you might be miserable pacing the floor, you'd be a lot more miserable out there in the teeth of the storm.

14 THUNDER-STORMS? NO SWEAT
by Ave E. Ator

THUNDERSTORMS don't faze me. Not any more. They *concern* me, naturally; they make me think and scheme and employ all of the tricks I've learned. They make me remember flights made years ago. But they don't excite me in the least. Not any more. When the map shows a line squall across the route . . . well, you can't have clear skies and smooth air every day, can you? When my terminal's sequences include such addenda as "TRW +," "FQT LTNG ALL QUADS," "SFC WND SHFT RPDLY," and the other phrases weathermen use to record storm conditions, my palms don't sweat—not like they used to.

No, I don't push my ship back into the hangar and await better reports. That is sometimes the best way to handle thunderstorm flying, but it just wouldn't work in my case. I am an airline pilot; I have to go, storms or not. Thunderstorms may delay us, but I cannot recall a trip of mine being canceled because of weather—thunderstorm or otherwise—in the last 10 years. But I'm getting ahead of my story.

First off, the byline on this article is obviously not my real name. For reasons that will become obvious, personalities (mine included) are best left out of this. Otherwise, the facts are quite true, and they are, I expect, almost identical to those that other airmen of similar experience can relate.

On October 6, 1941, at about three in the afternoon, I met my first thunderstorm. I can be specific because the date is right here on page eight of my first logbook, with a note by the entry that is, even today, an unnecessary reminder. On that day, I made my second solo cross-country, a round robin covering 400 miles. Armed with enough charts to fly from coast to coast, I was up, up, and away in my AT-6, my supreme confidence based upon 102/35, dual and solo, logged thus far. Outbound, everything went fine. Coming home, I had to duck under a deck of cloud that for some reason got lower the longer I flew. The situation resembled nothing the ground-school instructors had sketched on the blackboard. Pretty soon, I was flying at treetop level in driving rain, doggedly plowing on toward an ever-darkening sky. I was 15 minutes from base, according to my flight plan. A searing, evil bolt of lightning stabbed the earth not half a mile ahead. Carb heat on, cockpit lights full high. Well, I did something right.

I lost sight of the ground and pulled up to 500 feet having my hands full just staying right side up. I debated reversing course but doubted my ability to complete a turn in such turbulence. Solid instruments, incredible rain, continuous lightning, the acrid smell of ozone, and thunder clearly heard above the big Pratt's roar. I was one scared 19-year-old. Suddenly, unbelievably, I was free of the awful nightmare and in clear sky. The last drops of moisture streaked along the Plexiglas and vanished. Behind lay the sinister greenish clouds, ahead the runways of our base. God takes care of fools, drunkards, and young men learning to fly.

Overseas, it fell my lot to copilot C-47s, often for men recruited from the airlines, some of them old-timers who had begun flying on DH-4 mail runs. They came from Pan Am, TWA, American, and the smaller lines. It was the best possible training; I am probably alive today because of it.

The airline DC-3 gross load was 25,000 pounds. We routinely flew the military version at 27,500 and grossed out at 30,000 pounds or more on long overwater hops. Such weights sharply reduced the ship's reserve power and thus its ability to negotiate thunderstorm weather. We went anyway, cheered on at knowing that 15-percent losses would not raise eyebrows back at headquarters. Losses never rose to the "acceptable" rate, thanks to the expertise of our skippers and to good luck.

Each man handled the thunderstorm riddle in his own way, I noticed. Approaching a solid line, one would parallel it for miles, seeking just the right soft spot. Another would head without hesitation for a certain area and plow on through. The third might climb for penetration, the fourth descend to a hundred feet. There were as many theories as pilots to test them. Sometimes, we slipped through with a few bumps; other times, it was not so easy. When the guess was bad, we were pitched about like a cork in the ocean while the captain wrestled with the yoke and I tried to keep airspeed within 50 mph of a reasonable number. The sound and feel of a cargo of 2,000-pound engines straining against their tiedown ropes and slamming back onto the cabin floor were nerve-wracking. The option of diverting to another field rarely existed. The choice was not if, but how.

Douglas built stout ships, however, and our cups ran over with good luck. In my greenhorn's view, there was little skill to it.

Back home we went for new equipment, more schooling. Weather courses were thorough and explicit. There was a sensible, logical explanation for all weather phenomena. Thunderstorms were neither mysterious nor frightening, according to the composed experts who gave the lectures. They were, of course, nonflying officers, none of whom had ever researched an anvil head from the inside, but their pat solutions seemed to make sense. Thunderstorms—big deal. We were told, among other things, that cumulonimbus clouds occasionally built to 30,000 feet above the ground. More than once, a brash youngster who had been that high in a P-51 snorted at this, flatly stated that 30,000 feet was but waist high on storms he had seen. This would bring a patronizing stare from the instructor and a knowing look at the rest of us: you couldn't believe anything a peashooter pilot said. Besides, this conflicted with the data in the USAAF manual.

The gist of official wisdom was this: when thunderstorms could not be circumnavigated (avoidance was hardly overstressed), press on. Penetrate at this level or that, keep speed under control, expect heavy rain, sharp vertical drafts, and to come out the other side still flying. It was strongly intimated that any pilot afraid to tackle a line squall was chicken. Ever mindful of the rousing last line in our unofficial hymn ("Nothing can stop the Army Air Corps!"), we obediently tried to put theory into practice, not always with success. Thunderstorms in all theaters of operations, and with disquieting regularity, continued to receive intact airplanes on one side and spit out sheet metal on the other. Such incidents were laid to noncompliance with approved procedures. Pilot error. If they had only gone by the book . . .

My stateside assignments provided an unusual opportunity to fly a wide range of aircraft, most of them large bombers and transports. On my own, in command of my own ship (if not always of the situation), I quickly broadened my knowledge of thunderstorms. Some I tackled intentionally, others I blundered into. That was awhile ago, but certain recollections are still vivid. Once we were in a C-49 over the Rockies, struggling to maintain 14,000 feet in heavy snow. There was considerable chop, but nothing to be uneasy about. Without warning, we entered an updraft of unbelievable strength. With the gear down, full flaps and throttles at idle, we were swiftly lifted to 19,000 feet. It was rock-steady all the way up; you could have carried a tray of beer mugs through the cabin and not spilled a drop. It was eerie. When it ended, we cleaned up, restored cruise power, and flew out into the clear.

We were in a B-24 over Kansas at midnight—torrential rain, lightning all over, turbulence that tossed the ship from one steep bank to another and made it impossible to hold heading closer than 30 degrees either way. Then, the bottom fell out of everything and the altimeters went crazy. Takeoff power had little effect, even though the Liberator was empty. "A downdraft will never force an airplane into the ground," they had assured us in class, but at that instant, we were not convinced. The storm spewed us out at 3,000 feet. We landed at Forbes, where mechanics collected a sackful of sheared wing rivets and pointed to gas pouring from the ruptured bomb-bay tank. An hour later, a Navy fighter capable of absorbing nine-g loadings plunged out of the same line of storms. Its wings were found days later, a mile away.

Again—near Dayton this time, in a little Stinson L-5 liaison plane; in the back seat, a sergeant bumming a ride home. I tried to sneak between two merging thunderheads and made it, only to discover a third enormous black buildup. I reversed course, but it was too late: we were cut off from behind. Cursing an airplane that did everything at 80, I plunged into rain so intense the Lycoming drowned out and quit. Blinding shafts of lightning on all sides, the airplane almost beyond control. It was a terrifying situation. Finally, a light area appeared and I herded the unwieldy ship toward it. We emerged soaked to the skin and landed at Wright Field soon after. True to the code, the rider cleaned up the back seat before he left.

These were, of course, exceptions. Most of the time, it was possible to weave around the worst of it or completely avoid a troubled area (a trend I found myself adopting no matter what my copilots thought). A cautious penetration usually meant some rough air and a thorough wetting, but the trip would be completed, and that was enough for me.

I intend no criticism of the military flying philosophy of that day. For the most part, our training was superb. We were provided with a wealth of useful data, considering the time allotted for learning. That we were sometimes handed dangerously inaccurate information does not reflect upon the quality of instruction but on the state of the art. Flying was still a novel business when World War II began; weather-flying was an almost brand-new concept. Most prewar flying had been accomplished below 10,000 feet in reasonably good weather. What went on at 30,000 feet and higher was largely educated guesswork, some of which proved to be entirely wrong. When B-17 crews flying in Alaska reported headwinds that pushed them backward, they were considered candidates for Section Eight. In fact, they had discovered the jetstream.

Remember, we were engaged in a desperate war. Tens of thousands of boys, many of whom had never flown, had to be quickly trained to operate the most demandng aircraft ever built and imbued with a spirit that would see them through the most daring flights ever attempted. That the philosophy thus engendered was at times carried to unrealistic extremes is undeniable; but it is understandable, and even excusable, in the light of events. And don't laugh at the idea that thunderstorm flying was viewed as simply another problem that could be solved by writing regulations and passing out handbooks. Are we free of that sort of thinking in aviation today? Victory was the prime concern then, and victory we earned.

Then it was the airline job; now I would learn how the pros coped with thunderstorms. Surely the men who flew all-weather, all-season schedules had the answers. But again, there seemed to be as many solutions as captains. A surprising number steadfastly refused to penetrate a line unless clear sky could be seen on the far side. They would parallel a line squall for a hundred miles in search of a suitable corridor, then land and sit it out, shaking their heads at word from the dispatcher that another trip had made it "without any trouble." More than a few wouldn't even make the effort. "All captains have four stripes," said one. "I have five. The fifth runs down my back." He was not the only one. The majority gave it a real try. They usually got through, nearly always without incident.

It was the bold ones who scared me. This handful regarded thunderstorms as a personal challenge to their airmanship. Some of our routes lay right through "Tornado Alley," that midwestern region that sees more intense storms in an average year than any other large area on the continent. You cannot operate through the Alley for years without developing an eye for thunderstorms and seeing things that would escape airmen of lesser experience with such weather. Call it what you will—a knack, a sixth sense, a certain feel for the situation, a reading of the signs based upon long practice. It's a real and recognizable talent. The fearless few imagined they had it down cold, and I have to admit that their navigation through thunderstorm weather was uncanny. Time and again, we would slip through with minimum fuss, leaving the timid souls far behind as we wrapped up the run on time. They had it figured, most of the time.

But when they misread the clues! Reader, unless your logs list similar experiences, you would not believe the abuse a DC-3 or DC-4 could take. I have seen radios torn from their racks, entire galleys ripped from floor mounts, stewardesses thrown against the ceiling, and the turbulence so bad that the seat belt left bruises. The sounds of a twisting, creaking, shuddering airframe are sounds you will never forget. No one can tell me a story about thunderstorm flying that I won't believe.

The DC-6s and -7s, the Convairs, Connies, and Stratocruisers proved equally stout. All could absorb unbelievable punishment. There were limits to it, however. Any of them could be overstressed. A review of accidents since 1940 clearly proves that the airliner has not been built—piston, turboprop, or jet—that a severe thunderstorm cannot reduce to scrap metal. One can only wonder how many structural failures that seemingly had nothing to do with storms were actually caused by undetected damage previously sustained in storm-associated turbulence.

Again, no criticism of airline technique in the piston era is intended. No captain of that day deliberately flew into a cell. The trouble occurred when the signs were misread through ignorance or overconfidence. Considering the tools available and the level of weather knowledge, the wonder is that we weren't in jams more often and that the accident rate didn't soar.

My vow of long ago—that when I made captain, I would never again see a thunderstorm from the inside—has not been difficult to keep; not simply because of personal caution, but because of modern airline aircraft and weather aids. There is little excuse for today's airline pilot getting into a bad thunderstorm situation. In most cases, he flies a plane that can scramble to 30,000 feet or more in half an hour and cruise at 550 mph. He departs with a wealth of weather data, has radar on board and the use of ground radar stations. He has instant, static-free communications with other airmen in flight, airways controllers, and his own company headquarters. He has the speed and fuel for wide diversions. And he has his recollections of how it used to be.

Unfortunately, all this has led to some erroneous conclusions outside our ranks. Many private airmen are convinced that airline pilots regularly penetrate thunderstorms, using tricks we have acquired over the years. After all, they watch us arrive and leave in all kinds of weather; our trips depart on time, even when the sky turns blue-black and looks as if a welder's convention is in progress just beyond the horizon. Meanwhile, our flights from that direction are observed to touch down routinely. It adds up: the airlines have long since whipped the thunderstorm riddle.

"Write an article on thunderstorm flying," suggested the editor of a general-aviation magazine some years ago. When I said that airline thinking could be summed up in one word—"Don't"—he laughed. "Oh, come on! You fellows have schedules to keep. You don't ground yourselves every time a line squall blows through."

I was never able to convince him that 99 percent of us were scared silly of thunderstorms, and that we avoided them no matter what diversion or delay was required. We still do; believe me.

When an airline type talks like this to a private pilot, however, the man is apt to imagine he's being looked down upon. "This fellow *could* tell me, but he thinks the technique is beyond me . . ." That sort of thing. Too often, this line of reasoning leads to another: "If they can make it, I can." This is the raw material of which headlines are made.

Let's put it another way: if you imagine that airline pilots—or the men who fly high-performance military or business aircraft—are out there flying through thunderstorms, you have been badly misinformed. If you think they operate regularly in thunderstorm weather, you are right. The pilot of the average private ship can no more match their performance in this respect than he can match their speed or load. Suppose we are both in Denver, flight-planning trips to St. Louis—you in your 200-mph six-seater, me in my 707. A line of severe thunderstorms extends from the Texas Panhandle to Iowa, with reports showing it to be solid and building—a typical springtime setup, in other words. You don't think we're going to fly *through* that mess, do you? Not bloody likely, as a BOAC captain might put it. Oh, we'll file the regular airway across Kansas City and ask for Flight Level 410 all right, but don't be misled by that. We will look the situation over from 100 miles away—visually, by radar, and through whatever radio information we can get. Squall lines are overrated as often as not, and we might find a gap to slip through. If not, we'll head south to Dalhart or Lubbock or as far as it takes to get around the south end of it. This will mean a late arrival, of course, but we operate the airline with three points in mind: safety, passenger comfort, and schedule—in that order. It's right there in the book.

You have a totally different problem. You have about the same performance and capability we knew in the DC-3 days. The diversion that cost us minutes will cost you hours, for it will entail a stop for fuel. On the other hand, you may find a suitable gap down low that was not open at our level, in which case you will make your flight plan. "You cannot be too cautious" is not the motto of airline pilots, not in its literal sense. You can be too cautious. If every pilot grounded his trip every time a "TRW + " appeared on his route, his line would go broke in six months. You have to go out and give it an honest try. In most instances, a way through or around the problem will reveal itself en route.

Some observations, for what they may be worth: a thunderstorm is perhaps the most interesting of weather phenomena, being a thing of spectacular beauty, monstrous dimensions, incredible strength. We understand why thunderstorms come into existence, how they develop, and what goes on within them during their birth/growth/death cycle. We type them and sketch their insides and an airman must absorb all this knowledge. And he must always remind himself that all of it is *general* in nature. If such intelligence were rewritten with people as the topic, it would sagely observe that they are to be found in a limitless variety of sizes and shapes. Some are quiet; some are not. Two hundred pages of this, with sketches, would say a lot—and say nothing.

Thunderstorms are as individual, changeable, and unpredictable as humans. They defy specific analysis. The CB observed from 150 miles away is a brand-new fellow when you square off with him, eyeball to eyeball, half an hour later. The nature of storms, like the nature of nations, varies with regions of the globe. Fearsome lines in some tropical areas often prove to be more bark than bite and (usually) can be penetrated with no more inconvenience than heavy rain and light chop. Clouds a third their size over Kansas will (usually) shake your fillings loose. The general rules that (usually) apply over western Pennsylvania (usually) will not serve over west Texas. The pilot who thinks that his research and experience have brought him to the point where he can accurately size up storm situations every time is playing with fire. I don't want to ride with him.

Radar is an invaluable aid in thunderstorm weather—when it works. Like a checklist, it can get you into more trouble than it keeps you out of. It's not quite like TV, no matter how the ads read. I never met Walter Cronkite, but I know he's not green, so I can fiddle with the knobs until he comes up pink, and even give him a healthy tan when he's just back from Cape Kennedy. I know that my image of Walt is pretty close, but I cannot always say the same for the picture painted on my radar screen. When radar is good, it's a dream. When it's faulty, it can sucker the unwary into bad trouble. One major airline is spending millions to modify its sets right now—proof that current installations are not entirely satisfactory. Radio, in my view, is at least as helpful in obtaining specific weather data. Nothing beats talking with a pilot who has just run the course.

Hail is among the hazards of thunderstorm flying. A few years ago, a summer gully-washer hit our town—high winds, torrential rain, lightning strikes, the whole bit. For some reason, my home was in a 2,000-square-foot area singled out for hail. It hit the front of the house, smashing every window, denting the woodwork, and ruining the roof. Several upstairs-bedroom back windowpanes had been shattered from inside (the broken glass was in the backyard). It took awhile to figure that out. Dents on the room's back wall gave the clue: hail, traveling horizontally at high speed, came in the front windows, flew the length of the room (24 feet), and left via the back windows. Weather that contains surprises like that is no place for my tin airplane.

Lightning does not scare me. Terrifies is a better word. Anything that can split a six-foot oak down the middle deserves attention, the way I see it. Yes, I've sat through the movies of models suspended in test chambers with technicians standing around looking like dentists with clipboards while electrical charges go in one wingtip and out the other. Very interesting, but how closely does it follow the book? "Lightning has never been known to do major damage to a metal airplane," they told us in the war. If that flat statement is still in the cadet syllabus, I pray they add, "Of course, numerous airplanes have exploded in thunderstorms for undetermined reasons." The credentials of anyone who makes a flat statement about any aspect of flying should be regarded with suspicion. I fervently hope I have seen my last lightning strike.

The building of aircraft is an exact science. The operation of aircraft, particularly in bad weather, is not. In time, I think the act of flying will be recognized for what it is—a craft based as much upon an awareness of, and respect for, the unknown as upon what is known. It refuses to be hammered into such shape that it can be neatly pigeonholed among other components of the game. Weather must likewise go into a slot of its own dimensions.

Each airman develops a personal philosophy about flying. It stems from his training, experiences, and temperament, much of it being carved by forces beyond his control. Common sense is the one factor he has, or lacks, from the beginning, and it alone is enough to tip the scales. We see among us the timid, the cautious (to whatever degree), and the bold. We even see old, bold pilots, contrary to the Air Corps adage.

I know many pilots who have flown professionally in airline, military, and private aircraft for well over half their lives. The vast majority of them are *extremely cautious* in this business of thunderstorm flying. Make that read 99.99 percent of them.

I think that's worth noting.

15 HOW TO WORK WEATHER RADAR

by Thomas H. Block

WE WERE AN experienced crew in a large transport airplane, snaking our way through a forest of towering cumulus. There were thunderstorms all around, and hail, heavy rain, and turbulence had been reported. Yet we were lucky because—to repeat the words handsomely stenciled on the fuselage near the passenger entrance—we were radar-equipped. Up front, we huddled like gypsies over our screen, trying to divine a future from the ghostly pulses.

"Let's try some up-tilt."

"Is all that ground clutter? Let's see if it'll contour."

"Move the gain around. Pick me a soft spot."

We tried to choose the smoothest path on our tube. Would it jibe with what we saw through the cockpit window? Did we know what we were doing? Did we know what we were supposed to be seeing, and what should be done about it?

Hardly. At first, we knew only what the manufacturer's brochures and a quick ground school had taught us. Then, slowly—and often painfully—we learned how to draw more information from that dish of energy bolted to the nose. For operating and interpreting radar, experience is the only teacher.

Recent technology has done much to improve weather radar. The new digital sets offer unprecedented ease of operation. The color units provide instant interpretation of rainfall gradients. Electronic innards that create the picture have been revised and refined. Antennas and viewing systems have changed. Memory circuits are a welcome development. Weight, size, and cost have been reduced, while efficiency and accuracy have been increased. It's a good deal all around. Yet one basic truth still remains unaltered: getting the screen to provide a good image is only half the battle; interpreting what you see can be the more difficult part.

Digital radar has been heralded as the most important advance since the frozen waffle. While this new equipment does represent a marked improvement over older sets, it *still* must be operated correctly and interpreted properly.

Newcomers to radar normally get a bum's-rush introduction: they read the handbook and blast off toward the first weather targets they can find, only to retreat confused and disappointed. It is difficult to make your screen look like the neat and ominous patterns in the handbook. Part of the problem comes from too much emphasis on some of the sets' secondary functions. Radar manufacturers candidly admit that it takes actual operating practice to nail down even basic tuning and interpretation, but their handbooks are crammed with instructions on how to ground-map or use the self-test.

Becoming familiar with the intricacies of terrain mapping is like learning how to explode eggs in a microwave oven: interesting but seldom practical. I do recall once locating Erie, Pennsylvania via radar down-tilt and thinking how nice it was that the displayed range and azimuth agreed with the aircraft's dual VORs, DME, and ADF. Only a nonpilot electronics expert would think of saying, as one brochure does, that radar "is often invaluable as a navigation cross-check." Some specialized operators may need terrain mapping, but for most of us, it is simply a toy to be played with when time permits.

Self-testing is a nice function, because it provides some basis for determining that the radar is working. But efforts to remember what malfunctions the various patterns indicate are pointless unless you intend to do your own troubleshooting. Most faults are pretty easy to recognize anyway, since the radar screen simply stays black.

Even turning on the radar is easier now, thanks to modern circuity. In older equipment, the warmup period was five minutes or longer. Crews discovered that they would sometimes inadvertently move the selector to "off" while operating the set; many harsh cockpit comments ensued when flights found themselves negotiating nasty weather with blackened, warming-up screens. Modern sets take only 60 seconds or so to come on, so you can leave them off until shortly before you need them. (Don't wait too long, though; the time to learn that your radar is inop is *not* with towering cu in all quadrants.)

The gain control, which varies the threshold of intensity that is necessary to provide an image on the screen, has been effectively done away with. On older sets, all returns except for targets that reflect enough energy to contour would appear as a single shade of green. (In the contour mode, contoured spots are shown as inverted video; their images are black rather than green.) In order to differentiate among the various levels of noncontouring rain, pilots would sometimes turn down their gain control and watch as the weaker areas disappeared. It was an effective technique, albeit one frowned on by many manufacturers, who felt that pilots might inadvertenty leave their radars set at a lowered sensitvity.

Digital radar consoles still have a gain selector, but it's used primarily for ground mapping. The normal display now features three or four shades, and these color variations tell pilots all they need to know. The new radars discriminate among rain intensities automatically: light green is the lightest return, black still means contour, and the darker shades of green indicate intermediate intensities. For ultimate ease of interpretation, fly behind one of the new Bendix or RCA sets that use *different* colors (green, yellow, and red) to show the various intensities of the weather returns. There's nothing to fiddle with, and you get a color-coded message to boot. There's no mistaking the colors' meanings, and since interpretation is based on color rather than intensity, every return on the screen can be displayed at full brightness.

Contour circuits have been talked about a lot lately, yet they haven't changed much. The principle remains the same: the displayed video is inverted after the intensity of the return has reached a present level. By adding a mode that alternates normal and contour sweeps, and providing a switch that causes all but contoured returns to disappear, manufacturers have come up with ways to attract more attention to problem areas.

Cyclic contouring, weather identification, and other features have their good and not-so-good moments. While the pilot is playing with these special functions, he or she still hasn't acquired any additional data—and these extra chores can be distracting.

Another new function that has been given a heavy credit line is the "freeze" or "hold" capability on digital sets. Having it aboard is an advantage—but not because, as the brochures state, a pilot can "evaluate the significance of storm-cell movement." Except for some ability to watch targets drift sideways across the scope (which is as easily done without "hold"), determining which way cells are moving—while the airplane and wind alike are also in motion—is no small feat.

The hold function is worthwhile because it gives the pilot a chance to take in what's ahead minus the distractions of updating or additional ground returns. Saying "freeze" to the radar is like telling a fast-talking salesman to shut up: it gives you a chance to figure out what already has been said.

The most annoying trait of older sets was the fact that the images were present for only a moment. Pilots had to be patient, since the quadrant they cared about was visible only when the sweep line (representing the antenna's motion) was swinging by. Studying a specific area meant wait-wait-wait-look, wait-wait-wait-look, until you finally received enough meaningful hints to get the idea.

With digital radar, the whole viewing area stays alive at all times, like a TV screen. Most digital sets still have a sweep line, which assures the pilot that the image is being updated, but there is no erase rate. This makes the task of reading radar a lot simpler.

A slight drawback of digital displays is the boxiness of their images. The screen is lit in tiny squares, so round-edged targets (like clouds) have a strange staircase appearance. Time spent working a new set is the solution: a pilot soon begins believing that storm clouds are actually shaped this way. The sharp edges make them look even nastier.

Achieving finesse with the brightness knob merely means finding the setting that's comfortable for your eyes. The antenna-tilt knob is not so cooperative, however, and to work radar is to work antenna tilt. All other factors being equal, what you see is what you're pointing the dish at. But terrain complicates the picture by showing itself too often because of its high radar reflectivity. Run the tilt too low, and the radar screen will fill with terra firma.

Playing with the tilt control will demonstrate how much movement is necessary to get a change of image. Practice by attempting to pick out shorelines and lakes from the clutter. (Water reflects less of the radar's energy than hard soil.) Antenna tilt is easiest to adjust when you're sitting at the run-up block or (with prior permission) on the runway, since all the weather targets are backgrounded by nothing but sky. Move the tilt earthward enough to pick up the first hints of ground, then work it skyward.

Radar can become a cockpit distraction, especially during the maneuvers after takeoff. Unless you have antenna stabilization—and even that has limitations—any pitch or bank changes will show up as alterations in ground-clutter and weather targets. Pick a path around the echoes *before* pushing up the engine power. Fly using the clock and headings (''one minute on a 260 heading, then northbound''). The time to coordinate a pilot's wishes with those of departure control is before takeoff, not after the pilot is holding a fistful of airplane and another of microphone. Ask the tower to coordinate your intentions with departure control before accepting the takeoff clearance.

Don't try to get too much information from radar. It's nearly useless close-in, because it paints everything all too well, and you won't have time enough to discriminate among all the blobs, contours, and ground targets you see.

On many sets, the long-range settings are only wishful thinking. Intermediate ranges often work best, with an occasional sally to a longer or shorter increment. Unless there is some special reason for playing with range, the 50- to 80-mile scan normally gives a good array of information. Playing with range is sometimes a waste of time; adjusting tilt *never* is.

The airplane is the viewing platform. Any change in its even keel will be reflected on the radar. The antenna tilt is used to adjust for alterations in flight attitude. Normally, the tilt should be aimed low enough to skim the ground at the top edge of the scope. By keeping a touch of clutter at the outside edges, you'll be assured of not going excessively nose-up and searching out weather on Mars or Venus.

Airborne radar cannot turn a bumpy cloud into a smooth one. All it can do is locate the areas of greatest rainfall, which sometimes will correlate with the areas of greatest turbulence.

Remember to keep an eye on conditions outside the window. If it looks lousy ahead, it very well might be, regardless of what the scope is showing. Cloud formations and colors are sometimes a better indicator of where the rough ride lies than sophisticated electronic pinpointing of the heaviest rain. Radar is just another tool to help the pilot, and it shouldn't be used as the end-all implement of the cockpit.

16 WEATHER INFO: WHAT WE HAVE AND WHAT WE NEED

by Richard L. Collins

"GOOD MORNING, I'll be IFR from Asheville to Little Rock, flying over Chattanooga, Huntsville, and Memphis, below 10,000 feet."

"Okay, I see no big problems there. You do have a couple of lows, one in Kentucky and one in the northwest Gulf, with a front between the two. That front is occluded at the north end and cold at the south.

"Asheville is 500 scattered, 1,000 broken, eight miles, temperature 37; Chattanooga, 2,100 scattered, 13,000 overcast, eight miles, 42 and 39 on the temperature and dew point; Huntsville is 400 scattered, 800 overcast, three miles, fog, 43 and 41; Memphis is 1,500 broken, 2,900 overcast, eight miles visibility, 38 and 33; Little Rock shows 6,500 broken and seven miles, temperature is 38, and the dew point 33.

"Little Rock forecast calls for 10,000 scattered this morning, clear after noon. Winds in this area at six will be 260 at 36 and a plus two on the temperature; at nine, 270 at 48 and a minus two. Over around Huntsville, six will be 270 at 40, plus two; nine will be 270 at 52 and a minus two. There is nothing on the radar summary chart. Looks to me like you won't have any trouble. It's forecast to start clearing from Memphis on by the time you get there. Leaving here, you should be between layers anywhere from six to eight. Tops of all clouds should be about 10."

"Thanks, are you ready to copy an IFR flight plan?"

Thus begins another flight. The briefing had been above average. The FSS had delivered the pertinent information quickly and concisely with a minimum of extraneous data. I wrote it all down and felt that I had a good picture of the weather along my route. I wondered about icing, but the forecast of above-freezing temperatures at 6,000 feet was comforting. The warm surface temperatures were also a positive sign. The only cruddy en-route weather was the fog and low ceiling at Huntsville.

Based on the best available weather information, my mental picture was of a flight that initially would top all clouds, then I'd be between layers for a while, with the higher layer dissipating between Huntsville and Memphis and the lower layer disappearing around Memphis. The only "bad" thing was the wind, a blistering one on the nose.

The flight unfolded in a somewhat different manner than I had anticipated. There were layers at first, but they merged. I was flying at 8,000 in an area where that is the minimum en-route altitude, and when ice started forming, my only option was to climb.

Before requesting a higher altitude, I sought additional information:

"Center, 030. You have any top reports in here?"

"Negative, 030, no top reports. You are the first airplane to go through there this morning."

"Okay, let me try 10,000. If that doesn't work, I'll look at 12,000."

When I started the climb to 10,000, the only information I had on cloud tops was the FSSman's "should be" from a time that now seemed ancient history. His prediction of layered clouds had augered in; if his forecast of tops were equally in error, I would have to switch to plan B—an immediate diversion south, where the MEAs are lower.

My airplane made it to 10,000 easily even with a collection of ice. Happy days. That tops prediction was at least partially correct—I was clear of cloud most of the time. There were buildups all around, and I'd occasionally bust through one, but this did not create enough additional ice to cause a problem. But the tops were getting higher and the buildups more frequent. A solution lay just ahead where the MEA was lower and I could sample that advertised warm air at 6,000.

I failed to find that toasty air at six, and ice was plentiful. The next move was to try 4,000 as soon as the MEA allowed. I had stayed on top until almost to the 4,000-foot MEA, so the wait was short. The freezing level was at 5,000, and the ice dutifully slid away with appropriate bangs on the tail. Then ice started forming at 4,000. Down to 3,000 feet. Then back to 4,000 because of traffic. As soon as the traffic was past, down to 2,600 feet, the MEA, to shed ice collected at 4,000. Finally up to 6,000. The tops were lower by this time, and the icing problem was behind me.

Now, in terms of the status of our weather-information system, let's analyze the situation. The briefing had seemed good, yet unanticipated icing conditions had been encountered. Someone's fault? Does our weather-information system stink, and was this a hint of the odor?

No, nobody had been at *fault* up to that point, and the situation was not really an unfavorable reflection on the information that I had been given before takeoff. Weather forecasting is a very inexact science—meteorologists are the first to admit that—and the errors this morning could be measured in a few degrees of temperature, a few thousand feet of cloud and some merging layers.

It is tempting to use such a situation to condemn flight service, especially during and right after a flight, but as we examine our weather-information system, we must accept the inevitable errors in forecasting. And as we take the product of the system and go fly, our minds must be open enough to accept the actual weather, analyze it, and operate accordingly regardless of what might have been forecast.

One place where I yearned for up-to-date facts was along that high-MEA airway. There I was cast adrift; I was on my own. There was no way to find out precisely what lay ahead without flying through it. The center hadn't had an airplane in the area at low altitude, so they couldn't help. And when you think about it, any pilot report that I gave was probably of value for only a short time. There was convergence in the area, tops were building, and a fellow coming along an hour later might have found tops well above the ceiling of a normally aspirated airplane.

Moving along on this flight, en-route weather was not a problem once the ice had been left behind. The tops were relatively low and flat, and there was no indication of a weather system ahead to push them back up. So my primary need was surface weather at the destination. The last time I had checked, Memphis had been below forecast. If Little Rock was not clear as forecast, I would want to recalculate fuel—perhaps stop and top the tanks.

I set out to gather information from flight watch, the highly touted weather service for en-route pilots on 122.0. We are supposed to find an FSS person poised on this frequency, with all information at hand. All I got was great screeching and squalling. Flight watch is a one-frequency deal, and as pilots over a wide area seek information, it becomes a hopeless jumble. Jets with strong transmitters at high altitude pose a special problem: they can block the frequency over a huge portion of the country. When I finally did get a flight-watch briefer to answer, my request for weather was greeted with "Stand by." When he found the weather and started reading it, a jet blocked the frequency halfway through his discourse. I finally got what I needed by listening to the Memphis transcribed weather broadcast on low frequency. The TWEBs have been around for years and can be faithful servants if the tape is kept up to date.

Happy conclusion. The forecast of clear weather at Little Rock materialized. Another plus: the winds were not as strong as forecast. The flight averaged a 27-knot loss where one in the 40s might have been expected from the forecast.

To sum up the day's activities, there was the pesky ice problem, not mentioned in the briefing, that persisted for a couple of hours. The en-route surface weather was somewhat worse than forecast, and the winds-aloft predictions were only approximately accurate. (Filed time en-route was five hours; actual was four plus 30. A mistake in that direction is okay. The other way, and I'd have been nervous about the gas.) The ice came as the temperature aloft forecast proved inaccurate.

This was a rather typical encounter with today's weather-information system. The forecasting was about as good as it ever is when weather is not cut-and-dried, and the facts that I was able to gather were tolerably accurate.

There are a lot of questions on the horizon, though. The FAA does not plan to expand its FSS work force, yet aviation will grow. And the FAA apparently has moved away from a program to consolidate all flight service stations into 22 superstations colocated with air-traffic-control centers. Now, the FSS plan for the foreseeable future uses the Band-Aid technique—patch up what we have and go on. This approach raises three urgent questions: how can the FAA disseminate information as demand increases and their count of warm bodies remains stable? Will we be able to get more and better information in the future? How can pilots survive a situation that is bound to deteriorate before it gets better?

The briefing that I got was as good as this type of briefing ever will be. It did not take a lot of time, and it provided the information that was available for my route of flight. It was better than average because the briefer refrained from attempting what the FAA calls a "plain language" briefing, which amounts to a lot of bull and few facts. Using "plain language," the briefer talks as if he or she knows what the weather actually is along a route of flight. This is quite misleading because all the briefer really has is information on actual weather at a few places and an educated guess as to what it should be like along the way.

My man gave me the real stuff, with important items like temperature and dew point. As concise and efficient as it appeared, this one-to-one briefing— one government person giving his total attention to the unique needs of one user for a few minutes—is a luxury that won't be afforded the general-aviation pilot forever if there are to be more of us and the same number of them.

The FAA's big nut to crack is more efficient dissemination of information. They are starting from square one, too, because if you had set out to design an archaic method of spreading the word, you couldn't come up with anything more wasteful than the FSS system.

About 4,500 people working in flight service stations must satisfy the weather-information needs of general-aviation pilots. They also do some airline and military work, but not a whole lot. How many people do they serve? FAA figures suggest that nearly 50,000 pilot briefings are conducted daily in the United States.

Is that good? Consider that the National Weather Service caters to the weather needs of our entire population every day, with about 5,000 people in field offices comparable to flight service stations. The NWS serves 200,000,000 people a day with 5,000; the FAA serves 50,000 a day with 4,500. Before protesting that aviation weather needs are specialized, consider that the NWS does a lot of special chores for farmers, boaters, and people about to be flooded, as well as providing forecasts and other information for aviation users.

The rub is the dissemination of information. The NWS knows how to do it; the FAA is still trying to invent the wheel, and it thinks a square is the proper shape.

Hark back to my briefing once again. A person had to be at the FSS, waiting for my call, ready to go to work for me after he found out where I wanted to go. He woke up in a new world when I said "Asheville to Little Rock"; suddenly he had to learn all that geography. That is why our weather-information system is inefficient.

Unfortunately, the current FAA plan does very little to improve the dissemination system. The present "big" effort will increase briefer productivity by doing away with much of the paper in the flight service stations, which, according to the FAA, will free people to answer phones. The FAA has proposed two systems to accomplish this. You ready for the acronyms? MAPS and Awans. I will say only that both utilize cathode-ray tubes and allow the briefer to call up instantly current weather information for the pilot on the phone. You say "Asheville to Little Rock," he or she presses the appropriate buttons, the pertinent information appears on the scope, and the briefer reads it to you.

As a user, you will probably not notice much difference when your FSS gets one of these systems. A sharp FSS person usually can whip off a briefing as quickly from teletype paper, and the information will be the same. I once called an FSS with Awans (Atlanta), got a briefing, then called a plain old FSS, and got the same information. The Awans report took longer.

These systems *will* provide a few more briefers to answer telephones at busy stations. One FAA person estimated the potential productivity increase as 20 percent. That is certainly better than nothing, but when the inefficiency of the basic one-to-one system is considered, 20 percent is like paying off two bits on the national debt.

Another move will also produce more people to answer heretofore busy telephones without actually increasing the number of FSS personnel. The FAA is going to part-time flight-service-station operation at some low-activity locations such as Truth or Consequences, New Mexico. The personnel freed by this cutback will go to busier stations.

The vast majority of the flight service stations truly exceed our needs, and part-timing stations is a step in the right direction. It is very political, though. The majority of our congressmen probably have a station in their districts. Thus, FAA appropriation bills have specifically prohibited the closing of stations, and only the latest bill allowed part-timing. We surely can't blame the inefficiency of the small stations on the FAA. Congress, spurred by locals who want an FSS "right here in Podunk" even though information from a bigger station would work just as well, has created and preserved this archaic mess. The pressures are fantastic when a station is slated for shutdown—to hear some people talk, you'd think it would place every pilot in the county in immediate and grave danger. The irony is that the people who shout loudest are usually quite conservative on other government spending matters.

The part-timing is part of a process that should lead slowly into an FAA plan to upgrade the 43 busiest flight service stations. These 43 stations might eventually get MAPS or Awans equipment, plus some other good things, and should become the best sources of information. Smaller stations will still exist, but any upgrading in facilities there will be limited.

Some improvements slated for these busy stations constitute genuine good news. They should all have radar. In many instances, it will be a repeater system operating off the long-range air-traffic-control radar. These will show raw radar data before it goes through the air-traffic-control computer circuitry that minimizes precipitation return. This won't show precipitation quite so well as a pure weather radar, but it is certainly adequate for general-aviation purposes. This radar display is already operating at some locations, and specialists have found they can identify the radar return of an aircraft on the scope and give the pilot weather information in relation to his or her position. This function could add an important new capability to the flight service station. Do note, however, that no vectoring or traffic-advisory role is envisioned for the FSS.

Another neat addition to the bigger stations will be pictures from weather satellites. These are produced every 30 minutes and show cloud cover. Much information can be derived from them. Thunderstorm areas show with special clarity. The FSS has never had much information on weather between stations, and the satellite pictures will cover this important area. They come through at night, too, in an infrared version. Unfortunately, the night picture poorly displays clouds such as Gulf stratus because they are the same temperature as the land mass.

There is an interesting sidelight on the use of satellite pictures. These are available at National Weather Service offices now, and they have been put to effective use in search-and-rescue operations. If the takeoff time and approximate route of a missing airplane is known, the appropriate satellite picture can be consulted for the areas of worst weather along the route. Missing airplanes have been located by searching the indicated areas of bad weather; and if the pictures can be used in that manner, they surely could alert pilots to those very problem areas.

These new good things can't help us much on the other side of a busy telephone signal, though. Are they going to fix this?

Having each pilot actually talk to a government person before a flight is almost holy to the thinking of some FAA people. It will be difficult for them to face reality—to replace the one-to-one with mass dissemination for the majority of weather briefings, handling only the exceptional needs by phone call. It will be difficult because the FAA is not prepared to admit that the user is mature enough to take full advantage of information. A strong contributing factor is a laziness in the pilot population. For years we've called and gotten tailored briefings. Many pilots feel this should continue indefinitely. There is mention of mass dissemination in the plan to improve things, but I don't get the feeling that the FAA's heart is in it.

The FAA does have the shell of a mass-dissemination system in place, and a lot of us do use the transcribed weather broadcasts on low-frequency beacons and VOR stations. The low-frequency TWEBs are potentially most useful to a person on the ground because of their good range, and because the information covers a fairly wide area. They are also the most neglected. The equipment is old and antiquated, the FAA's effort to install new equipment has been a shameful exercise in procrastination. The equipment is nothing more than a recording and broadcasting machine, but the difficulty that has surrounded procurement would do justice to something as complex as a moonship. I've heard "next year" so many times that a year in FAA parlance must be at least a decade. As an illustration of how the FAA thinks, consider this: they have installed machines that play a recorded message saying all the briefers are busy, while doing nothing to update the TWEB equipment.

An FAA procedure has coupled with the old equipment to make the quality of TWEBs poor. The preparation of the broadcast is at the bottom of the priority list of things to get done in a flight service station. Listen to some of the TWEBs, and it becomes apparent that this is taken literally. The TWEB is done only if the briefers can't find something else to do. Outdated and incomplete reports are frequently broadcast, and occasionally the TWEB doesn't play at all.

I called about an inoperative TWEB once and was told that it was off the air because they couldn't spare a man to cut the tape. They were all busy answering the phone. Nobody in the FAA seems to realize that a TWEB of consistently excellent quality eventually would attract a wide audience and cut down on the number of phone calls. The user need only know that the recording contains the most current information available, and that nothing further would be gained with a phone call. Some FAA people suggest that nobody listens to the TWEBs now. Make them good, and people *will* listen.

The TWEBs on VOR stations give information on a smaller geographical area than the LF TWEBs and are designed to cut down on the number of air-to-ground weather calls. These are quite effective, and if they are expanded and fine-tuned, they might eventually eliminate a very high percentage of the FSS radio chatter. A broadcast of all the hot poop would cut the traffic on flight watch in most areas of the country, too. Again, the quality must be consistently excellent for the service to become popular with pilots.

The National Weather Service has an expanding TWEB network far superior to the best the FAA envisions. The NWS system isn't for aviation, unfortunately, but the FAA could surely look at it with copying in mind. The NWS uses 162.55 MHz plus two other frequencies; 90 percent of the population will be in range of one of their weather broadcasts when the system is completed. A wide variety of radios have this frequency, so almost everyone will be able to have up-to-the-minute (hour, really, since that's how often the tapes are made) general weather information from the NWS. The broadcasts include radar summaries when appropriate, so they will have some aviation value. There is an experiment in Baltimore that piggy-backs aviation weather on 162.55, but unfortunately this isn't expected to become a normal procedure.

The FAA's partial answer to this VHF broadcast might be the pilot automatic telephone-weather answering service. Patwas is expanding, and is often excellent. The good tapes are made hourly and include current weather. In some areas, Patwas is available for local operations as well as for selected routes; in others, they are putting the low-frequency TWEB tape on a phone recording. But wouldn't it be nice to also have a VHF TWEB system like the one the NWS has developed? Only one person at a time can listen to a phone recording—thousands can listen to a broadcast. You could cuddle up under the covers and get the latest aviation weather.

Any FAA VHF broadcast would have to be a new program. The TWEB on a VOR frequency doesn't serve a pilot before flight because VOR stations have very poor ground coverage. The NWS puts their VHF antennas on tall buildings. A VHF TWEB would also eliminate static, a common complaint about the low-frequency TWEB during thunderstorm season.

Certainly if any portion of the FAA's plan needs expansion and emphasis, it is the transcribed broadcasts. A goal should be set to reach 90 percent of the pilots with top-quality transcribed weather information. A complete reevaluation of the information in the broadcast is also necessary in order to maximize the TWEBs' effectiveness. Study what pilots ask for in phone calls to the FSS and then put that information on the TWEBs. And certainly the TWEBs will have to move to the top of the FSS priority list.

Radar at more flight service stations and satellite pictures have been mentioned as important new information sources. Many people consider a better system of pilot reporting of equal importance. This has had some attention, and at one time there was an experimental program in which pireps were recorded and rebroadcast for everybody. Pilot reports of weather must always be taken with a whole shaker of salt, though. Weather is dynamic, especially when it is causing problems. In many situations, a pirep's useful life might be just a few minutes—far less than the time it would take to disseminate it.

There is still much value in pireps. The airplane is a fine weather sensor, and it should be used as such. For example, the information that you now get on temperatures aloft is almost always a forecast. (Two actual upper air soundings are made daily at NWS locations, but the results of these are not generally used by FSS personnel. If you do ask for the sounding, its value might be sharply diluted by age.) Every airplane has a thermometer, though, and actual temperatures-aloft information is available to the Government only for the asking.

Winds aloft are all forecasts, too, yet in every airplane flying, you can determine actual winds. (That is, if you remember how to use a computer.) This is a ripe area for pireps and, like other information, it can be disseminated best through an efficient system of transcribed broadcasts over LF, VHF, and the telephone system.

We pilots must make work whatever system the FAA proffers. It is interesting that many FAA personnel honestly feel that they make the system hum. That is a misconception. They only make it work for themselves. Any value a pilot derives is a result of his or her individual effort in seeking and using the available data.

The basic staples to use in learning about conditions include actual weather reports, weather maps, radar reports, pireps, and forecasts. The new things that are coming—satellite pictures and expanded radar service—will enhance that information but won't really change the basics and the way that we have to use the information to survive. And where there might be a security-blanket feeling about the personal phone call, rest assured that it matters not whether we receive the information that way or from a broadcast. What counts is that we get it.

The weather map will always be the basis of any weather briefing. A map can be described easily in words: "There is a low in the Texas panhandle with a cold front extending southwest from the low . . ." If a pilot has rudimentary knowledge of weather principles, he or she can envision easily enough the likely distribution of inclement weather around map features. We should never fly without understanding the current weather map. The ones on TV and in the newspaper don't count because they are often forecast maps.

The actual weather reports go a step further by telling us what effect the map features are actually having on the weather. Reports are only slightly less important than the map and are easily covered in a broadcast. Any reduction in the number of flight service stations will undoubtedly result in fewer actual weather reports, but contract observers or automatic weather-reporting equipment can and should provide reports at most locations.

Radar reports tell us about precipitation and are a vital adjunct to reports from stations even though they are not precise. The current procedure is to outline a large geographic area and give a percentage of coverage within that area. Lines of storms are more precisely defined. Satellite pictures will make thunderstorm information much more current and precise. Radar equipment in the flight service station will do the same. These items will be of great pre- and inflight interest and must not be hidden behind a busy signal.

Pilot reports are often thought of as a cure-all, but they are of value only if very recent and reported by an aircraft of similar size and capability concerning your precise area of interest. Only brand-new reports need to be included in broadcasts.

Forecasts are probably least important because they do not constitute facts. Forecasts are estimations of what will probably happen if all the weather systems behave as expected. The criteria for grading forecasts as incorrect are so broad that a forecast can be a serious operational consideration and still be graded correct. They should have relatively low priority in broadcasts. It is important to have forecasts, but it is equally important almost to expect them to be incorrect. The pilot's fondest wish has been for better forecasts, but it is not likely to come true.

There we have it. The necessary information is simple. What's available now is adequate, and improvements from satellite pictures and radar expansion in flight service stations will make things even better. The rub is getting the information. Today it is generally available, primarily by phone. There are to be a few more people to answer phones but not many, so the present system is destined to collapse. The FAA needs to put more emphasis on mass dissemination and less on adhesive gauze applied to the present outmoded system. If this isn't done now, and if we do not learn to use the information when it is presented in this manner, we'll have a true crisis in weather information within five years. We *must* replace that recording that says, "All the briefers are busy, do not hang up" with a system of recorded current information that tells us what we need to know.

PART II: IFR

1 THE WORLD OF INSTRUMENT FLYING

by Len Morgan

INSTRUMENT flying. You sit in your metal cocoon, relaxed but attentive, scanning the multicolored display of needles that confirms the fact of flight. Beyond the cockpit, a sullen whiteness shuts out reality, making sinister suggestions: you are motionless in space; you are drifting far from course; you are slowly settling back to earth. But these evil whispers fall on deaf ears.

The gauges don't lie. They say you are on course and comfortably close to flight plan. They verify that the remaining fuel will take you home and on to an alternate if home socks in. You are a hundred miles ahead of the airplane, visualizing the approach with the ILS page already on the clipboard. You're fat. That's instrument flying.

Or you sit there with everything going from bad to worse. What the forecaster guessed would be scattered showers has turned into a full-blown line squall, solid and menacing. The headset is filled with voices asking for vectors, new altitudes, the latest sequences; you can't get a word in edgewise. Inexplicably, the vortac ahead has gone off the air. Headwinds and turbulence have eaten into fuel reserves. What to do now? That, too, is instrument flying.

Instrument flying is almost any weather-flying situation you can imagine, good or bad. It can mean no more than climbing through 500 feet of stratus to get at the clear skies above. It can mean six hours without a glimpse of the ground—six hours of rough air, rain, ice, diversions, holding, and a tight approach to minimums to cap a long and wearying trip. For the airman with the right blend of skill, experience, and self-discipline, it's a routine way of doing business, whatever the circumstances. It is more demanding, but not more exciting or dangerous, than flight with continuous visual reference to the ground—and it is far more satisfying.

An airplane unequipped for instrument flight is like a car without headlights. A pilot without an instrument ticket is like a swimmer with one foot on the bottom. Are you ready for the deep end of the pool?

Before you become immersed in a welter of details (a fair definition of instrument flight), stand back and get the whole picture. What's it all about? Well, you say, it's the technique of flying from here to there when the ground cannot be seen en route. The problem could be a thin layer of haze or fog at the departure point. The gear is hardly up when you pop out on top. No sweat. Or you might have to enter an overcast at 12,000 feet and ride down to 3,000 on the gauges. Unless ice is involved, no big deal. An active front along or across the route is another matter altogether. So is the quiet coastal airfield that lies ahead, sparkling and windless in the evening, only to disappear from view in ground fog as you turn onto base leg.

Weather is the riddle to solve, the enemy to conquer. Question: how good is the advance information about weather given to a pilot planning an instrument flight? Answer: it's not nearly good enough. No criticism of the National Weather Service is intended; all things considered, the NWS does a magnificent job. It's just that it is still impossible to make consistently accurate forecasts. Today's forecaster can predict the trend over Kansas and Nebraska and hit it almost every time, but he cannot be absolutely sure about the ceiling at Topeka when you arrive overhead. He cannot tell you what the winds are doing at 15,000 feet over Wichita, but what they were doing there 3 to 12 hours ago when the last observation was made. You are interested in specific route conditions and the exact situation to be expected at a distant pinpoint hours from now, and these no weatherman can guarantee. To launch a flight into weather without all available information and advice is sheer folly; to forget that weathermen deal in probabilities, not certainties, is equally hazardous. There probably isn't a forecaster alive who hasn't pulled off his snow tires a week too soon.

A second point to bear in mind, now and for as long as you fly instruments: there is almost no weather you cannot tackle—providing you have the right training and the right tools. It is imperative that you continuously reappraise your own experience and equipment, matching them against the problem at hand with scrupulous honesty.

The average handyman can accomplish much with a hammer and saw, but there are many jobs beyond him. Conversely, the finest woodworking machinery is useless without a craftsman to operate it. A properly equipped Bonanza or 182 is, in the right hands, quite adequate in many instrument-flight situations, but there are limits beyond which such aircraft should not be pushed. With a deiced and radar-equipped small twin, you can take on more, and a Learjet or DC-3 will see you flying when the lightplane crowd is—or should be—waiting for better reports. Yet even the pilots of these types will sit out certain route/weather situations that are considered a routine in a DC-8. Finally, there are hours each year on most air routes when nothing moves—no DC-8s, no 747s, no "all-weather" fighters. A weather map depicts a hundred different riddles, each requiring its own level of skill and its own minimum list of tools. When you have learned when *not* to go, you'll always know *when* to go.

Instrument flying should remain a challenge to a pilot's skill. Too often, it becomes a challenge to his pride. It is not easy to sit and wait while other pilots in other planes are flying in questionable weather, apparently without incident. The temptation is to rationalize. Am I not as good an instrument pilot as they? Is my ship not as well equipped as theirs? Unchecked, this line of thought may lead to a new look at the situation, this one clouded by a tendency to minimize the problem and inflate your ability to handle it. You end up talking yourself into something you would never have tackled had you been locked up to think it through. Forget what the others are doing. It's your skill, your airplane, your judgment, your neck. Do your own thing and your own thinking; fly when you like the prospects, and only when.

Now, to the actual technique of flying an airplane on instruments: if you come far enough to be in the market for an instrument rating, you can go the rest of the way—and you should. This is not a field reserved for super airmen. It's a brand of flying that makes you think more and work harder than VFR flying, but its tricks are well within the grasp of any good VFR pilot. Why not try it on for size? All you need is an instrument-equipped airplane and an instrument-rated safety pilot to keep you clear of other traffic. A turn-and-bank, airspeed indicator, and altimeter are enough tools for the first experiments. Get away from all airports, off all airways, stuff a small pillow above the glare shield to block forward vision and . . . you're on instruments. Now you must relearn how to fly, starting with the straight-and-level bit and working at it until you can maintain altitude and heading without conscious effort.

A beginner's tendency is to stare at a gauge until he has stabilized its reading, then transfer his attention to another. This won't work. The idea is to *scan* the panel continuously, never spending more than a second or two on any instrument. Scan: you will hear that word many times from your instructor later on. The skill of scanning comes with practice, though you may doubt it at first. You will develop a scan that works for you—a "flow pattern" and speed that are comfortable.

When straight-and-level flight is being kept within reasonable limits, standard-rate turns can be tried, then turns while climbing and descending. Two or three sessions under the hood should see these basics shaping up. Now try speed control. Let's say you are in a Cessna 150, cruising at 4,000, and indicating 90 knots; ease the power off, carefully maintaining heading and altitude, and nail it on 60 indicated, retrimming as necessary. Then increase power and return to cruising speed, maintaining altitude and heading through. When you can hack this without straying off headings more than five degrees either way or altitude by more than 50 feet up or down (in smooth air, at least), you're ready to complicate things a bit.

Try this one: slow to 60 knots in straight-and-level flight, then go to full power and start a climbing left turn, keeping speed right on 60. After completing a 180-degree turn, reduce power and enter a right descending turn back to the original heading and altitude, flying right on 60 all the way around, then stablize in straight-and-level flight at cruising speed. Your safety pilot can suggest similar exercises or you can invent your own.

Look at what you are doing after two or three hours of such experimentation: you are flying a precise straight-and-level flight path, changing course and altitude at will and keeping airspeed at a selected figure, all without a look outside. Not bad, eh? And far more satisfying than slopping across the landscape with heading and altitude wandering all over, isn't it? You may not be ready to make an approach to an RVR 1800 runway, but you are well on your way. Nor is your toying around with instrument-flight time wasted. Until you can do the basic airwork, keeping all readings within close tolerances, and manage all of it as a matter of second nature, it's useless to clutter your thinking with navigation problems and radiowork. They will occupy your full attention later on, leaving you little time to consider the mechanics involved.

Having come this far, the chances are that you will want to sign up for the full course and get that rating. You have already learned a few things about flying instruments and they intrigue you. Obviously, it's a mental workout from start to finish. The proper state of mind in this end of things is based solely upon your knowledge and preparation—your knowledge of weather as a science and as it will affect today's trip; your knowledge of the route and airports involved; your knowledge of instrument-flying rules, procedures, and techniques; your awareness of your plane's capabilities and limitations; your recognition of your own degree of proficiency; and the thoroughness of your preflight preparations. When this background is right—and it is entirely up to you to make it right—then the actual execution of a flight into weather is almost an anticlimax.

Old-timers talk about the days "when we flew by the seat of our pants." Well, we still do, even on solid instruments. The human body is sensitive to changes in direction and rates of change. When the DH-4 airmail pilot felt his bottom floating off his parachute, or being rammed into it, he took a quick look at the horizon. The same sensations make a DC-10's pilots glance at the panel. Obviously, you cannot depend on the senses except for backup information. Alone, they won't keep you straight and level in a cloud for 60 seconds. Relied upon, they will kill you. The point is this: during training, you will be told to believe your instruments, which you must, and to ignore the seat of your pants, which you can't. It's quite impossible to throw your physical feelings into neutral in an airplane or anywhere else.

Being aware of this beforehand makes the transition from VFR to IFR flying easier. The initial awkwardness and confusion will quickly pass, and you will smoothly maneuver your ship through all regimes of flight, comfortably assured by the seat of your pants that all is well. "It's a matter of getting your brain into gear with your butt," said one pilot. Few pilots appreciate the reliance they place on physical sensations until they fly a synthetic training device. A 30-degree banked turn in a simple Link or a million-dollar simulator just doesn't feel right because it does not press you firmly into the seat. The average airline pilot much prefers to take a check ride in the airplane itself.

The shortcomings of any sort of education become apparent upon the first attempt to apply its teachings to real-life problems. Looking back on formal schooling or specialized training, we can see the fallacies in them and the wide gulf they leave between theory and practice. Education does not provide answers as much as it teaches us how to find answers on our own. In this respect, an instrument-training program is not unlike a high-school chemistry course. The bare facts learned in a school lab hardly convert a student into a chemical engineer, but they do put him in the proper frame of mind to absorb the advanced intelligence that chemical engineers must possess. Haven't we all entered a new school or job, certificates of previous accomplishment in hand, only to hear, "Forget everything you've learned; *this* is the way we do it!"

The beginning student in any field has no choice but to accept without question whatever his teacher tells him. He studies, passes the tests, and moves on to something else. Should the course in fact be weak, outdated, and incomplete, it will be recognized for what it is in due time. In most areas, an incompetent teacher does little permanent damage to his charges; at worst, he wastes their time.

In flying, the situation is quite different. Having taken his course and passed his tests, a student may immediately stake his life and the lives of others upon what he has learned. If his instruction failed to give him a firm grasp of things, well . . . the dreadful possibilities do not have to be spelled out. Our trade is fortunate in having some government controls that go a long way toward ensuring that instructors know what they are talking about and that students demonstrate certain minimum standards of performance. While these controls place our game head and shoulders above those played by most of our nonflying friends, they don't guarantee perfection.

Pick the best instrument school you can find and hope you get an instructor. Oh, the pilot you are assigned to will have "instructor" typed on his license, but he may not be one. Instrument courses are often "taught" by check pilots.

A check pilot says, "You're 200 feet low," or "You're off your heading again," or "that's not the way to enter this holding pattern." An instructor analyzes your thinking and tells you *why* you are out of tolerance and *how* to stay within it. He has not forgotten how it was to be inept. The art of really teaching a subject is not widespread in any field.

There are as many ways to skin this instrument cat as there are airmen with instrument ratings. Your man will teach you his way, stressing the points that gave him trouble and perhaps hurrying over those that he found easy. This is human nature. Accept at face value everything he dishes up, for his way will put the rating in your pocket. Later on, you can and should make any minor adjustments in thinking that will make the chore easiest for you. Remember that the manuals you study and lectures you hear—and magazine articles you read—reflect the ideas of those who put them together. Those authorities won't mislead you any more than your instructor will. They understand the rigid framework within which all instrument flight must be conducted; they want you to operate safely and efficiently in their fascinating world of weather flying. Their ways are proven and they will fit you—once the sleeves are taken up a bit.

Lucky is the newly rated instrument airman who starts his weather flying as a copilot. I spent 16 years in the right seat. To paraphrase Will Rogers, I never flew with a captain who didn't teach me something. Some of what they taught was about instrument flying. My FAA-designated physician, a man who has 35 years of experience with pilots, once told me, "I don't want to ride behind any pilot who is not convinced, deep inside, that he's the best in the world." He was talking about self-confidence based on knowledge and experience, not conceit. I never pulled the gear for a man who did not firmly believe his way was the right way. Some, of course, imagined that their way was the only way, and a month on schedules with such an unbending type never seemed to end. But even these static characters had something to offer an apprentice. Copiloting with a good assortment of qualified instrument airmen is by far the best way to learn the ropes.

If it won't be this way for you, if you will be alone and on your own once you are rated, do your homework now. Learn not only what is taught during the formal training itself, but by reading everything you can get your hands on and by practicing whenever you can get a plane and a safety pilot to ride with you. Accept all available knowledge and advice, but regard all of it as a guide to use in honing your own technique to personal sharpness. Be as suspicious of the weatherman as you are of the lineman who topped your tanks. Neither is going along for the ride.

Most of all, be suspicious of yourself. Double-checking takes minutes but saves lifetimes. A story in my local newspaper this week says it all. A professional man with his family aboard took off and disappeared into a

600-foot overcast. Two minutes later, his $40,000 airplane fell out of the clouds and crashed. Six fatalities. The pilot was considered competent and was instrument-rated. His plane was adequately equipped to climb above the cloud and fly in the clear to its VFR destination. What should have been a lead-pipe cinch ended in disaster. When the scraps were collected, the fuel selector was found to be set on the only one of the ship's four tanks with no fuel in it.

During the overall process of training yourself to be an instrument-flying expert, learn something about how corporate and airline pilots handle weather flying—not so you can copy them but to understand why you cannot. Compared with the task facing a Cessna 182 pilot who plans a trip from, say, Atlanta to Chicago when a line squall lies across the route, a 707 crew has it easy. A 707 can scramble to 41,000 in half an hour, and up there, it is usually possible to sneak through a saddleback. Otherwise, a detour can be made to skirt the end of the line, even if such a diversion takes it 300 or more miles off the regular route. Along the way, the airline types are in immediate contact with as many other airliners or business jets as they wish to consult about conditions up ahead. At their disposal is a corps of company dispatchers and weathermen that usually provides advice even before it is requested. While the old man flies (with the good help of an autopilot that keeps altitude nailed down), his copilot handles the radio chores and his engineer keeps a watch on fuel and engine readings. And the three of them are flying an airplane they know better than their own cars on a route they have covered hundreds of times before, often under almost identical conditions. They make it look easy because it is easy, for them.

By comparison, the single-engine man who must navigate the same distance without radar or autopilot or professional cockpit assistance, making his own contacts and decisions with little of the expert help so readily available to the airline fellow—well, he has his afternoon's work cut out for him. His problem is immeasurably more complex and difficult to solve. He should recognize it as such, manage it as such, and take real pride in pulling it off as safely as the big-plane boys, and with but a fraction of the equipment, assistance, and experience at their command. An instrument pilot isn't measured by the number of expensive devices stuffed into his cockpit.

It should not be necessary to mention this at all, but any analysis of private-aircraft accidents involving weather suggests that a dangerous illusion is shared by some newly rated pilots. They appear to believe that expertise is a transferable commodity. Just how this notion gets started in a pilot's thinking is beyond me, but proponents of the theory continue to make the headlines with tragic regularity. Having proven himself at some difficult profession or skill that is obviously beyond the capacity of the average pilot, an individual decides that instrument flying is a minor challenge that can quickly be mastered by one of his demonstrated intelligence. Sure enough, the instruction doesn't faze him; he breezes through and wins his rating. He's a bright fellow, but not smart, and this is suddenly revealed when he piles his friends into his expensive six-seater and roars off into the wild black yonder. The six of them rate front-page attention the following day. Friend, your success on the ground, no matter how remarkable, does not automatically follow you into the clouds. Don't prove me right by assuming that if us ordinary-looking, run-of-the-mill pilots can make it, you can, too. We can tell you tales about weather flying that will curl your hair.

There's a grandmother I've been trying to take flying since she was a schoolgirl. No luck. The idea of it, even aboard a huge airliner, terrified her. Recently, she went up for the first time and told me all about it. It seems that she rode a Bonanza across 200 miles of mountainous terrain at night in foul winter weather. I expressed an interest in her pilot, who is her neighbor. "Oh, Ted is a wonderful flier," she gushed. "Why, he has a college degree." Good luck, Ted.

Get that rating, even if you have no inclination to launch into extended weather situations. It's great insurance and will get you back to base should a scattered undercast suddenly become broken. It will get you going on mornings when a haze layer separates you from clear skies. It will improve your VFR flying by increasing your speed and range while cutting fuel burn per mile, since you'll make it a habit to keep those gauges on the money. It will make all of your flying more precise, more expert—if you exploit the tricks your instructor taught. It's good stuff to know, and you will be a better pilot for learning it. It's also great fun.

2 GETTING YOUR INSTRUMENT RATING

by John W. Olcott

"THE MOST IMPORTANT thing an instrument-flight instructor can teach his IFR students," said the airline captain who moonlighted as manager of the flight school, "is *not to fly instruments*. Get them their rating, but convince them not to use it. Teach your students that IFR flying is for the pros, and that it is too sophisticated and demanding for general-aviation pilots."

Perhaps too many instrument-flight instructors get such misguided and counterproductive advice. Such an attitude may be why nearly one out of five pilots is IFR-rated, but only a fraction of the nonprofessionals fly instruments on a regular basis.

IFR flying definitely requires a precise, professional approach toward aviation. With proper instruction, the many benefits of instrument flying are well within the grasp of the average general-aviation pilot. IFR instruction should supply him with the necessary knowledge, skill, and guarded self-confidence to use those benefits. Instrument training that does not provide the pilot with the ability and desire to participate in the actual IFR system is deceptive and a waste of time.

The objective of instrument instruction should be safe and effective flight according to instrument rules. It's as simple as that; and good instruction will support that objective. There is much more to flying IFR than pilotage of your plane solely by reference to instruments; all applicants for a private license have to demonstrate that capability. Maintaining heading, altitude, and airspeed is just the tip of the iceberg, and if that's all you're getting from IFR instruction, you should demand your money back.

Your training must prepare you to handle all the real-world tasks you will ultimately encounter while flying IFR. Some of those tasks will involve normal ATC procedures and characteristics: copying a clearance in flight, entering a holding pattern from any orientation, capturing an ILS after approach control has vectored you onto the final approach course right on top of the outer marker, etc., etc. Phony instrument instruction protects you from these annoying little problems by advising not to file actual instruments, or by having the flight instructor act as a supersmooth copilot. In addition to coping with garden-variety ATC tasks, the pilot must make and execute decisions involving en-route, destination, and alternate weather conditions. There is more to a weather decision than saying, "I'll cancel and go tomorrow." IFR training that doesn't prepare you to make a realistic appraisal of the weather leaves you no better off than you were before you spent time and money for the rating. Then there is the question of emergencies, such as instrument malfunction or radio failure. Those things do happen, and you had better have been there with an instructor prior to facing such experiences by yourself.

Landing out of a circling approach at published minimums, or a missed approach executed when reentering instrument conditions after initial visual contact, are some of the more challenging IFR tasks. Weather conditions do change, so you must be prepared to handle those necessary piloting tasks just as well as you fly a regular omni-tracking problem. Real instrument training prepares you to handle realistic situations, by yourself, without relying on the instructor to tell you what to do.

The pilot who feels that these IFR tasks don't concern him, but are solely for the person training for a corporate job, is misled. Even if you plan to limit your instrument flying to simple situations, the realistic training is indispensable. Maybe the IFR rating is solely to augment your VFR flying. You don't really plan to compete with the airlines for honors when the runway visual range is 1,600 feet. You just want to file to on-top conditions and let down at your destination. Perhaps you want the IFR rating as a safety measure for bad weather, or possibly you wish to improve your overall flying proficiency. All these scenarios provide valid reasons for seeking an IFR rating. It's the mark of a good pilot to recognize limitations and back off accordingly. We wholeheartedly support a realistic approach to what a businessman-pilot flying by himself can and cannot do *vis* à *vis* instruments. But you haven't satisfied your objective unless your training prepares you to fly instruments in the IFR environment as it actually exists.

To use your rating safely and effectively, you have to file, and that means living with the ATC system. Communicating with ATC and following their directions represent fundamental IFR tasks: it is part of the regime to handle them effectively by yourself. You share the frequency and the airspace with others who use the ATC system, so you have a responsibility to be proficient for their sake as well as your own. If receiving a new routing, or copying clearance in the air, or executing whatever approach is given to you are common demands of the ATC environment, how can you consider yourself

IFR-qualified unless you can handle those situations? The answer is simple: you can't!

Take ILS approaches as an example. IFR flying involves cross-country operations; and eventually, if you plan to fly to any major city, you will be asked to make an ILS approach. Even though the FAA hasn't revised Part 61 to require ILS approaches on all IFR flight tests yet, it is rather naive to be unprepared for radar vectoring to the ILS final approach course. I'd estimate that better than half my IFR approaches have been ILSes, and I suggest that an instrument instructor who doesn't prepare his student for ILS approaches under actual or simulated high-traffic conditions isn't doing a complete job.

The properly trained IFR pilot can handle the idiosyncrasies of ATC. There is no guarantee that you will always receive an on-top altitude, or that your assigned altitude will allow you to remain clear of the clouds. Likewise, there is no absolute assurance that the en-route and destination weather will be exactly as forecast. Even if you don't plan to tempt weather, you must be prepared for its contingencies. Even a simple thing like pitot ice can be a fatal challenge for the poorly trained instrument pilot.

You must be prepared for instrument and radio malfunctions. It may be VFR under the clouds or CAVU at 1,000 feet above your assigned altitude; but if you are on actual instruments, the loss of an artificial horizon or the failure of your communications equipment requires an effective level of IFR knowledge to handle the situation satisfactorily.

The implication is not that an IFR flight can't progress exactly according to plan. Many do, but a slight deviation from your original plan of action isn't unusual, and it shouldn't put you off stride. Contingencies do arise, and you should be able to handle them effectively and certainly without panic. Your flight planning needs to consider all aspects of the IFR environment— characteristic anomalies and contingencies as well as normal situations—and your learning experience should prepare you to meet all those possible challenges. That is what we mean by a professional approach to instrument flying. Anything less is not to your advantage.

The concept of a working IFR proficiency applies to the pilot who obtains his rating solely for weather insurance. He will find his policy may have been worthless to start with unless he learned how to handle himself in real IFR situations, and it may have lapsed if he did not maintain his proficiency.

Was the cynical airline captain right after all? Is instrument flying too demanding for the general-aviation pilot? Once rated, should he limit the use of his newly acquired knowledge solely to impressing people at cocktail parties? In each case, the answer is no, provided the pilot receives training that prepares him realistically for the actual conditions of instrument flight.

IFR flying is not difficult. With proper training, any person with reasonable flying ability and background can learn to use the IFR system safely and effectively. Becoming IFR-qualified takes a strong desire to learn, and a qualified flight instructor who wants his student to use his rating once he receives it. Properly familiarized with the system, the pilot can acquire the necessary knowledge to manage himself and his aircraft on instruments. Equipped with 40 or more hours of realistic instrument instruction and drill, which must include actual IFR exposure under typical conditions as much as possible, the newly rated pilot will be able to conduct successful IFR operations.

Instrument flying is well worth the effort it takes to become IFR-qualified. It is a lot easier and considerably safer to file IFR than to sneak under the clouds during marginal VFR conditions. Also, the money invested in the rating can expand utilization of an aircraft, provided that you use the rating after receiving it. The organization and self-discipline that accrue to the pilot who really learns to fly actual instruments pay dividends in all facets of his flight activities, whether they be VFR or IFR.

Here are a few things to look for in a realistic instrument-training program. Select a flight instructor who flies instruments on a fairly regular basis. A pilot who flies instruments often has made his peace with the ATC system. He has learned to cope with its demands, and he has determined how to use its characteristics to support his objectives. He can teach you how to use the IFR system to support your needs, safely and realistically. Be sure your flight instructor is willing to make actual IFR flights with you during your training program. One or two IFR cross-countries is not enough. Insist on an instructor who will give you as many flights as you need under typical instrument weather and traffic-density conditions to become competent.

Naturally, this criterion implies that you must own an IFR-equipped plane or find a school that rents them. Every hour of flight time need not be in an aircraft with all the latest whistles and bells, but you'd better have equipment available for serious IFR flying. A boom mike is one piece of equipment that is worth its weight in gold when it comes to instrument flying. Picking up and putting down a regular microphone is too awkward and time-consuming when you must copy clearances in flight, particularly for a person learning to fly instruments.

Insist on a training curriculum that includes all the maneuvers and tasks you might experience in actual IFR operations. In the area of airwork, it is important to have good airspeed control over the entire range of flight conditions and configurations, and to be able to transition smoothly from one airspeed/ configuration to another. It will help you when being radar-vectored to a final approach course and making the transition from the IFR approach to the VFR landing. In fact, good airwork technique is a prerequisite for all precision IFR tasks, so don't bypass good training in this area in favor of more glamorous topics like approaches and cross-country trips.

Your radiowork must encompass all the various facilities and approaches available to the IFR pilot. This is not to suggest that you find an obsolete LF range and practice listening to the build and fade, but a recommendation that you learn to fly an ILS even though the nearest facility might be 100 miles away. Your IFR rating won't be limited to the local area, so there is no reason that your training and experience should be. You can easily plan IFR flights to an airport that has an ILS, and conduct all the demonstrations and drill necessary for ILS proficiency. Learning to fly an omni precisely is excellent practice for flying the localizer, so it's possible to transfer some of your local omni-tracking and approach skills to ILS work. It is entirely practical to develop workable ILS techniques regardless of where your training base is located, provided ILS work is an integral part of the curriculum.

The bulk of your IFR training should be under actual filed conditions, where you do all the work including clearance copying, radio navigation, decision making—the whole thing. Once you have the fundamentals of airwork, radio tracking, and approaches under sufficient control to tackle putting it all together, your curriculum should concentrate on the principal objective of IFR training—the cross-country flight that fully utilizes the ATC system. A lot of the mandatory material can be taught with ground-based training aids and simulators, but short of airline-type training devices, there is nothing available that can substitute for actual ATC experience.

Initially, your instructor will and should assist you in the management and housekeeping functions of IFR flying. His role is to teach you the proper techniques and to demonstrate how to apply them under realistic conditions. He will supervise your drill and make sure you don't hurt yourself in the process. He can test your progress and give his evaluation. The transference of duties and responsibility must eventually be made, however, and it must be complete: it just isn't good enough to have the instructor constantly bail you out on new routings or clearances or estimates. You must learn to be independent of him, so you will be competent on solo IFR flights.

A major part of your IFR training is the development of the required techniques to get the job done safely and effectively without undue stress or panic. That is a tall order, but it is possible, and the piloting composure you develop in the process is invaluable. The best—and perhaps only—way you can develop an IFR technique that is effective is to work at it in the actual ATC system. When you consider IFR training, make it clear to your instructor that you are not interested in a phony instrument rating. Tell him you want the real thing, and that you plan to use it.

③ ME AND MY RATING

by Gordon Baxter

INSTRUMENT FLYING, I had concluded, is an unnatural act, probably punishable by God. Even if it isn't, the whole idea of flying an airplane when you can't see out the window seems self-defeating.

I knew a few instrument pilots. Technicians. The type who would read the marriage manual on their wedding night, or sip Coca-Cola at a party when everyone else was being thrown into the pool. I have a friend—a doctor, a little steel-rimmed guy with bald pate and the strongest, coldest fingers, like forceps. He flies his Mooney just exactly like in the book. His approaches leave a crisp dotted line in the clouds. He bugs me.

I respected instrument pilots, understand. Awed by them and their sudden popping out of the lead belly of a cloud and taxiing up to where we stood, huddled like fluffed-up birds under the dripping eaves. Getting out with their little secret chart books. Who is this Jeppesen, anyway? I peeked at those flimsy little chicken-track charts. No rivers, no railroads, no drive-in theaters, all written in Egyptian. No rosy colors to tingle a pilot's imagination.

I put aside any idea of my ever flying instruments. I'm going to be a low-time pilot all my life. Flying since 1957, yet the total hours stay below the magic thousand. A very sparse ticket, lots of white space around those words, "private—SEL." That's for SELdom. I had heard that instrument flying was such a demanding skill that unless a pilot went out and flew in the clouds often enough, his rating would turn to ashes in his wallet. Why should I be an instrument pilot? Would you trust a part-time brain surgeon?

Anyway, being a VFR pilot kept things exciting. Residents of small Texas towns still wonder about that madman circling their water tanks in rainstorms. Or the guy who flew a Champ down Main Street to read the name on the post office. (Had to. All it said on the school bus was School Bus.) Come to think of it, the IH (interstate highway) approach system has never been openly credited. Think of its merits: you are too low to be hit by other airplanes, too high for the cars; no TV towers grow in the middle of the interstate; and if you stay with it, it's bound to take you *somewhere*.

Flying VFR all these thrilling years has kept me in touch with America's heartland. The farmer on his tractor, shaking his fist. The big, shiny roaches in the No-Tell Motel during unexpected little overnight stays. The thrill of eventually finding my way home again and that phone call to my widow—I mean wife—"Cheated death again!"

I admit it: I was nervous about some things. Like being around the Feds, for instance; and omni navigation. Remember now, those gadgets crept into airplanes long after I started flying. (I always meant to ask why one day you fly toward the needle, other days away from it.) But with a good pipeline right-of-way underneath me, I could usually get it sorted out after a while.

I really hated going into big, busy airports. All those chaps snapping at each other in strange languages; and something about approach frequencies, which I decided not to fool with. I'd just fly up low, real close to the tower and say, friendly like, "Hi, this is ol' Cessna Two-Niner Betsy, can I land with y'all?" I don't know why that seemed to get everybody's noses out of joint.

Well, as you have no doubt noticed, instrument pilots are evangelistic. "Bax, you really ought to get your ticket."

"Yeah, sure. I already got one." And I'd whip out my blue card with the hole in it, and hold it up against the sky and read what it says on it: "When color of card matches color of sky, fly." I believed in that. Like Pappy Sheffield always said to his students out at our little grass airport, "Keep one wing in the sunshine and keep smiling."

That's true, too. You look at the face of a pilot who has both wings in the cloud. He looks positively gloomy.

Finally, one of those evangelists got to me. Jerry Griffin; runs a Piper outfit over at the cement airport. Calls it Professional Aviation. They wear coats and ties, shave every day, and keep those little stubby airplanes waxed so slick that a fly landing on one would break a leg. I respect and admire Jerry. He's the blue-eyed, blond-haired, all-American kid who grew up at the airport fence. His operation is his dream come true.

Jerry is a "nonstress" instructor, as Bob Blodget used to say. After I got used to the funny way the Pipers ran down the cement and popped into the air, I began to look forward to the hours under the hood beside him. He purred, blending with the purr of the low-winger. I remember the enchantment of those first hours with Jerry. He gave me the confidence that I could defy all the senses that all my ancestors had developed since they came down out of the trees and began to walk on their hind legs. I was fascinated with the little science-fiction world that grew in the green-and-white glow of the needles and numbers. I had my own horizon, my own gravity. We flew into the sunset, into the nights, into the clouds. It made no difference, for I had my own little universe swimming before my eyes.

Gradually, the instruments came to translate into attitudes of flight; they became an intellectualized extension of my own senses. Drunk with professionalism, I went right out and bought a black attaché case and some bifocals. It was the first time I had ever needed a place to keep papers while flying or needed to know what all those little close-together marks on the dials meant.

Somewhere along in here, Jerry sent my license off to Oklahoma City and it came back with a blue seal upon it. Now, at least, I would not be an automatic dead man in 120 seconds if I flew into a cloud.

Griffin said the instrument-flight test would come easy enough, in time, but we were both worried about that glass mountain ahead: the written exam. In real life, I am the morning mouth on local radio. On a disc jockey's schedule, that meant early to bed, early to rise, attend a night school, and look more dead than alive. So Jerry did a noble thing; he referred me to the Bob Marsh Aviation School, on Houston Hobby Field. After that, I'd pass the written in a matter of days, I was assured.

My first thought was that Marsh operated a diploma mill. His school teaches the FAA exams, and I never heard of anybody failing to pass a written when Bob Marsh sent them reeling off to the Fed's exam room nearby. I went to Marsh expecting three days of cram, wham, bam, thank you, ma'am, but I was staggered by the mass of knowledge needed and the relentless digging into a massive pile of government publications to get to it.

Marsh gives the student a dummy exam, the student researches his answers, wrong ones are circled on the graded paper, and the student has to display a complete working knowledge of every subject missed during a *mano a mano* oral review with an instructor. The process simply goes on and on until the student has learned this new vocabulary and all its attendant rules and can snap out the correct answer to every squiggly blur on a weather-depiction map and every bird footprint on a Jepp chart.

Marsh says he builds his study course around the FAA written-exam format so that the student will be familiar with it when he sees the real thing. He has mastered a workable technique of imprinting "government-speak" upon the clay of a student's mind.

After three days of this, all I wanted was out. I kept gazing off toward distant blue skies, longing to find a place not boxed in, so I could lay the Stearman over on its back and shout "Ya-a-a-ah!" I was never born to fly with a bunch of books in my lap. Marsh, who learned to fly from Moses, and had retired as chief pilot for Howard Hughes, shook his fine silver head and looked wonderfully sad. "There's only one learning obstacle I cannot overcome in a student—lack of motivation." We parted with regrets, but as friends.

A few years rolled by. *Flying* published its *Guide for Instrument Flying 1973*. The demand exceeded the supply, and the magazine soon became a collector's item. Plans were made for a sequel. The New York office searched its roster of far-flung editor-writers. "Who is our most hard-case VFR pilot?"

"How about Bax? He doesn't even like to sit indoors in a cabin . . . He's nearly 50, a part-time flyer, set in his ways . . . if he can get a ticket, anybody can."

"We can really make a case for instrument flying . . ." And so the Word went forth.

I kicked. I screamed. Bled real blood. Offered to fly around the world in a Champ with one hand tied behind my back. Said I'd grow a moustache and write about a talking seagull. Anything but this. Didn't they realize it would be the ruination of one of America's last unspoiled pasture pilots? Anyway, I never got a chance to read *Guide for Instrument Flying*. They never got as far as the grass airport in Texas.

New York sent a copy. I read it, and it like to have scared me to death. All this nonchalant tossing off of such phrases as: "Of course, anyone can keep the airplane right side up." Hell, I couldn't.

But if this cup cannot be passed, then let us drink the most bitter portion first. In November of 1972, I once more appeared before Bob Marsh, hat in hand, asking for a go at the written exam. His only comment was a bear hug, and that he had been expecting me.

We mounted an all-out campaign, and this I recommend to any busy man. Get your affairs in order; take about a week off; present yourself to a reputable, well-organized ground school; and concentrate on dat ol' debbil, the written exam. The flying part of it will make more sense to you later.

In order for me to live with this ground-school monster for a week, a most strange business alliance was worked out, involving the owners of Radio KLVI and a 19-year-old Houston bride. We moved the radio station from Beaumont, Texas to Houston and installed it in the front bedroom of my newly wedded daughter's home, which is just blocks from Marsh's school on Hobby Field.

The housewives listening to my morning show got music, news, and witty recitations of FAR Part 91. I was totally saturated with this new world of words; I couldn't shut it off. I went to sleep at night dreaming that I had my hands around the throat of whichever government lawyer had written all that stuff. It couldn't have been written by a human pilot. It obviously was never intended to support an airplane in flight, only its own bulk in court.

I who had known the freedom of the hawk became a studying automaton. A sponge of the assorted knowledge that a man must arm himself with to fly without outside eyes . . . true course is magnetic course, plus or minus variation . . . magnetic heading plus or minus deviation equals compass heading . . . the following documents must be displayed inside the aircraft . . . Zulu minus six equals Central . . . pressure altitude . . . density altitude . . . indicated airspeed . . . calibrated airspeed . . . true airspeed . . . the hemoglobin absorbs carbon monoxide 200 times faster than oxygen . . . My mind staggered. It reeled. It separated and rolled up into little balls. I had slit open the belly of the body of aviation and was swishing around in the innards, and I began to wish I'd never seen them. If it was going to take all of this just to fly by instruments, then ATRs must be gods. They should be allowed to ascend the flight deck adorned in silken garments, with private pilots as acolytes bearing candles.

For me, there was no way. I felt like a savage brought out of the jungle to be taught the whole body of civilization in one week. By the third day, I was silly, numb. My head felt as though it had been injected through the ear with numbered putty. Bob unchained me and sent me home. "Look at your handwriting. It's gone all to pieces. Go home, fix a drink, go to bed. Tomorrow will be worse."

On the last day before the exam, they heaped me with books. *AIM,* FAR, ZAP—all the stuff I needed to know, and all of it scattered through a million subparagraphs. If a lawyer had to go through this many slithering books and back-flopping pages to research a case, nothing would ever get to the court-room. They would settle it out on the lawn with short-barreled pistols. Why doesn't somebody codify all this junk? I rebelled. I told Bob the hell with all this, teach me with words. Let me sit at your feet, tell it to me in real words as the ancient masters did. No man can read an FAR.

But inside me, the growling glow had begun. I was ready, and I knew it. I was stiff with it. For a week now, I felt as though I had been shackled with chains inside a wooden box only four feet high and four feet wide, new government documents and a pan of water being shoved in through the trap door. After seven days and seven nights, I was ready. I sprang out into the daylight, all covered with hair, howling and gnashing to get to that real exam.

That FAA exam room was a place Hitler could have loved. Rectangular, white-walled, sterile, windowless except for the one-way-mirrored door. The matron, in a seersucker uniform with black name tag, ran her Geiger counter over my lunch bag. (Man, it's a six-hour exam.) The Geiger counter found my little transistorized calculator and shrilled its alarm.

Feds in dark coats with narrow ties came running softly. I explained that ever since the fourth grade, I had flunked anything to do with math, and what a shame to blow questions on weight and balance and time and distance just because the towers of figures toppled for me. A hurried conference over the exam rules found nothing that said such a device could not be used, "So long as it does not have a memory circuit," they warned. It didn't, and that little handful of transistors saved me about an hour and at least a half dozen wrong answers.

I entered into the stark chamber, my footfalls crashing among the earlier inmates, who looked up, glaring, until I had rattled around and built my nest.

From that time, the only sound was an occasional deep sigh—more like a drawn-out sob. Once some guy's stomach rumbled and I almost bolted and fled; thought the building was caving in.

Marsh had commanded, "Use those Mickey Mouse things I showed you. Place a strip of paper over the question and read it line for line. Move your lips if you have to. I've watched you read, you gulp a whole paragraph. Do that in there and you'll swallow the hook that's buried in the question for smart guys."

I drew up two columns on a scratch pad, headed "possible" and "out." Bob had said that every question would have four answers. "Two will be dogs, one will be the 'deceiver,' one is the correct answer. Put the dog numbers in the 'out' column, then you just have two answers to consider. Then think, man, because if you don't have a full knowledge of the intent of the question, the deceiver will appear to be the most believable answer."

A part of Bob's Last Rites and Benediction ceremony is to challenge the applicant to predict his score and write it down in secret. Bob does the same, and a two-bit bet is made on the outcome. I gave myself a four-point spread; 68 to 72. Bob had written down 78. When that ominous letter arrived from Oklahoma City, my final score was exactly what that crafty, mean old man had predicted. He insisted on payment by check.

Rejoicing, I started for the door, free at last to go fly the airplane, the easy part. But another manacle had been slipped over my cuff. This time, the chain led to what you had better not call a Link Trainer, for it is in truth a Cessna 172 Flight Simulator—a one-of-a-kind modification of the Link. It will do anything a 172 will do, and some of it much more suddenly.

Had I but known what the next 19.2 hours in the green box would bring, I would have given up on the rating and put in for pit crew at *Car and Driver*. "The airplane," said Professor Marsh, "is a very poor classroom. And if you cannot do the required stuff in the simulator, then you will have plenty of problems out there." So I climbed in and they shut the coffin lid.

If I said "Ahhh" or "This is . . .," then I had to start over. Me, with 28 years of broadcast radio—I began to get mike fright. They badgered me, and how I longed to get one of those instructors into my control room for just a day and make an ass out of him. But soon I was using this new-found skill with the radio as I commuted into Hobby by Cessna each day. Some hillbilly, like I used to be last week, would call up with 17 planes on the frequency and say, "Hobby Tower, Hobby Tower, this is Cessna November 1234 Alfa, and ahhh, we are . . . about . . . over downtown, ah, and we plan to land at Hobby, ah, what's y'all's active runway? Over."

Gritting my teeth like the rest during this soliloquy, I would wait for a breath of silence in the chatter and hit them with my carefully rehearsed "Hobby, Cessna 8529 Bravo, Fry Intersection with Juliet." And like magic came,

"Two-Niner Bravo, report right base 30." Just like that. In seven seconds, each of us knew all we needed to know about the other. I felt pretty smug, although it had taken me a week to find Fry Intersection.

But the confines of that little green box were paced off by Satan himself, and his footprints are called holding patterns. Paul Carrington taught me how to find my way into those little private cells of airspace. Direct entry, parallel entry, or teardrop; what a delicious name for it: teardrop. Also sweatdrop. It got so bad that each day, I used a different brand of antiperspirant. And each day, I surfaced from the box dripping like a fish. The laundry began to reject my shirts.

Big Brother's metallic voice became inescapable: "You are flying through your fix . . . you are 100 feet above assigned altitude . . . what is the reciprocal of 134? C'mon, c'mon!" And more nagging about radio technique: "Six-One Xray, say altitude."

"Twenty-two hundred."

"Never heard of it."

"Two thousand two hundred."

"That's better."

Time stopped in the green box. No yesterday, no tomorrow, just those walls closing in. Bunched up, I tried to cheat. I muttered, cursed, got vertigo, spun, crashed, and burned on the VOR approach to Galveston. Once, I even stalled and crashed on takeoff. When that thing shuddered and broke and those needles started unwinding, I felt a shock of real fear in my chest. Slammed in power, dived for airspeed, actually died when the altimeter hit field elevation. Only the raucous laughter and beating on the box from outside, and the fact that my smoking hole was now 400 feet below sea level, brought my senses back. Still, no one will ever make me believe that thing is just bolted to the floor.

I, an airplane pilot, was suffering the indignity of having to jiggle this infernal bungee-cord-and-bellows box into level flight and keep it there among slithering needles. My red tracings on the plastic overlay outside looked like the homeward route of a drunk boll weevil. I alibied that the mechanical phoniness of the thing was impossible—that I could fly a *real* airplane. Then the immaculate Tony Koenninger stepped up on the running board, reached in with one hand, and flew a perfect approach. For that, I could hate him. "The trouble with you, Bax, is that you have never flown a real airplane like we are asking you to here. Going home tonight in that 172, see if you can fly within five degrees of heading and 20 feet of altitude."

This was the beginning of the next great turnaround in my thinking about flying. I discovered what a sloppy pilot I had always been. I had been wandering along like a strutted cow coming in at milking time. I began to really try to nail those numbers. Soon it got to be a habit, and a second bud of pride—of wanting to be professional in flying—broke through the thick bark.

But it was still slow going in the box. I regressed for a time, and they put me back to basics. Depression led to savagery. I slammed it around, rejoicing as the bellows hissed and groaned. New clearances to copy piled upon corrections not yet made. In my great fury, I would have leaped out of that thing and punched my instructor, except that the years Tom Adams had spent in that flying trapeze called a helicopter had developed his biceps powerfully. Always, his voice would come in: "You can fight it, but you can never win. Make it come to you."

The urge to kill born in the simulator was not just mine. Tommy Reedy, a high-time ag pilot and part owner of Reedy-West Air Service, in Angleton, Texas, was going through the same torture an hour ahead of me. Tommy was in the school because "It's getting so you can't go anywhere VFR anymore." The quiet, taciturn Reedy once sprang from the infernal box and made Marsh a cash offer for the thing so he could take it home and chop it up with an ax.

Lorene Holmes, who was going the IFR route so she could be as good as her Beechcraft, could sometimes be heard softly weeping inside the infernal device. She suggested that a drain plug be installed in the cockpit floor so that a finished student could be poured out into a pan and set in the window to cool.

An hour or two in the box would dissolve me into utter stupidity. Once when turning right again instead of left for a teardrop entry, a ghostly voice whispered through the fuselage, "Left. Turn left." I later learned that there is a hidden peephole in the thing, so they can see how white your knuckles are, but I never found out who my prompting angel of mercy was.

Later, in the real airplane, I was being eaten up alive on a back-course approach into Hobby. I had forgotten about reverse sensing. Approach radar couldn't even tell whether I was heading for Hobby, Andrau, or Ellington. A shaking cup of coffee in my hand, I told my instructor at debriefing that I didn't see how I had botched it up so badly; all the other times had been right down the slot. He gently reminded me that this had been my first one; all the "other times" had been in the simulator. That's how real the experience had seemed to me.

The campaign to get the instrument ticket had begun in early November. It was now approaching New Year's and the winter gunk had moved in over the Gulf Coast. I soon learned that if Hobby and Beaumont, my home base, were reporting VFR minimums, the clouds would be hugging the ground between the two, over the spooky Trinity River bottomlands over which I crept and cheated along Interstate 10. During this time, a Cherokee Six pilot trying the same stuff at Baytown got vertigo and slow-rolled in at the threshold of Humphrey Field, dispatching himself and his passengers.

There was only one tall obstruction along this course—the 570-foot-tall obelisk of the San Jacinto Battlefield Monument. My game was to find the white stone shaft in the haze and mist without hitting it. In wilder moments, I thought of what a fine way to go this would be:

"Say, whatever happened to your old man—the one who used to fly and write about it?"

"Oh, he hit the San Jacinto Monument in the fog one day. You can still see the mark, about halfway up to the star."

Tour guides would add this event to their spiel. A niche in Texas history. Oh, how I longed for that instrument rating, and how I hoped to live long enough to get it.

And now it was 1973, early January, and we were walking out to the real airplane for the first time. Bob Marsh's instrument trainer is a Cessna 172 bristling with antennas for almost every navaid except radar and inertial. It sat crushed low on its spats with its burden of electronics and long-range fuel tanks.

Each day began with one of those preflights that everyone ought to do but nobody does unless under the eye of the Feds or a flight instructor. As I inspected those little cotter keys that hold the aileron pins in place, as I have done on every Cessna for the last 15 years, I got to wondering where this custom started. Has there ever been a report of the ailerons falling off a Cessna? Does Dwane Wallace ever worry about those little bent wires? Later, in the course of 31.5 hours in that thing with Jim Shelton threatening to tape my hands if I didn't stop wigwagging those ailerons and zooming my turns all over the sky, I began to consider that while the cotter keys may never drop out, it did seem likely that I might *wear* them out.

Shelton and I droned through January together in the growing understanding and friendship that develops between men who fly side by side for long hours. The flight syllabus was a repetition of the pattern learned in the simulator. Radar vectors to a holding pattern, then to Galveston VOR, another hold, then the Galvestor VOR approach, missed approach, and back to Hobby for whatever instrument landing was in use. About two hours under the hood, with the worst saved for last, as it would be for real.

With the elastic headband of the hood slowly squeezing off the blood supply to my brain, I entertained Mr. Shelton with such comedy as attempting 180-degree intercepts to a radial and getting lost over the Galveston VOR. I was mildly surprised to learn that actual flight was even more demanding than those dark hours in the womb of the simulator. I say mildly, for by this time, I didn't really expect anything to be easy or turn out right. Shelton helped: he turned all the radios up to full blare to get me used to this new element of chaos. During the first few hours, he worked all communications; I had my hands full just keeping us right side up.

"Scan, man, don't stare at any one instrument . . . take little bites at that radial intercept, then center those needles with the rudders . . . hold that heading; in this business, flying a heading is everything . . . pay attention . . . concentrate . . . this is an exercise in self-discipline . . . quit flopping those ailerons; you want a five-degree correction, so skid into it."

And more: "I let you get lost over that VOR. Everybody does it sooner or later. You must keep a mental image of where you are relative to what those needles are telling you. If you ever lose the mental picture, you are *really* lost. There's two attitudes—yours and the airplane's . . . relax, man, your legs are like iron on those rudders . . . relax . . . don't fight it. . . make the airplane come to you . . ."

Holding at a VOR was the worst for me. As the needles became more fidgety, I would cramp in more and more rudder until I was unwittingly holding reverse aileron. "Jim," I said, "this is a crooked airplane. It's kinked up somewhere and will not fly straight."

"Turn it loose a minute." I did, and it jumped sideways so far I missed the whole VOR again. Shelton laughed out loud.

Missed approaches were not too bad, except for my fatal fixation on heading to the detriment of altitude. I was also still plagued by occasional vertigo. Once on an approach to Galveston, Scholes Field did a slow roll right inside my head. Jim had said, "If I ever yell 'I got it!' I mean I got it. I'm not interested in finding out which of us is the strongest." He stopped the barrel roll and said, "If you ever get disoriented on approach, declare a missed approach and get the hell out. That other will kill you. A missed approach is for reorientation. Never be too proud to go around."

I'll probably remember Shelton's ILS technique forever. "Your road is narrowing down to two and a half degrees on either side. Say those limits out loud to yourself if it helps, but leave those ailerons alone. Use rudders. Pedal it like a bicycle. And that glideslope is just like sliding down a slanted wire. If you're too low, level out and gently fly into it again; if you're high, ease off a little power and let her sink a little."

I'd come down that chute babbling to myself, sweat trickling cold down my ribs, "Two left, just two, that's all you get, wait for it, and you're low, hold level, and there it comes . . . centers at 36 . . . hold 36 . . . and now you're high . . ." When the hair started to crawl on the back of my neck as the altimeter went below 250, Jim would flip up the hood and there it all lay, right where it was supposed to be, like the pathway to the Pearly Gates.

At this instant of transition to contact flying, Jim would lean forward, peer into my face and ask, "You all right?" I don't know what he was looking for, but I felt great. He varied between being a most gentle and sensitive instructor and the most infuriating stress teacher, but a word of praise from him was worth the Medal of Honor to me. Once he flipped the hood and I found us in a bucketing crosswind. I painted one on and made the first intersection turnoff.

He may have smiled, and I thought I heard him mutter, "Ain't no hill for a stepper."

After other flights, we would taxi to the ramp in sullen silence, and once he said, "You were so bad today I don't even want to talk to you about it." I regressed badly at about the two-thirds point, as I had in the simulator. The blowup came on an ILS approach when I reached for the mike to report "outer marker, inbound." While I reached and talked, I swerved right, pegging the needle. Jim shouted, "Now what the hell did you do that for?"

Furious with myself as well as him, I yelled back, "Because I'm just a gahdamn student, that's why!" I overhead him telling this to Marsh later; both of them whooped a laugh.

These regressions brought another pride-crushing back-to-basics session. We flew the course without the hood, "so you can see what you have been visualizing." Then without a word, Jim took over the controls and began the most astonishing demonstration of routine flight that I have ever seen. The sly old master unwrapped his hidden stock of tricks and opened up my eyes to why professional pilots seem to be doing so little at the controls. He painted the needles in the center with almost imperceptible rudder movements. Went from approach speed to cruise speed and through altitude changes apparently using only the throttle. The last thing he ever touched was the trim wheel, and that just once. "Now you do it."

I copied his techniques and felt the tingle of having broken through to a subtle management of the airplane that I never knew was possible.

"Jim, why in hell didn't you show me all this from the beginning?"

"You weren't ready for it. Not deep enough into instrument flying to know that you needed it until now."

"Yessir."

It's funny; I always call him Jim on the ground, but calling him sir in the airplane seemed natural, too. Now I understood why. One of the unexpected dividends awaiting you when you finally decide to don the plastic hood and find out what *all* the knobs on the panel are for is that you will enter into a level of flying proficiency that most VFR pilots don't even know exists.

January dissolved into February, and now I was so anxious to get it that I was overtrying. "Bax, you are worrying this airplane to death. Relax, man; you got a lot of bad habits to unlearn. I can force you to fly that panel with your eyes, but you are still listening VFR. You're flying partial panel. And quit ducking down to pick up those charts in a turn or one of these days you'll go snap-roll vertigo doing that . . . and think ahead, man. Plan. Always consider your alternatives; so the VOR just flipped from 'to' to 'from' and the needle never centered? Well, what are you going to do now? Listen carefully, Baxter: When you are sitting up here in the gunk all by yourself, if you don't have a cold, deadly reason for doing something—then don't do nothin'."

Fatigue, frustration, and humiliation began to move in on me. Even if I ever learned to fly the insane perfection that these guys wanted, would I have the cods to actually bicycle down that beam someday, alone and blind, scared witless, knowing that the altimeter was going through 400 and there was solid earth down there in one minute more of this? Would I have it to go on down to the published 250? That one I still don't know, though I did it every day under the hood. I was pretty discouraged. Fed up, in fact.

A particularly shaggy approach at Galveston and the start of the long, butt-weary flight back to Hobby, concentrating on those slippery needles. Jim could sense how it was. Lit a cigarette and stuck it up under the hood on my lip. I slipped off my boots, wiggled my toes, unclamped my hands and decided he was probably the greatest guy in the world. The airplane flew better, too. But as for me, I was probably unteachable past a certain point we had already reached some time back.

Then one day I heard a chilling exchange over Houston Approach that wiped out any lingering doubts I may have had over the loss of "freedom of the skies" and my own sinking motivation. A Braniff captain on the ILS into Hobby reported in resigned tones, "There's a 150 playing around just under the clouds out here."

But for the random chance of a few seconds either way, that Boeing would have popped out of the cloud base on gauges and speared that happy VFR pilot in his little tin fish. The hell of it was they were both legal. The VFR pilot was clear of clouds and all that, and nothing required him to have an approach plate that would show him that he was playing in the middle of the busiest street in Houston, or be tuned to the approach frequency, or even have his radio on.

I realized then that flying may be a great game, but it can only be survived by everybody knowing and playing by the same rules. Yet eight out of ten of the players are required to know only about one-fourth of the rules. Freedom of the skies be damned; this is madness.

I may become insufferable about this, like a reformed drunk busting into saloons and slapping the drinks out of everybody's hands, but for the first time in my life, I don't feel too guilty about calling myself a pilot in front of other pilots. With the ink still wet on my ticket, I'm no instrument pilot. (They said, "Here—now you have a license to go out and kill yourself.") But I'm working at it. And I have never worked so hard for, or been so proud of, anything in my life.

Getting there took from November 10, 1972 to February 10, 1973, and it didn't finish with any flourish of trumpets or garland of roses. At the low point of the sullen rides with Shelton, I was smitten with pneumonia. Two weeks of skulking around wondering if I was well or sick, and then I presented myself to Marsh's on a Saturday. I wasn't well, wasn't still sick, but any flying would be better than no flying. It seemed natural that the Old Man himself came gruffing out to ride with me; Saturday was Shelton's day off. "I don't like to shout in airplanes," was Marsh's only comment as we set sail to go crashing around the course.

On the home leg, he stirred himself out of his corner to pester me with a lot of irrelevant questions about what I'd do if this or that happened. He only stiffened up once when I almost pegged the needle on the localizer coming home. As I bicycled out of that one, I heard us both emit a long, shaky sigh.

Taxiing back to the ramp, he asked: "How do you think you did?"

"No better, no worse than usual, sir."

"Well, you just passed the check ride."

4 THE FIRST CLOUD
by Nina Galen

"YOU KNOW what an instrument ticket looks like?" Bill Gates, the chubby ex-military chopper pilot asked me as he tied his gear onto the back of his motorcycle. I shook my head, looking at him curiously. At my grass airfield in southern France, Americans were almost as rare as IFR-rated pilots. "Well," he continued, "it's bright blue. You take it and hold it up against the sky. When the color of the sky matches the color of the card, *that's* when you file IFR."

It wasn't the first time I'd gotten such echoed advice from a pilot. At first, I'd listened to them patronizingly. In my opinion, 180 horsepower and a single prop were enough to achieve a neat, safe, and relatively professional ascension. But now my early self-confidence was fading. It was weeks since I'd passed my instrument-rating flight test, but I had yet to file IFR. And this in spite of owning a well-equipped Rallye Commodore, which was sitting handily nearby in a hangar.

Still, the reasons for this reticence were sensible. Violent thunderstorms were making even VFR flight a misery all across Europe that spring. The IFR levels over the Alps, which almost invariably lay in my path, were simply too high for my plane, and to circumnavigate the mountains on airways was too time-consuming. Yes, these reasons were irreproachable, but there was another that was even more significant and less easily defensible. The sad truth was, I was scared to death of actually flying into cloud.

I'd done it only once, by accident, on one of my first VFR cross-country flights. Without even a turn indicator on my instrument panel to help orient me, I realized immediately that I'd flown into a kind of preliminary death, and that unless I was very lucky in the next few seconds, I'd be obliged to remain. Refusing to panic, I performed the gentle movements with control stick and pedals that had heretofore always resulted in a level 180-degree turn, and a moment later, the world was back.

Since that day, hundreds of flight hours ago, I'd carefully avoided clouds. During instrument training, a hood shaded my eyes from the bright Southern California sunshine while the earth below turned yellow from drought. We never even saw—much less entered—a cloud. Since returning to my home in France, I had become less and less clear about the things I'd learned. If I didn't do something about it pretty soon, my IFR adventure would end before it began.

As Bill Gates and his motorcycle disappeared up the road, his words lingered in my mind. I thought ruefully of the little card in the plastic window of my wallet. "Commercial pilot. Airplane single-engine land." And the still almost-mystical word, "instrument." Turning, I started walking back toward the hangars. Then I realized that except for a little blue circle in the right-hand corner, the instrument ticket issued to me was white as a cloud.

The next day dawned clear and sunny. I was due to leave for Perpignan and planned first to cross the frontier and land in Spain to detax the gas for the trip, a worthwhile saving on a tankful of fuel in spite of the detour. Today, there would be no Alps in the way, and the IFR route over the water was even shorter and easier than the VFR route along the shore. I studied the weather charts at Cannes, where I went to file my flight plan. As usual, the highs and lows spread across the map of Europe like some vast and obscurely controlled game of checkers. The forecast was favorable for the trip, and in any case, I had no intention of flying into actual instrument conditions. So I filed IFR.

Runway 18 at Cannes heads out over the sea, and as I climbed toward the St. Tropez VOR, the view was magnificent. Snow capped the Alps to the north. On the left, the distant horizon of the Mediterranean traced the shadowy shapes of Corsica and Sardinia, their high peaks hidden under wreaths of cumulus. It was hard to look away from all this to the instruments, and as a result, each seemed to go its separate way in a depressingly sloppy manner. Feeling lazy because of the lack of traffic and the perfect visibility, I nevertheless began to make some effort to bring all the needles to their proper positions.

And then I saw it. A cloud. My heart gave a terrific thump. The cloud seemed to be directly in my path, between me and the STP VOR. I guessed it was a very small cumulus, not hard and bubbly, rather a bit of white fluff lying in the air. All reason indicated that the cloud was as inert as a wad of cotton, yet I was firmly convinced in my soul that many such clouds contained red brick walls.

What to do? I looked at the altimeter. Should I request a higher altitude? Eight thousand was a much better en-route altitude; what had ever possessed me to ask for six? On the other hand, the cloud looked hardly larger than a house, so conceivably, one could just change course slightly and fly around it, or above or below it. In any case, it probably wasn't directly in my path.

But it was. I knew it was. And I further knew that for better or worse, if I were to have any self-respect as an instrument pilot, I would have to fly through it.

Looked at objectively, the cloud presented no problem. It was too small to stall out in, too innocuous to flip the plane onto its back. I flew straight at it, my heart pounding. Larger and larger it loomed, rushing toward me with terrible speed. I stared at it wide-eyed as it reared up and flung itself upon me. And then, as the plane plunged into the smothering white embrace and disappeared along with the entire universe, the most unexpected and miraculous thing happened: my eyes went directly to the instruments and my body relaxed comfortably back into the seat. In the few seconds I was inside the cloud, an instrument pilot was born.

That first IFR trip was a personal success. It didn't really matter that the people sitting around the radar screen somewhere near Marseille were operating a betting pool on my wildly inaccurate ETAs to the FIR boundary, or that a line of mountain thunderstorms along the Spanish coast caused the cancellation of my IFR flight plan and an ignominious retreat to my alternate. My appetite for instrument flight had been whetted.

More was in store, for my next IFR experience occurred a short time later. I was spending the weekend with friends in their country home outside Paris, when I heard a depressing tale. One of the guests, a young, handsome marquis, had had his driver's license suspended for a traffic violation and now could not drive his yellow Maserati around England on a business trip. The poor fellow was in despair, until our host suggested that perhaps the trip could be made in my plane. I paused to reflect, as all pilots do before committing themselves.

"I will pay all your expenses," said the marquis.

"What time do you want to take off?" said I.

The next day, I filed a VFR flight plan from Toussus-le-Noble to Ashford, on the English coast. The ceiling en route was reported to be rather low, and I did not feel that someone who had logged only two seconds of actual had the moral right to carry a passenger into cloud. As we flew northward, the ceiling became progressively lower until we had to abandon the land and fly out over the sea to have some headroom. We flew at 500 feet with the blue sky just a few hundred feet above. VFR, I decided that day, was strictly for the fish.

Though this was his first voyage in a lightplane, it soon became obvious that my passenger had no fear of flying, a courage undoubtedly based on a carefree ignorance, but a courage nonetheless. Upon arrival, he cheerfully volunteered to help me tie down the plane for the night, and I watched with fascination as he attached the Rallye firmly to earth by a rope looped around its pitot tube.

But such people, charming as they are, present a dangerous influence on the normally wary pilot, and I determined that I wasn't going to allow my fearless friend to push me into taking any risk in the hours of flight before us.

It was two days later, when we were about to take off for Exeter, that the marquis demanded why I wanted to waste time going up to the control tower for weather information. No weather, he assured me, was going to keep us from reaching Exeter that morning. I stared at him blankly, surprised at my own inclination to heed his words. Then habit took over; I asked him to humor me this once and ran at full speed across the tarmac to the tower.

The weather reported seemed reasonably good for VFR, though there would be a weak front on our route. But as we passed the Biggin Hill VOR and headed toward Bournemouth, I noticed the ceiling was descending fast. Bournemouth reported 400 feet. Low hills up ahead, their tops almost hidden in cloud, convinced me that continuing like this would be impossible. Either we would have to turn back or get on top.

A decision had to be made. They say that he who hesitates is lost, but this seems like pretty bad advice when applied to flying. Was I, I asked myself, being pushed into a situation beyond my abilities to handle? Had my pilot's conscience and judgment been "bought" in the last 48 hours by the money that had been spent on me, the lovely hotel room, the meals in some of London's finest restaurants? I hadn't a clue what it was going to be like inside that dark, wet, relatively active mass, yet I was expected without further ado to commit two lives to its mercy. Was this a classic instance of undue influence? No. It was not. I *had* to be able to handle this stratocumulus with the training I had received. This was part of what instrument flying was all about. I would climb on top.

I made a 360 in the murk while waiting for the Exeter weather. It was reported good. The front, the cheerful English voice on the radio told me, was very weak, and tops were at about 4,000 feet. I asked if I might be allowed to climb to VFR on top, and he immediately replied that I should first climb to 2,000 feet and head direct to the Bournemouth VOR. I made the necessary power adjustments, and a moment later, we were inside the dark gray cloud, seeing occasional flashes of the earth below. Again I felt that curious sensation of relaxation.

As I flew along, scanning all the instruments, listening to the traffic and waiting for those reassuring words that I was in radar contact, my passenger was tapping me on the shoulder. What did he want? I couldn't be bothered right now. I didn't smell smoke, all the gauges were stable, and since we were in cloud, he couldn't be trying to tell me that we were on a collision course with another aircraft. I decided to concentrate on keeping the plane straight and level. A few minutes later, our blip was observed crawling onto the Bournemouth radar screen, and we were given clearance to climb.

A short time later, we were suddenly on top under the blue sky with the overcast now an undercast. I was grinning a wide, self-congratulatory grin as I leaned toward my passenger to ask what he'd been trying to say.

"I just wanted to tell you," he shouted in my ear over the engine's road, "that there's absolutely nothing to it. You just pull back on the stick."

The marquis was startled a second later to receive his first boom-mike kiss.

Since the day I spoke with Bill Gates, my instrument ticket has been held up to many skies—blue, white and all the colors ranging to black. It has fluttered in mistral winds sweeping down the Rhône Valley, in mountain wave, sirocco, and tramontane. It has taken me to new worlds and shown me strange and wonderful sights. I have seen the runway lights of invisible airports floating on the blinding late-afternoon haze. I have visited in the most intimate recesses of clouds, the birthplace of the rain. And I have journeyed in a world of trust and communication out of sight of the entire universe, where experience seems ever fresh and therefore unreliable, and the only thing between you and some ultimate, fatal decision is self-honesty and calm nerves.

It is no wonder that to some pilots, flying out of an hour-long cloud into the limitless visibility of a beautiful day, the very sight of earth and sky is disappointing, appearing to them as nothing more than a cheap and unnecessary thrill.

5 BEFORE YOU GO...

by Peter Garrison

SINCE IFR FLIGHTS are much more tightly structured than VFR trips, they require more advance planning—or, to put it another way, they require more advance awareness. After all, a pilot with years of experience may put less time into planning an IFR flight through the Chicago-Boston-Washington Golden Triangle than a novice puts into planning his first VFR cross-country, but however little time he actually spends preparing for his flight, the pro has as thorough an awareness of the requirements, restrictions, and alternatives that will surround him as if he had studied the flight for a week in advance.

There is, furthermore, almost always a difference in reality between the planning procedure as set forth by a conservative teacher, commentator, or examiner who doesn't want to omit a single technicality, however nugatory, and the actual practice of real pilots undertaking actual flights. There are always more variables than generalizations can possibly cover: different types of terrain and weather, different types of airplanes, different degrees of pilot skill and currency. The flight you set up on the instrument written exam is planned on a big table under the benign eye of a lazy clock with hours to spare; real-life flights are planned in motel rooms and airport limousines, filed on the run in terminal phone booths or under the basilisk eye of a lonesome FSS man in a small, rainy town. Plans take shape in the nebulous uncertainty of often-contradictory weather forecasts from sources as disparate as the *Today* show, the National Weather Service, and a wife or FBO looking out the window at the destination. The solidity of the flight's substantives is undermined by the particular modifiers of time, place, and party: the pilot is tired or rushed, the glideslope receiver is unreliable, the weather is improving more quickly than forecast, the route is familiar, the engine sounds funny.

It would be hard to program a portable computer to handle all the variables—especially the nebulous ones the precise nature and importance of which are not clear even to the pilot himself. The mind of the instrument pilot serves as such a computer, however, and he must make his decisions out of the sum of the raw material given—sometimes dredging intuitive results out of a trackless murk below the minima of reason. Sometimes, a past experience, good or bad, will make you take a chance at which you might otherwise balk—or, alternatively, make you refuse a routing that others are accepting.

Every pilot's planning procedure becomes, with passing years of experience, more individual and more his own, incapable of transmission to others. But what are the common strands that tie them together, and what are the unstated questions to which answers must be found before the flight can take shape?

The first answer to be found, as with every flight, is to the question, "Go or no go?" It's usually easy, but not always. Punching through hundreds of miles of benign warm-front clouds in the Midwest gives no one pause; but what about a narrow but powerful cold front over the Rockies, where you might have to deal with high terrain and a low icing level? What about a forecast of strong upper winds, which, if wrong by 25 percent to your disadvantage, might stretch your reserves thin, combined with an alternate that's legal but unstable? The trouble with rules of thumb to cover such situations—such as "If in doubt, don't"—is that nobody obeys them. Maybe that's a hidden merit. If you don't try your luck and get your feet (or at least your palms) wet a few times, you will never start to develop a feel for your own limitations, and for the mercies and cruelties of real-life weather. If you don't wander into icing once or twice, how will you ever learn a true respect for its possibilities? The fallacy of "If in doubt, don't" is that in the beginning, everybody is somewhat in doubt about his own capabilities as he sits down, shivering slightly, on a new instrument ticket in a chilly cockpit and wipes the raindrops off his map bag. You just have to sit through a few cases of the jitters, and a few recurrences now and then when things don't go your way, to get the seasoning that makes the difference between book learning and savvy.

The best rule of thumb is: "Don't fly into funnels." If you fly, you know what I mean: no matter how many little obstacles and difficulties may plague you, you are never so completely in trouble as when the course you are taking is the only one left.

Assuming that the flight is practical and that you have the equipment for it—meaning the necessary radio gear and charts, and an airplane that can still climb decently at the highest MEA on the route and can make the trip with more than legal fuel reserves—how do you start your material preparations?

I'll assume that you have Jeppesen charts, since most pilots do. (If you're using the NOS charts, you may also want a copy of FAR 91.117, which gives landing-minimums changes for partial equipment.) Let's say that you've already gotten a general forecast the night before, or from the television or newspapers; you have a rough idea of the destination and en-route weather.

You take out the low-altitude flight-planning chart and find your points of origin and destination and a suitable route of flight. Sometimes the selection of a route is straightforward: it is the only one, or the shortest one, or both, or it is an ATC preferred route. Sometimes, particularly in the eastern half of the country, the choice of a route is more complicated.

Staying for a moment in the West, where the route structure is much sparser than in the East, there are many flights for which airways do not serve very well; for instance, a flight from Douglas, Arizona to Gallup, New Mexico would have to go about twice as far if it stuck to airways as if it did not. The requirement for direct routings in controlled airspace is that direct routes lie between omnis not more than 80 nautical miles apart; in uncontrolled airspace, there is no such limitation, and where there are gaping holes in the victor-airways structures, the holes are generally occupied by uncontrolled airspace. Our Gallup-bound pilot would simply file direct from Douglas to San Simon, from there direct across 129 nm, most of it uncontrolled, to St. Johns, and thence direct Zuni V421 Gallup.

When you fly off airways, you have to provide your own obstacle clearance; to do so, you need the relevant sectionals or WAC charts, or else you need to fly high enough to clear everything in that part of the country—about 17,000 feet in the Rocky Mountain states if you want to maintain the 2,000-foot terrain clearance that is standard on airways in mountainous regions.

In parts of the country where there is little or no uncontrolled airspace and no shortage of airways, route choice is determined by the pilot's whim, the preferred routes recommended by the FAA, or by considerations of altitude and weather. The designation of preferred routes arises from internal reasons at ATC; preferred routes are not always the shortest or simplest ones from the pilot's point of view, but since they facilitate traffic flow for the controllers, they are preferred. Commonly, if you file other than a preferred route for a trip for which one exists, your clearance will come back modified to fit the preferred route; or you may get cleared as filed and then receive an amended clearance later that takes you via the preferred route. One reason is to feed you into a standard terminal arrival route (STAR), which you will probably end up using anyway, unless you are flying during a lull in traffic. In short, if there is a preferred route, use it.

Other considerations may override that one, however. Suppose you want to go from Van Nuys to Oakland, in California, during a period of winter frontal weather with the freezing level at 8,000 feet at Los Angeles and dropping to 6,000 feet around San Francisco. Tops are above FL 200. If you are flying a powerful airplane—say, a lightly loaded twin that handles ice well—you might be willing to take either of the preferred routes, both of which take you into forecast icing conditions over the southern part of the state. If, on the other hand, you are flying a 180-hp single with some passengers, you would do badly to flirt with ice. In this case, a routing via V186 Fillmore, V12 Henderson, V27 San Luis Obispo, V113 Los Banos, and V107 (the preferred route) thereafter would be safer, while unorthodox, and would be accepted by ATC, particularly if filed with a notation that you did not want an altitude assignment above the icing level.

In order to know the icing level, you have to have called weather; but before calling weather, you have to have familiarized yourself with the route structure in the areas where you'll be flying. Therefore, the next step after assessing the options on the planning chart and settling on a provisional choice is to call weather and get the reports for the locations that concern you. It is usually a good idea, if you're not in reach of a synoptic chart, to ask the weatherman for a thumbnail description of the overall weather system that is producing the conditions you'll be flying in; the big picture might help you make your own weather judgments and certainly would be an aid in choosing an alternate. Local weather disturbances such as thunderstorms, squall lines, and tornadoes will influence your choice of routing as much as ice would.

Depending on how quick a thinker you are, you can now hang up and complete your flight planning or keep the FSS on the phone and file an ad-lib flight plan. You ought to take several quick steps before filing, at any rate: check the revision notices in the avigation and approach sections of your chart binders for late news that may be pertinent; alternatively, ask the FSS for notams. Also, you must often select an alternate.

The requirements for selection of an alternate are outlined in the FARs; you need an alternate any time the weather at your destination is much less than VFR—the precise limits are in FAR 91.83(b)—and the weather at the alternate must be forecast to be above published alternate minimums. If the alternate does not have published alternate approach minimums, the forecast weather must be VFR from the MEA on down.

The importance of the alternate in initial flight planning is that when low ceilings or poor visibilities prevail over a large area, and the forecast weather at the destination is marginal, it may not be possible to find an alternate with suitable minimums—since alternate minimums are much more conservative than normal ones—without adding a great deal to the length of the flight. Conceivably, you could have to file the airport of departure as the alternate; and sometimes, even that won't meet the requirements.

The alternate can actually be the probable destination—if, for instance, you depart for a destination where conditions are worsening in the knowledge that you will most likely have to make a missed approach and go to your alternate. You would rather go to the filed destination, but you are pessimistic; so you just drop by for a look. In that case, the choice of an alternate should be influenced by an educated guess about the weather; you'll probably be happier if you start off in the morning from your RON stop in better weather rather than worse.

It is obviously advisable, if there is any doubt about destination weather, to keep posted on it while en route; but you should not be too complacent about the alternate. Keep an eye on it, too.

The minimum legal fuel requirement for IFR flights (FAR 91.23) is enough to go to the destination, then to the alternate, and then to have sufficient fuel for 45 minutes of cruising flight remaining. This seems to me rather slim, especially in the light of some peculiarities of the IFR system. For one thing, since 15 minutes' worth of fuel is barely enough to hold off panic unless a runway is right in front of you, I think the 45-minute reserve should be one hour. Next, you may file at an optimum altitude and then end up cruising somewhere lower down; or you may be troubled with carburetor icing and use carb heat a lot, which increases fuel consumption per mile unless you relean with each change in carb heat, which most people don't do. Also, in any tense situation, there seems to be a tendency to be very aware of normal engine roughness, and to enrich the mixture to deal with it—precisely the wrong direction, in many cases. For any of these diverse reasons, or any combination of them, you might use fuel at one or two gallons per hour more than your optimum VFR rate.

Finally, since you might file one route and find yourself following another that is more circuitous than the first, and since you might be vectored or held for one reason or another, or might fly two or more missed approaches before settling for the alternate, the picture of a flight straight to the destination and thence straight to the alternate with a safe reserve is not very realistic. Personally, I would like a minimum reserve of 90 minutes at the alternate, assuming no delays en route, with this figure subject to variation depending on the weather conditions and trends and the degree to which the fuel consumption of the airplane can be carefully controlled and the fuel remaining exactly measured. In a fuel-injected airplane with an EGT indicator and reliable fuel gauges, fuel margins can reasonably be more precise than in a carbureted airplane with sloppy gauges and no precise way of controlling the mixture.

When you have picked a route, you have to pick an altitude. The choice of an altitude is somewhat provisional, since it is subject to change by ATC; it is often just a way of expressing your preferences as dictated by the performance of your airplane or by conditions aloft—winds or icing, mainly, or sometimes cloud tops. There are a number of obvious considerations involved: MEAs, freezing level, winds, whether or not you have oxygen aboard, the climb performance of your airplane, and other factors. Sometimes, when warm air overlies cold, it is worth penetrating a layer of icing in order to take advantage of winds aloft, or perhaps to avoid encounters with freezing rain at lower levels. Light singles are usually most efficient between 7,000 and 12,000 feet, and depending on the physical condition of the pilot, oxygen may be required starting anywhere between 9,000 and 12,000 feet. Often a general familiarity with the geography of the area to be overflown makes a detailed check of all the MEAs en route unnecessary; at any rate, the stated altitude for the flight is an initial one, and if a higher MEA is required down the line, you will find out from center soon enough. MEAs take on overriding importance in planning only in mountainous areas, and where the icing level is so low as to make it a question whether the flight can be conducted safely at all. If only a single cardinal altitude is available between the MOCA and an icing layer, and that altitude is the wrong one for the direction you want to go, don't forget that ATC can route you eastward at an even altitude, say, if you request it and traffic permits. Also, of course, the whole question of an icing layer shrinks in importance as one gets into more powerful and better-equipped airplanes; moderate rime ice may be insignificant on a twin with prop deice and serious on a weak single with nothing but pitot heat. If you do plan to penetrate areas where some icing is reported, try to get a tops report; icing is usually worst near the tops of clouds.

When you are in the process of filing, you can make things easier for the man taking down your route if you phrase it as tersely as his shorthand requires. Avoid chatty phrases like: ". . . and from there I'd like to swing down to Soandso Intersection, and, let's see, then over to Etcetera Omni . . ." The clearance will be filed as an unpunctuated list of fixes and routes, with the route name omitted if direct; except for inserting the word "direct" where appropriate, you should phrase your request in the same manner.

Depending on the terminal area and the traffic, ATC may give you a standard instrument departure (SID). If you don't have SID charts, say so in the remarks box of the flight plan; the consecrated phrase is "No SIDs/STARs." On the other hand, if you *do* have them, you may as well request them if you think they may be used. Otherwise, specify your departure route in sufficient detail to make it clear that you want to go that way rather than by an SID. If you're in an area where every move you make is covered by an SID, preferred route, and STAR, you might as well file the whole works.

Sometimes—say you're filing westward out of Denver in good weather— you might want to file "VFR on top" for the first part of the flight to give yourself a chance to clear the mountains while climbing visually, rather than climb in random directions, under vectoring, over the plain until reaching the MEA for the mountains. Once you get to the MEA, you can request a "hard altitude," meaning one for which IFR separation is assured (except, of course, that in VFR conditions it is incumbent on the IFR pilot to keep a watch for possible conflicts with aircraft not on IFR flight plans).

You should file not more than two hours nor less than 30 minutes before your proposed takeoff. In practice this means either just before leaving for the airport or just after arriving there. There are some exceptions; for instance, on a towers-en-route flight plan, you can usually make the clearance request on first contact with ground control, since your release is likely to be timed to fit into unpredictable local traffic conditions.

Before starting up and calling ground control at any of the larger airports, check the basic approach plate for the notation "Cpt" in the communications-frequencies box. It refers to clearance pretaxi, and occurs alongside a frequency on which your IFR clearance can be obtained up to 10 minutes before takeoff time.

The basic requisite for relaxed IFR flying is familiarity not only with the techniques and the equipment but also with the route to be followed and the departures and approaches to be used. When nothing is left to chance, the flight unfolds routinely and effortlessly. The flight-planning phase is the time to get all the unknowns into the light and all uncertainties under control. If you take off uncertain of the details of your route or the severity of the weather conditions you will encounter, you're already in the mouth of the funnel.

⑥ THE IFR PREFLIGHT
by Peter Garrison

FOR THE VFR PILOT, "preflight" means those minutes he spends checking the prop for nicks and the fuel tanks for water. For the IFR pilot, it means that and a lot more.

The IFR preflight begins in the pilot's home, office, motel room, or dispatch shack, and really ends only when he calls the tower and says he's ready for takeoff. In many ways, IFR flight is like stepping off the high board: it's hard to turn back once you've launched. The consequences of a careless preflight before a good-weather mission can be serious, of course, but usually, they're nothing more than embarrassing: wax on the static ports? Well, lookee here—the ASI says we're going minus 280 mph and we're obviously climbing out at something around our normal 80. Forget the right chart? What a pain—have to go back and pick it up. Pitot-heat circuit breaker popped? Must remember to mention it to maintenance next week.

Even something like the ultimate catastrophe—a solid slug of water in the fuel system—means only that you work very hard to keep your cool and set up a nice approach to that empty supermarket parking lot, worrying all the while about how you're going to explain away your carelessness.

Now, imagine any of the above occurring during departure into a cold, 300-foot overcast, and you'll get an idea of the importance of the IFR preflight.

The proper IFR preflight checks not only the airplane but its pilot, his equipment, and his preparations for the flight at hand.

Let's start with the pilot. Are you ready for the mission at hand? Sober, rested, unmedicated, emotionally fit? Instrument flight is the most physically and intellectually demanding kind of aviating that any of us are likely to indulge in. If you had a hard day at the office or didn't get as much sleep as you'd like, you might not have any trouble climbing to on top through a benign overcast and shooting an easy approach to an 800-and-five situation at your destination—but you'd better be in absolutely perfect shape if you plan to tackle a solid-IFR mission with possible ice or thunderstorms, en-route weather or alternate decisions to make, turbulence, or an approach to minimums. Check your health as the very first step in your preflight: a *minimum* of 8 to 12 hours since the last drink, more like 24 hours if you did anything more than very moderate boozing. Well-rested, no head colds or other illnesses that might affect your motor reactions or balance mechanisms, and *no* medications.

If any of the above conditions aren't met, wait to fly another day. Or work it the other way around: if you must fly on the appointed day, make sure that the above conditions are satisfied.

Equipment comes next if the pilot is ready and able, and a good place to start the preflight is the chartbag. Do you have the *right* charts in that bag? That may sound like a silly question, but after living with east-of-the-Mississippi Jeppesen coverage for a couple of years and finding it entirely satisfactory, you may file for Dallas one day and discover that your en-route charts cover the flight but your approach plates die at the Arkansas-Texas border. I know: I did it, and only discovered my stupidity when I opened my approach-plate books to get out the pages of the destination.

Luckily, the weather was good—which was why I hadn't checked the approach plates for minimums in the first place—so there was no problem. If you do find yourself airborne in cloud with the wrong charts or plates, don't feel totally embarrassed to ask for headings, frequencies, and whatever else you need, for you won't be the first pilot to open his briefcase and discover that the family dog has digested all of Indiana, or that number-one son took the plates for Memphis to show-and-tell class.

After you've flown a little IFR and used radio-navigation charts, you'll be tempted to give up those clumsy VFR-only sectionals. The low-altitude radio charts have all the frequencies you need, they show all the airports you'll ever be going to, you're paying a lot of money for them, and besides, it impresses the fair-weather pilots down at the airport when you spread out your Jepps on the countertop.

Maybe so—but resist the temptation to leave sectionals out of your flight kit. They could be a lifesaver, and your IFR preflight should confirm that you have all the sectionals covering the areas in which you'll be traveling. Why? So that if you have a sudden fuel, mechanical, or radio problem and have to get down *now,* you know you can find that little airport marked only by a little green circle on the featureless Jepps by looking just south of the river, near the edge of town, hard by the Interstate. Or so that if you have trouble maintaining altitude with one engine feathered or a load of ice someday, you'll know exactly where the safest and lowest terrain is.

Another basic part of any IFR pilot's flight kit should be a set of headphones—or a light earplug headset—and either a boom mike or a standard microphone. Many experienced pilots consider the boom mike the most important single item in that assortment, since it frees one's hand for more important and less distracting chores (flying, for one); others find that the advantages of the rig are sometimes canceled out by the rat's nest of wiring required unless the airplane has a push-to-talk switch for a boom microphone built into the control yoke. In any case, earphones are a necessity because they can make garbled, long-range transmission almost as clear as a telephone conversation and can offer passengers comfort that pilots inured to screeching fire-sale loudspeakers are unaware of. Most important, however, they are a backup against the possibility of speaker failure—remote, indeed, but wouldn't it be ridiculous if every part of your $10,000 radio rig worked except for the $5 speakers, and you therefore had to fly as though you had a full-scale IFR radio failure?

The same reasoning applies to the spare mike: if the regular microphone fails (I once had one come unscrewed in my hand, with little hairsprings and plastic parts springing to every corner of the cockpit, and innumerable other pilots can tell similar, if less spectacular, stories of mike failures), how do you continue to fly IFR? The easiest—though not the only—way is to use your spare mike.

Is there a good flashlight in your bag? Never mind that you plan to be home by three in the afternoon; carrying a flashlight on *every* IFR flight is the best way to ensure that you'll have it during the one flight when you need it. A pack of matches might be enough to help you find the tower frequency if you have a cockpit-lighting failure on a moonlit VFR night, but it'll take a lot more than that to read a full panel, charts, and an approach plate if you ever lose the lights in cloud.

If you use a regular dry-cell flashlight, storing it in the refrigerator between flights can prolong battery life greatly, but an even better idea is to use one of the many rechargable lights on the market today.

Two last equipment items that are almost too small to mention—until you forget them: do you have a pad of clearance-copying paper, and a couple of spare pencils?

The pilot and his gear are ready; is the airplane? The aircraft preflight should revolve strongly about those items that, if they failed, could catastrophically affect an instrument flight. The only good preflight is a complete one, but if you must cut corners somewhere, scrimp on those tired old items that perhaps had meaning back in the days of Jennies but hardly do any more—brow-furrowing examination of the aileron hinges, for example. Spend your time worrying about the engine and its bodily fluids, the flight instruments, and whatever deicing or high-altitude equipment you might optimistically be intending to rely upon.

Uncowl your engine and examine it for leaks and drippages, loose parts, and any apparent anomalies; you'd be surprised how little one has to know about powerplants to be able to spot a loose wire, an oil puddle, a cracked mount, or any number of other problems.

Check the oil level, of course, and if you have a belt-driven generator or alternator, spend a moment checking the belt and its pulleys and idler for condition and tension; remember, you'll be leaning heavily upon your electrical system, and the belt drive is one of its weaker links.

Drain your fuel sumps for water, then drain them some more, just to make sure. Do this not just once a day but every time you refuel if you're departing into clouds—and check the color of the fuel as well for proper octane. Make absolutely sure the tank caps are on securely, too.

Reassure yourself that your pitot and static vents are apparently clear, and turn on the pitot heat long enough for the palm of your hand to determine that it works. Prop heat and deicer boots can usually only be checked with the engine running, but remember to do so before leaping off into possible icing conditions; the prop can be checked by noting the appropriate amperage drop, and boots can be cycled once to ensure that they pulse properly.

(Be aware that the preflight and runup checklists found in many owner's manuals are written by the advertising and public-relations departments of the manufacturer in question, or an equally authoritative source. Feel no compunctions about making up your own versions—including such usually forgotten items as pitot heat in the preflight and prop deice in the runup—and putting everything in a more logical order. For example, many manuals still insist that you go through the engine runup and prop check first, then check everything from trim setting to DG caging while the engine loads up from the prolonged postrunup idling; in fact, mag and prop checks should be the very last items done before calling ready for takeoff.)

While you have the master switch on for the pitot heat, give the strobes, beacon, and position lights a quick check; they'll be especially valuable to you on those marginal-VFR days when you'll be filing but a lot of other pilots will be stumbling around in the murk depending upon their eyeballs.

Take a look at your oxygen supply if there's any likelihood that you'll be depending on it—which could well be interpreted to mean "anytime the weather's IFR." Wouldn't it be a shame to be unable to climb above some weather because your oxygen bottle was empty? Another incidental benefit many pilots claim for oxygen is that 10 minutes with the mask on at 5,000 to 10,000 feet can do wonders in helping to prepare one for a tough approach.

One other invariably neglected item: are you and your airplane legal for IFR? We'll assume that you've flown the required six hours of actual or simulated instrument flight during the last six months, because if you haven't, it's doubtful that you're going to cancel the flight and go buy some dual or simulator time; but what about your airplane? Have the static pressure system and altimeter been checked sometime during the preceding 24 months? Is the altimeter accurate to within 75 feet? Have the VOR receivers been operationally checked within the past 10 flight hours *and* 10 days?

Since the biannual static-system check is the sort of thing that is usually taken care of during an annual, that leaves you with altimeter and VOR accuracy to worry about. The altimeter's easy: set it to field elevation, and when you get the pressure setting from ground control or flight service, determine that the reading varies no more than 75 feet up or down from its original setting. If it does, you can't legally take off or fly IFR. As for the VORs, accuracy must be within plus or minus four degrees if an on-airport checkpoint is used, or six degrees if the check is made while airborne; if you check one VOR against the other, the maximum difference allowed is four degrees.

It would be discouraging to the bureaucrats if they knew how many aircraft were currently in violation of the carefully framed VOR rule, but that's no reason not to make a habit of running a VOR check at least once a week. Look in the *Airman's Information Manual,* Part One, page eight, for the specifics of how to do it. Record the results right on the panel with a soft-lead pencil, and a glance at this spot will soon become an indispensable part of your IFR preflight.

You've copied your clearance, your engine is running, everything's in the green, the mags are healthy, and all the strictures of the typical VFR runup and pretakeoff check have been satisfied. But there's more, and your IFR preflight really isn't finished yet.

You'd be amazed at how many instrument pilots taxi into position for takeoff and only then realize that they haven't done a thing with the clearance but write it down. Their departure isn't set up, they have little idea where the first nav fix is, and about all they're sure of is that they're to "maintain runway heading for radar vectors." An essential step in the IFR preflight, therefore, should be thorough study of the departure clearance or SID, and a check that the radios are set up to do the job that route requires. (That includes dialing in the correct transponder code, incidentally—a common omission.)

Next, clean up the cockpit and organize the needed charts—and *only* those charts: SID or area chart on top, if applicable; then the first en-route chart properly folded; and then the plate for the approach in use at your *departure* airport, in case you have to make an emergency return. Everything else should be filed away and *out* of the way, until you need it. It's a bummer to grab the taxiway diagram for your destination when what you desperately want is an instant look at the ILS 15 plate to go back home again because you have a burning bird's nest.

You're almost ready to roll—but not quite. If the weather's bad enough that you'd actually have to make an approach to return to your departure airport, set up one of your nav radios to do the job; if you'd also need the ADF and/or marker beacons for the approach, check that they're on, too.

Turn on the pitot heat if you're heading into cloud and it's anything but midsummer, switch the transponder from standby to on, set the little red marker hands of your panel clock to mark takeoff time (or record the time however you wish), and guess what? Your preflight's complete, and you're ready for takeoff, IFR.

The IFR Preflight Checklist

You might want to add the following items to your normal checklist, if they aren't part of it already.

Equipment

—Proper IFR charts and VFR sectionals
—Headphones and spare microphone
—Flashlight
—Pencils and paper

Aircraft Walkaround

—Check engine(s) visually
—Check oil and fuel levels
—Drain fuel sumps thoroughly
—Check fuel caps for security
—Check alternator/generator drive(s)
—Check pitot and static vents
—Check pitot heat
—Check beacon, strobes, and position lights
—Check oxygen supply
—VORs legal?

Cockpit Pretakeoff

—Altimeter legal?
—Check deicer boots and prop heat
—Set radios and transponder for departure
—Organize charts
—Prepare for possible return approach
—Pitot heat on
—Transponder on
—Takeoff time recorded

7 INSTRUMENT FLYING MADE SIMPLE

by Stephan Wilkinson

WHEN I BEGAN flying, I spent a lot of time in the right seat of a turbo Twin Comanche flown by an excellent pilot who had started his IFR career in fabric Pipers. At the time, I thought it remarkable that his sole instrument-flight tools, other than the panel, his Jepps, and the radios, were a ballpoint pen and a lined yellow legal pad. It took me years—and an almost endless series of flight-planning forms, plotters, lighted kneeboards, computers, protractors, and matched sets of luggage in which to carry them all—before I realized that his little lap pad was all that I'd ever need.

That's what flying IFR is all about: uncomplicating an essentially dog-simple procedure that has often been made overcomplex in the interest of thorough training. Especially today, when radar, transponders, and DMEs have so simplified instrument flying. "Cleared as filed" is about all you'll ever have to copy, and though we should all stand ready to give dead-accurate position reports when the radar fails, grope our way via ADFs and beacon bonfires, and cut precise little teardrop holding-pattern entries . . . take heart, because if one of those procedures is ever required, no doubt you'll find yourself milling around with a lot of ATPs who are just as rusty as you are.

Certainly we shouldn't deliver ourselves blindly into the all-knowing hands of radar and the heading bug on the autopilot, for there's no substitute for knowing where you are at all times; but if you can uncomplicate everything *else* about filing IFR, you'll have plenty of time to keep track of your position, your fuel, and your options.

Flight Planning

The first step is planning the route, and you can most conveniently determine that by working from both ends toward the middle rather than drawing a line along the straightest victor airways from the liftoff to the destination runway. Pull out the approach plate for wherever you're going, and see what the initial approach fix is or find the terminal routes that bring you to the approach itself. If the IAF is a vortac 15 miles south of the field, or the terminal route comes up from the south, why file for an airway that takes you 20 miles *north* of the airport?

Then crank your thinking back to the departure airport. If it has standard instrument departures, pick the one that sends you off in the direction you want to go, then note where it drops you off (where it says, ". . . to So-and-So Vortac, then via assigned route"). You should flight-plan along the most convenient airways between that first en-route point and the initial approach fix.

If your departure airport doesn't have SIDs, take a look at the en-route chart and pick the nearest convenient intersection or VOR in the general direction you want to go—don't try to file direct to the one 100 miles west just because the VOR 20 to the southwest is a little out of the way—and make that your first en-route point.

With your basic route in mind from the flight-planning chart, spread out the relevant en-route chart(s) next to your faithful yellow pad and write the whole route at the top of the pad—in plain English or whatever abbreviations you fancy. There's nothing worse than using an official designator that you later can't figure out; MXE is Modena, not Moxie, so you'd be better off spelling it out if you're new in town.

Insert the mileage between fixes in parentheses as you go. "Br Mtn 6 Dep Hugo (45) V126 Lk Hen (40) V58 Wmsport (59) Pburg (61) V35 Tyr (19) Jons (34) V113 Alleg Co direct (19)" is how my basic sketchout for a flight from Westchester County Airport to Pittsburgh would look. On the actual flight-plan form, I'd write it, "BRMT 6 HUO V126 LHY V58 PSB V35 JST V113 AGC D," and I'd add up the numbers in parens and come up with a total distance of 313 nautical, which will give me my time en route once I get an idea of the winds aloft.

The best reasons for writing out the route in this fashion are that you've probably just copied part of your clearance (and on the same part of the yellow pad where you'll be copying the actual clearance), the route description takes only a couple of lines, and it saves having to add the en-route mileage in your head. Some pilots will put the route down on a complex kneeboard form—a vertical list of checkpoints, nav frequencies, distances, estimates and actual times, and "to" and "from" courses—which is a lot of extra work and one more sheet of paper to worry about. Besides, it repeats perfectly legible numbers that are right on the en-route chart and contributes to head-in-the-cockpit flying while you're doing all the necessary bookkeeping en route.

Getting the Weather

There are people who have minds like maps, but I'm bad enough at geography that I need one other item of paperwork besides my yellow pad: a homemade Xeroxed sheet upon which I have reproduced an outline map of the part of the U.S. within which I do 99 percent of my flying. A group of compartmentalized boxes below the map provides space to write out reported and forecast weather for a number of departure, en-route, and alternate airports, plus room for winds aloft for several stations, relevant pilot reports or sigmets, freezing and icing levels, and tops reports. Sounds complicated, but it's not. It gives me everything I need to know (and reminds me to ask the briefer for it) on one simple, graphic sheet of paper that I can fold and stick in the back of the yellow pad.

The secret is to use a full 8½-by-11 sheet for the form; commercial versions are available (though I've never seen one with a map), but invariably, they are miniaturized forms that require you to print the entire weather forecast for a station in a square the size of a postage stamp.

I draw my route on the map, which shows me the en-route and alternate weather stations that I should list and concentrate upon. With the map in one hand and the telephone in the other, I can sketch weather systems and movements that otherwise would be only a blur of words (". . . well, you've got this trough stretching from northeastern South Carolina to eastern West Virginia moving southeast . . .").

The map also solves a problem I had for years: where do you go when you need an alternate? With the weather pattern right in front of me on a map that shows all the major terminals, I can see immediately that if the destination is *that* bad, my only real option is to continue westward to where the front has already passed, or whatever.

The IFR Preflight

How do you do an IFR preflight? Like the man said when somebody asked him how porcupines make love, "very carefully." You do everything you'd normally do, plus you open the cowl, if possible, and *look* at the engine, paying special attention to wires, hoses, and the source of your electrical energy—especially if it's an alternator with a drive belt. Check the static ports and pitot tube, which you probably haven't done since basic training, then turn the pitot heat on and go out and feel it (again, very carefully, for those things can get surprisingly hot).

One of the most important parts of the IFR preflight, however, takes place in the cockpit, *after* you've kicked the tires and counted the wings: sitting down and getting your act together. Put your Jepp books where you can reach them. Make sure your sunglasses are available for when you bust out on top. Know that your pen is at the ready and that there's a backup pencil if the ballpoint runs dry. Wind the clock. Make sure the DME is channeled to the correct nav radio. Set departure control into the number-one com radio. (You can use number two for clearance delivery, ground, and tower, because none of that is nearly as important as having departure ready to talk the instant you need it.)

Then insert into your yellow pad the charts and plates for the first half-hour of your trip. At the most, they'll consist of: 1/the plate for the approach in use at the departure airport, in case you left the oil cap off and have to come back; 2/the SID that you filed; 3/an area chart folded to show the area over which you'll be heading; and 4/the first en-route chart that you'll need, similarly folded.

I put all these charts and plates in the back of my faithful yellow pad, piled in the sequence in which I probably will need them.

Now look at the SID or likely departure route and set up the nav radios to fly the first part of it. If the weather's really bad, set up the number-one nav radio for a return to the airport and use the number-two for your initial navigation (since you may well have ILS frequencies and a glideslope only on number one). Don't forget to tune the ADF as well if it's part of the approach.

In short, arrange everything in the cockpit you possibly can for an orderly departure *before* you turn on the master, which is when things begin to get hectic and chores are too easily forgotten.

Ready to Copy?

Copying a clearance is as simple as stopping at a gas station and asking directions—easier, in fact, for few of us are smart enough to have a pen and paper handy when the man says, "Go down two lights, take a left, two blocks and you'll come to a big church, go right, and you'll see the sign on your left."

But that's all a clearance is: a condensed set of directions on how to get There from Here. You can write it all down in the time it takes the controller to read it off, and if it's so complex that you *can't* write it down, an airline captain probably couldn't do any better unless he'd flown the route before. Copy what you can and then ask the controller to run it by you once more. And don't get hung up trying to copy a complex route so that you miss all the simple things—transponder code, initial cruise altitude—because your pen spun out back on line two.

Clearance-copying shorthands are great, but only if you're comfortable with the one you use. Forget about trying to memorize somebody else's shorthand and concentrate on writing down words and symbols that are meaningful to *you*. The book may say that "right turn to intercept the Carmel

175-degree radial" should be copied "⌐ X CMK 175r," but "R turn icpt CMK 175" may work better for you.

One of the easiest things you can do to simplify clearance copying is to write out as much of the clearance as you can *before* you call ground. If you're IFR from Charleston to Jacksonville, you can write on your yellow pad (just below where you wrote the route while flight-planning):

ATC clears 201M Jax Airport.

Via _____ .

After takeoff _____ .

Climb/maintain _____ .

Dep freq 119.3 [It's right there on the airport plate, remember].

Squawk _____ .

Then all you have to do is fill in the blanks, and if you've left a line or two on the magic yellow pad for each part of the clearance, you have plenty of space to copy down the surprises—if, for instance, they don't clear you all the way to the airport but to a holding fix en route.

The Easy Part

Level at cruise, leaned out and cleaned up, the autopilot engaged, you're almost through for the day; what's left of the flight is pretty much the easy part. There are several kinds of bookkeeping you should do en route, though, and they can all be done on the pad on your lap: recording com frequencies, keeping track of fuel, and running an ongoing groundspeed check to get a feel for your progress and the winds aloft.

Some pilots will alternate their two com radios, keeping the previous frequency ready for recall in case there's no contact on the new one, but if you fly an airplane with one com that's far better than the other, or find yourself inadvertently wiping out the stored frequency on number two, list the center and approach frequencies used on the far right margin of your lap pad (just the last three or four digits, incidentally; after all, the first number is always "one").

Managing fuel is simply a matter of recording the time when you switch tanks in order to keep the airplane in reasonable trim and to avoid unnecessary back-and-forth switching; any time you change tanks, it's one more opportunity for the handle to come off in your hand or for the selector to malfunction. (This caveat should extend to the landing checklist as well. Just because it says "fuel on fullest tank" is no reason to switch from a two-thirds-full tank that has been performing admirably to a three-quarters-full tank that is still an unknown.)

Keeping track of groundspeed will give you a chance to play an interesting game called Hide the Computer. Get an idea of what your groundspeed is and then try doing estimates for en-route points in your head; you'll be amazed at how easy it is. The astound-your-passengers "secret" is that most of the airplanes we fly IFR go roughly two, two and a half, or three nautical miles per minute (120, 150, or 180 knots). If you're in a 172 with a bit of a tailwind component and it's 86 miles to the next fix, you'll be there in 43 minutes—two miles a minute. Your Aztec is doing 150 knots over the ground into a standard westbound headwind and you have 32 miles to go; this one may require pen and yellow pad, but it's more fun than finding and spinning the plastic wheel. You multiply the distance to go by .4, so two and a half miles a minute puts you there in 13 minutes. If you're doing 100 knots, use .6.

With a bit of interpolation between those salient 120/150/180-knot speeds, you'll consistently be within ATC's plus-or-minus-three-minutes requirement, and you won't have to do the old, "Uhhhhh, stand by one," when center says, "Radar contact lost, forward an estimate for Blind Intersection."

Now that you're so well organized, the best way to use all the extra time you'll have is to keep absolute track of the weather ahead and at your alternates, comparing what it's doing to the forecasts on your weather-briefing sheet. Call for the new weather every hour, and if trends are worse than predicted, *now* is the time to consider your options, not when they turn you over to approach control and you hear, "Sky obscured, visibility one-quarter mile in blowing snow, what are your intentions?"

Don't be shy about asking center if you can leave the frequency for a few minutes, but don't ask them to get the weather for you.

Down to Earth

Start to prepare for the approach as much as an hour out. There's nothing worse than realizing you're about to be turned over to approach control and you don't have the plate, the ATIS, or any idea of where you're going.

Study the approach plate with the intent of getting a broad, *general* picture of the approach; scanning the plate completely every 10 minutes or so should do it nicely. The salient things you want to know are:
—Initial approach altitude
—Inbound course
—Time or distance to the airport if it's a nonprecision approach
—Decision height or minimum descent altitude, msl
—*Basic* missed-approach procedure

Don't bother memorizing frequencies, crossing altitudes, procedure-turn courses, or the intricacies of the holding pattern. The relevant numbers are right on the plate, and you'll have plenty of time to check which radial to intercept *after* you've pushed the power up and missed the approach.

The same kind of basic arithmetic that brought you computerless estimates can now be applied to the task of descending comfortably to initial approach altitude. A standard descent is 500 feet per minute, so if you're level at 8,000 and the approach starts at 2,100 feet, you have roughly 6,000 feet to lose, and at 500 fpm, that'll take 12 minutes. If you're going 3 miles a minute (180 knots), you'll want to start down no less than 36 miles out (i.e., 3 times 12). If you want to drift down even more gently, pick a rate just a needle's width more than 300 fpm—330 fpm, which means it'll take you 3 minutes for every 1,000 feet of descent.

Once you're flying the actual approach, it's better *not* to have the distraction of the approach plate in your lap. Put the book on the right seat or give it to your passenger to hold. In your possession, it can only get in the way of the yoke, and even if you have one of those little approach-plate clips on the yoke hub, the sheet can do nothing more than take your eyes and mind away from the all-important instrument panel. After all, you should by now have all the radios and instruments tuned and set up, and you should know which direction you're going and how low to descend. You shouldn't need the approach plate again unless you miss the approach.

Lights in sight, runway dead ahead. Wasn't that a piece of cake?

As a reward for doing such a nice job, you get my favorite IFR tip: how to halve the work involved in updating your Jepp charts.

Pull out your approach-plate binders and first go through them, ripping out all the pages marked "destroy" on the first sheet of the revision letter. Then start at the beginning again and insert all the revised plates, ripping out the pages they replace one by one. The trick is *not* to open and reclose the rings of the binder every time you insert a new page; just stick the page in place loosely. When you get the entire revision assortment collated throughout your binders, open the rings, take out a handful of as many pages as you can handle, pat the loose sheets into place as though you were squaring up a deck of cards, then replace them *en masse*.

And you thought that doing the revisions was the hardest part of instrument flying.

⑧ WHEN YOU'RE ALL ALONE
by Peter Lert

THE PHRASE "single-pilot IFR" seems to strike fear into the heart of many an otherwise sanguine aviator. Perhaps it's the thought that the practice must be difficult if it merits the special designation (as opposed to "just plain IFR"); perhaps it's that the airlines don't fly without copilots, despite all their automation and IFR aids; perhaps it's just insecurity about all those terrible things that can happen when you're up there all by yourself. "No sooner am I up in the stratus," thinks the neophyte IFR soloist, "than the weather will deteriorate, engine and avionics gremlins will proliferate, and a sniggering controller will take a bite of his peanut-butter sandwich and reach for the mike to set a new words-per-minute record in the freestyle amended-clearance event."

Fortunately, this need not be the case. Sure, single-pilot IFR has its own requirements and pitfalls; but remember that the military flies single-pilot IFR missions as a matter of course—missions far more complex than the most rigorous general-aviation IFR flight, flown in extremely complex aircraft at very high speeds. Granted not all general-aviation pilots can maintain the level of proficiency of a military pilot; nonetheless, single-pilot IFR operations can be not only safe but actually very pleasant experiences. A well-planned and well-executed flight brings with it a deep sense of accomplishment, and the knowledge that one can handle any situation that might arise is satisfying indeed. It's practical, too; not everyone has a copilot available, and many of our smaller planes don't have all that much room for passengers when both of the front seats have drivers in them.

A very important key to painless solo IFR is preparation. Complete preparation of pilot, aircraft, and each individual flight are not only essential from a commonsense point of view but have an important psychological advantage as well: the pilot who is satisfied that everything pertaining to the flight is in optimum condition (himself included, of course) will be re-

laxed and confident in flight, and much more able to cope with any unfore-
seen circumstances than someone who's a bit insecure about the equipment,
the situation, or his own capabilities.

Taking pilot preparation first, it would seem obvious that the brand-new IFR
pilot, fresh from 40 hours under the hood in nice weather, is not the best
candidate for immediate "heavy" single IFR flights. Of course, the more
actual instrument-flight time, the better; lucky indeed are the pilots who were
able to take at least a part of their training under actual IFR conditions.

The way to get into solo IFR flying is to get your feet wet slowly; start with
flights of the "IFR to VFR" variety—a simple IFR climbout to VFR on top, with
VFR destination, is ideal—then progress gradually to VFR cruising on top with
IFR letdown to VFR underneath at the destination (this is the way most IFR
flights are anyway) before going on to approaches closer and closer to
minimums. Don't trust the process; be completely secure at each level of
difficulty before going on to bigger and, presumably, better things. Ego-
bruising though it may be, a couple of hours of dual instruction and checkup
at each stage can be very valuable here in terms of confidence as well as
proficiency. Currency is important, too; the new FAA requirement—six hours
of instrument time, including six approaches, every six months—is minimal at
best, but the recent increase in the number of simulators makes it easier to stay
sharp. Remember, there'll be no one to check on your mistakes up there; get
them out of your system *before* you launch into a cloud.

Even superpilots won't be much good if the airplane isn't up to snuff. Those
little idiosyncrasies—the OBS knob that falls off in turbulence, the creeping
prop control—may not be particularly dangerous in themselves, but they add
to the stress of IFR flight and tend to leave the pilot with jangling nerves by the
time approach clearance rolls around. Not an ideal situation by any reckoning. Any equipment that fails will weigh more heavily on a single pilot than
otherwise, so it behooves the pilot to be sure that the possibility of such failure
is remote. It goes without saying that VORs and static systems should be
checked and current, yet a look at accident reports shows how often this isn't
the case.

No discussion of aircraft and equipment for solo IFR would be complete
without some thought on the subject of autopilots. There is no denying that a
good autopilot is invaluable, and that it can mean the difference between hard
work and a relaxing ride. By "good autopilot," though, I do not necessarily
mean the latest three-axis wonder with approach coupler, altitude hold, and
sundry other automatic gadgets; a *reliable* single-axis wing leveler is almost as
much of a boon. Autopilots malfunction in fairly direct proportion to their
complexity, and an unreliable unit is far worse than none at all. The real
function of an autopilot in single-pilot IFR flight is to hold the airplane more or
less level while the pilot's attention is *momentarily* diverted (for example,

looking at the chart or dropping one's flashlight); those who fly behind the snazzy, fully automated brains sometimes tend to be less intimately aware of the situation and less ready to take over if something occurs to break the normal flow of events. I hesitate to say that an autopilot is absolutely essential, but many single-pilot IFR practitioners, including a number of military and airline types, would not attempt any but the simplest of flights without one of these silent copilots along and working.

The final factor in the preparation trinity is preparation for each individual flight. Assuming that pilot and aircraft are in order, the same general rules apply as for the preparation of any IFR flight, but more stringently. Bear constantly in mind that the single pilot has somewhat less margin in workload before proficiency starts to suffer; anything that will make things easier on the pilot will increase safety and decrease fatigue. A cardinal rule here—even more important than for multipilot operations—always have a way out. Solo IFR is no time for paper alternates filed only to fulfill FARs; if at all possible, have a severe-clear alternate within easy range. If this isn't available, a bombproof alternate, of the sort where you'd be willing to bet large sums of money that it will stay *way* above minimums, is a must. Remember, if you have to use it, you may already be a bit off peak performance after one or two missed approaches at your primary destination, perhaps with a bit of holding thrown in for variety.

Exercise some foresight in your choice of primary, in fact; some pilots like the simplicity and relaxed pace of less frequented fields and shun large urban areas; others like the latter, with their constant surveillance and control, and less chance of wandering off course without being spotted.

It helps a great deal to have all one's materials and documents ready and in order before starting. Bucketing along solo in turbulence is no time to start thumbing through the Jepps for the approach plate or chart you need, or breaking a pencil and looking for a spare. It's not necessary to have a kneeboard with clips, drawers, and loops for every conceivable item, although some swear by (others, at) these intriguing devices; just stocking the SIDs, en-routes, STARs, and approach plates for primary and alternate in appropriate order in some easily reached spot can save many a muttered epithet and gnashed tooth. Having a Jepp binder snap open and dump *all* the plates on the floor while the bank approaches the vertical is a situation not conducive to safety.

Find a good place for the mike, too; aircraft manufacturers seem intent on putting it somewhere that almost guarantees instant vertigo if it has to be located by eye and instant dropping if located by fumble. Those pilots using headsets or boom mikes get an A-plus for this section of class; they are, in fact, mandatory for IFR operations in many European countries. The quantum jump in relaxation just from the use of one of these has to be experienced to be believed, especially in areas where the radio exchanges come thick and fast.

Finally, remember what your mother would say: get a good night's sleep and eat a good breakfast (or, as the case may be, lunch or dinner). Why start off tired or hungry? That will happen soon enough anyway.

Honed by all these preparations to a keen edge of readiness and objectively confident of our overwhelming competence, we approach the flight itself. Just as in the preparation phases, solo IFR is like any other IFR, but more so.

It is much easier to stay ahead of the situation than to regain the picture once it has been lost. This is, in fact, the great pitfall of single-pilot IFR: once something has changed the situation—not necessarily a problem, perhaps just an amended clearance or holding instruction for which the pilot was unprepared—it's not hard to get further and further behind. This is what can happen to the marginally prepared pilot who suddenly finds he's bitten off more than he can chew; there are, however, a few simple hints that can make the problem an unlikely one.

Again, preparation and practice are the keys. It requires a certain skill to always know where you are and where you're going; this skill can only be gained with practice. It's a type of practice, though, that can be gained in a simulator just about as well as in flight; and a good sense of orientation in relation to courses and checkpoints makes it much easier to devote the major part of one's attention to the primary task of flying the airplane. Simple though it sounds, one of the most effective ways of "keeping the picture" is to mark the chart or the approach plate with one's position at frequent intervals. In addition to providing ready reference, the act of putting a mark at a given point serves to underscore the pilot's awareness of where he is.

Another, similar crutch: be it ever so humble, write it down. Not just clearances but new headings, altitudes, times—anything. It's surprising how quickly such things can be forgotten or garbled; jotting everything down saves a lot of "Wha'd he say?" especially when you have no one to ask. For information that changes frequently, such as vector headings, some pilots use a child's "magic slate" or erasable grease pencil on a plastic overlay. All sorts of flight logs are available with myriads of tiny boxes for every conceivable item of relevant information; these are usually manufactured by the makers of the cunning kneeboards mentioned above and elicit a similar range of pilot preference. Some find the format of a plain pad of paper (Paper, aviator's, yellow, Mk. IV Mod. 7) somewhat more flexible. The end result is the same: if in doubt, you can look at your lap and find out what you're supposed to be doing.

Speaking of "if in doubt": contrary to popular belief, controllers are there to help you. Your pilot certificate will not be revoked if you ask to have a clearance reread, an unfamiliar intersection defined, or anything explained that you don't fully understand. ATC personnel will bend over backward to help; they'd much rather take a little extra time now than a lot of time later if you're not 100-percent sure of what's happening but forge ahead regardless for fear of embarrassment.

It takes courage to retreat, but it is sometimes necessary. One of the reasons for staying ahead of the situation is to be able to decide early enough on some alternate course of action. There is absolutely no stigma involved in admitting that a situation is showing signs that it will develop into something you don't feel like dealing with; unfortunately, far too many pilots have waited too long to make that admission. Solo IFR is the type of operation where the decision to turn back or land at an alternate must be made (and adhered to) earliest; with no one along to consult with en route, this requires a considerable degree of maturity and honest self-assessment of the solo IFR pilot.

Single-pilot IFR is not easy; it is not, in itself, harder than any other sort of flying. It does require its own commitments and abilities; but, given these, it can provide great enjoyment and satisfaction as well as practical transportation when the weather is less than CAVU. Even if you try it and decide it's not for you, it's a fine learning experience—one in which you can learn about yourself as well as about flying. When you're flying solo IFR, remember that you may be the only occupant of your airplane, but you won't be alone—there will be lots of others doing exactly the same thing you are.

⑨ THE IFR PASSENGER
by Peter Garrison

THE LONGER YOU FLY, the more you take for granted things that are completely alien to everyday experience. Instrument flight is such a thing. To be enclosed in a throbbing capsule adrift in a gray mist and suspended a couple of miles above the earth is an experience, until one grows accustomed to it, that borders upon the supernatural. It holds the key to the fears locked inside most human beings—fears that are without a clear object, springing from the very abstract core of insecurity. They well up unpredictably and sometimes turn into uncontrolled panic. Most people can contain their fears if they have to; sometimes only a slight outward edginess betrays an abysmal terror struggling to get out.

Pilots should understand fear, even though they pretend to have none. It is, after all, habit, not heroism, that makes pilots look indifferently downward over depths of empty air that would petrify an aborigine. Pilots should understand fear and what occasions it not only because the day may come when they will have to cope with it within themselves but also because, more commonly, they will be carrying fearful passengers.

The passenger who allows himself to be carried aloft only to go into an uncontrollable panic at 10,000 feet is a rarity. Most people with deep fears of heights or confined spaces know themselves well enough to avoid flying. There is an element of acrophobia and claustrophobia in almost everyone, however, and many people who for one reason or another find themselves traveling in light airplanes have to quell their fears by a more or less conscious effort. The better they hide their fear, the less likely the pilot is to be aware of it; and if he is not aware of it, he does nothing to alleviate it.

Airlines take pains to supply a confidence-inspiring atmosphere to set their passengers' minds at rest. The interiors of passenger transports are partitioned so that it is impossible to see the entire inside of the airplane from any place at once;

the potential claustrophobe always has a partition or doorway in sight to sustain the impression that there is somewhere else he can go from where he is. The crewmen are in uniform not because they fly better that way but because the ubiquitous uniforms create the impression of some obscure but highly organized officialdom infinitely competent to deal with the hazards and problems of flight. The passenger is encouraged to give himself over to this vast, impersonal network of computers, conveyor belts, radar beams, and uniforms with a docile faith in its infallibility, just as any country's citizens are encouraged to imagine that whatever the government does must be right, because it knows so much more than they do.

There is, of course, not so much that a nonairline pilot can do to sustain a confidence-inspiring mystique; still, he does have it in his power to make an IFR flight easier for a nervous passenger. He can do so both by what he tells the passenger and by the way he conducts himself during the trip.

It is important for the passenger to understand that what is happening on an instrument flight is routine, and that the pilot fully understands and expects every event of the flight. Thus, if you take off into a 100-foot overcast in heavy rainshowers, you have to be aware that for a passenger unaccustomed to the experience, plunging into cloud and the roar of pounding rain on the windshield may be very unnerving. A little advance warning is in order; if you tell the passenger before the takeoff that this will be an instrument flight, that this means that you will be operating under ground guidance as well as with electronic aids in the cockpit, and that although the ground will not always be visible, you will always be sure of your position, he may be more reassured than if you simply put on a *macho* grin and charge off.

Understand that a passenger might be acutely aware of things you habitually ignore, such as the sharp, loud splatter of rain on Plexiglas, or the sounds of power changes after takeoff. Remember that people who do not fully understand how airplanes manage to fly might easily think that rain can interfere with the process. If you explain that the overcast is simply fog, and that flying through it is no more difficult than driving a car in fog—less difficult, in fact—the passenger is less likely to experience claustrophobia because of the sense of being enclosed in an envelope of stationary grayness.

On the other hand, there are aspects of instrument flight about which some minor fibs might be advisable. For instance, if you're asked about the chance of running into another airplane in instrument conditions, it would be better simply to say that no non-IFR takeoffs are permitted in IFR weather than to explain over the din of the engine that in fact, some nut actually *could* be tooling around out there without being in touch with ATC. It is difficult to be convincing when you are explaining that the odds of running into such an aberrant nitwit are smaller than those of meeting a rhinoceros coming down the freeway in the wrong direction; odds are no comfort to someone who is beginning to feel in danger of losing his life.

Never make idle comments about minor malfunctions or equipment idiosyncracies; to an edgy passenger, the remark that the marker-beacon audio is out but the lights are okay, or that the DME groundspeed indicator seems stuck, can be tantamount to saying, "Do you smell smoke?" or "Boy, that's a healthy-looking crack in the wing out there."

When you are carrying an inexperienced passenger, it is important to conduct the flight in the calmest, most orderly way possible. If you are in the habit of throwing charts all around the cockpit and hunting for them when you need them, you should discipline yourself, for the passenger's sake, into keeping everything in perfect order so that you can do your bookwork in an unobtrusive way. A boom mike is advisible—it is advisible even for solo IFR flight, actually, but it is especially useful when you have passengers—because it spares you those clumsy moments when you have to do three things at once and answer an unexpected call from center as well.

Some passengers enjoy hearing radio talk, and begin to feel nervous if the radio remains silent for too long; others find the hard-edged radio voices unpleasant. With the boom mike, it is well to have an earplug speaker or a set of headphones; ask your passengers if they would like to follow the radio guidance or if they would prefer relative peace and quiet.

Passengers enjoy being made a part of the operation of the flight, and sometimes a front-seat passenger will be interested in an explanation of the instrument panel and of the principles of radio navigation. Keep it extremely simple, though; it's easy to lapse into a lot of hangar jargon that nonpilots won't understand. Show him or her how you take a radio fix, and point out on the map where you are. Err on the side of precision; it is more confidence-inspiring if you put your finger on the map and say, "We're here" than if you draw a circle with your fingertip and say "We're somewhere around here, I think—according to the radio, anyway."

You might turn over to a passenger the duty of finding charts and approach plates, as long as you leave him ample time in which to do so. During the en-route portion of the flight, there is no harm in letting an ambitious passenger take the controls and try to keep the artificial horizon and DG in the right position. On the other hand, it is good to warn a passenger before stages requiring serious concentration on the pilot's part—an ILS to minimums, say—not to chatter, question the pilot, or interfere with his activities.

Once people get used to the steady roar of the engine, they find it disturbing for that noise to change; when you are planning your descent, remember to give your passengers some advance warning, and perhaps an estimate of how many more minutes it will be before you land. It is wise on any flight, but most especially during those parts of IFR work when your maneuvers are somewhat out of your own control, to find out if any of the passengers has a cold, sinus trouble, or any likelihood of difficulty in clearing his ears during altitude changes. Pilots who have never experienced the pain that even a gradual descent can inflict upon a clogged sinus cannot imagine how severe it is.

If you find yourself aloft with a person with sinus trouble who complains of pain on the way down, there is nothing to do but keep the descent as gradual as possible and assure the sufferer that the pain gradually goes away. Ear clearing is a skill that must be learned. Yawning and chewing gum are popular remedies for an "unpopped" ear, though holding one's nose and blowing against it works a lot better. Once a person gets the hang of it, it is possible to clear the ears by a voluntary movement of muscles in the neighborhood of the inner ear, but the method is impossible to teach. If you have to try to help a passenger with ear or sinus trouble, remember that when you are coming down, the problem is that air is trapped inside the cavities of the head at low pressure, and high pressure outside is squeezing on sensitive membranes, causing pain. In the case of ears, blowing against a closed nose helps because it raises the air pressure on the inside end of the eustachian tubes. When the problem is a clogged sinus, however (which may be recognized by a sensation resembling that of a red-hot needle being twirled into one's skull), blowing makes it worse; inhaling against a closed nose may give temporary relief, but the only permanent relief comes from the slow seepage of air or mucus into the blocked cavity. This sometimes takes 10 to 20 minutes after landing. The ear problem can be especially serious in IFR flight, since clearances for unusually steep descents are possible, and they are difficult to alter.

Since airsickness has not yet disappeared from the earth, and IFR flight may involve the turbulence and anxiety that are especially conducive to it, it would be foolish not to take off with a supply of sick sacks if there is any doubt about the passengers' reactions. There is a prescription anti-motion-sickness pill called Bucladin, which is supposed to work immediately, unlike Dramamine, which can be bought over the counter but which has to be taken half an hour before the flight. If you can obtain a bottle of Bucladin, it might be good to have along for an emergency. Consult your physician to find out his opinion about offering Bucladin to an airsick passenger; if he thinks there would be no harm in it, he may supply you with the prescription as well.

Apart from giving his passengers a comfortable familiarity with the airplane and the workings of IFR flight, the pilot has to deal with the difficult question of advance briefing for possible emergencies. The airlines are required to brief passengers on the use of emergency oxygen and, on overwater flights, life vests. Printed cards give details on the use of emergency exits and escape slides. Most of this is just part of the confidence-inspiring hocus-pocus, since few passengers read the escape directions or pay much attention to the spiel about oxygen masks and life vests.

The situation in light airplanes is different, since the occupants are close enough together to permit the pilot to supervise evacuation of the airplane. There is always the possibility of the pilot being incapacitated in a forced landing or ditching and the passengers having to figure things out for themselves; but it is doubtful that giving a long list of instructions for all imaginable eventualities would be much help. Passengers should understand how to open the doors by themselves, however, and in airplanes that, like several Beech and Aero Commander models, have emergency window exits, their presence and function should be pointed out. The importance of wearing seat belts should be stressed, and since it's hardly possible to get up and stroll around in a light airplane, the pilot needn't feel guilty about asking the passengers to keep their belts on all the time. If a shoulder harness is installed, it should be used during the entire flight—or, at the very least, during takeoff, climb, descent, and landing. Remember that in the excitement of an engine failure duing the climb, for instance, a panicky passenger might take much longer to reconnect the shoulder harness than usual.

Some light-aircraft cockpits are noisier than others, but all are noisier than they should be. The noise can be fatiguing and irritating, especially to a passenger who would just as soon be somewhere else in the first place. The solution, for VFR as well as IFR flying, is to provide ear protection for the passengers, either in the form of sets of rubber earplugs, which you can boil to sterilize after each use, or fine glass wool specially made for ear protection. Hearing-aid stores are good sources for this type of equipment, or for advice on where to obtain it. Earplugs do not interfere with conversation, but they make the cockpit a much more comfortable place in which to pass one's time.

Fear of the unknown is a challenge every pilot has to face, both in himself and in the people who entrust themselves to his skill. Knowledge is an effective antidote to apprehension, as is the sense, conveyed by the pilot's own apparent confidence, that everything is all right even if it doesn't seem that way. The most useful tool the pilot has for making an IFR flight comfortable for inexperienced passengers, however, is his own awareness and imagination. They permit him to put himself in the anxious passenger's place and to anticipate his needs. A pilot who does not allow his own experience to prevent him from understanding other people's reactions to IFR flight can make even a difficult instrument flight seem novel and enjoyable rather than mysterious and menacing.

Let's turn now to the passenger who is more accustomed to lightplane travel, the wives, husbands, sons, daughters, and friends of pilots who fly.

I presuppose in this class of passenger, perhaps optimistically, a certain incidence of courage, adventuresomeness, curiosity, or intolerance of thumb twiddling that might motivate learning something about IFR; I presuppose that those habitual front-seat passengers who have picked up this book—or have had it thrust upon them—have shown an interest in taking part in the game. I presuppose, in short, that you would like to help, if you only knew how.

You will be happy to know that no less awesome a behemoth than the Federal Aviation Administration has, in a sense, come out in your favor in considering, and in some instances enforcing, rules requiring two pilots for IFR flights. It has been decided in Washington or Oklahoma City or somewhere that the plain, ordinary, noncommercial instrument flight requires but one pilot; but the fact that two are required on certain commercial flights in light aircraft not far different from the one you may fly in suggests that in some situations, that lone pilot may be almost overburdened. Your help may be needed and welcome.

You may have to be subtle about offering it, of course—depending upon the size and fragility of the pilot's ego. Some pilots like to surround themselves with mystery, and expect mute passivity from their charges. For these, the role of the amateur copilot may have to be confined to service as a bookshelf.

They also serve, however, who only sit and hold books. The books (and the like) in question are the charts by which the flight is guided, and they usually consist of one or two brown leather binders, rather thick, filled with delicate pages of "approach plates" and pockets containing "en-route charts." There are also in-between maps called "area charts" for the environs of large cities, and these may be filed, according to the taste of the pilot, in a group near the front of the book or among the approach plates for the city in question.

Generally, the flight begins on an approach plate, then goes to an area chart, one or more en-route charts, possibly an area chart used in transit over some busy city, then an area chart for the destination, and finally an approach plate. Various exceptions may be made to this pattern. Most commonly, no area charts may be involved; or you may depart or arrive in VFR—good weather—conditions and omit one or more area charts or approach plates; or you may make use of the "SIDs" and "STARs"—standard departure and arrival procedures for major airports, which are filed with the approach plates for the city. The overall pattern, in any case, is logical, using large-scale, detailed maps at the beginning and end of the flight and smaller-scale, less complete ones en route.

If you end up with the chart book in your lap, ask the pilot if he or she minds if you look through it. You'll quickly recognize the different types of charts, and easily see how you proceed from one to the next. When you are familiar with the organization of the charts, you can provide the pilot with the ones he needs when he needs them, and file the ones he is done with in the right places. If the landing is made in VFR conditions, offer to relieve the pilot of whatever books and papers may be in his lap; most pilots prefer to land unencumbered.

Some pilots carry sectional or WAC charts with them on IFR flights in case they lose radio contact with ground stations but are able to proceed by visual navigation. These charts are much larger and more colorful than IFR charts; they are not carried in books; and they contain detailed topographical information. If these charts are aboard, you can keep track of your position on them by picking out salient landmarks or radiobeacons that are also shown on the IFR charts.

You can also do useful work by keeping abreast of your position, noting the time you pass each successive radiobeacon and calculating your position at any moment on the basis of your groundspeed. Speed can be determined from the time needed to cover the distance between any two points a known distance apart, and corrected later if necessary. Ask your pilot to show you how to use his time/speed/distance computer, and you'll be able to make the necessary calculations in seconds.

During an instrument approach, you can assist by noting the time of passing the final approach fix and figuring out, on the basis of information supplied on the approach plate, how many minutes and seconds it should take to get to the runway threshold. You can also do what real copilots do: look out for the first glimpse of the ground, and tell the pilot when you see ground features and how well you can see them. When you first glimpse the ground, it may be visible only downward and not ahead; as the airplane descends, the forward visibility improves until the runway becomes visible ahead. At the moment you can see the runway or the lights, say "runway in sight" or "approach lights in sight" to signal the pilot to look up from the instruments and complete the landing visually. This procedure, while not difficult, requires a little practice, since someone who has never seen a runway or an approach light may not recognize it the first time. After you have ridden on a few approaches, you will get the idea of what the pilot needs to see to land.

If you have enough passenger experience to be familiar with the operation of the radios, you can lighten the pilot's workload by doing some frequency changing for him, setting up VOR bearings, and checking DME distances and speed. While en route, it is perhaps better to let the pilot do his own frequency changing; it is just as distracting for him to confirm what you have done as for him to do it. During an approach using the ADF for both outer- and middle-marker passage, however, the pilot will appreciate someone else taking care of the frequency change. If he likes, you can also do the work on the frequency changes from approach control to tower at the outer marker, and to ground control after landing.

There are a lot of little chores that one often forgets about: turning off the transponder after landing, winding the clock, watching the oil pressure, switching tanks on time, and so on. Some of them are on checklists, and the pilot might like you to read off the checklist to him so that he does not have to take his eyes off the instruments. Some are not, and it takes a trained pilot to know them all. If you have no other chores to take care of, however, you can remind yourself to check the oil pressure, fuel pressure, engine temperatures, and fuel quantity and tank selected from time to time.

It is always important for the pilot to know what is going on in the airplane; even if a passenger is himself a licensed pilot, he should not lend a hand in conducting the flight without the consent of the pilot. On the other hand, though some pilots would doubtless object to the slightest outside intervention in their work, most realize that there is nothing about flying that a layman can't quickly learn; the only thing that takes long to learn is doing everything at once.

If you fly very often with a spouse or friend, you might try your hand at flying the airplane by instruments. My own constant companion insists on flying by reference to the instruments even on a clear day, which suggests that it can't be terribly hard to do. Really, there isn't much to it for short periods, and if the airplane in which you ride doesn't have an autopilot, you can quickly learn to take the place of one. It is easiest to practice in the left seat on a VFR flight, but since you will end up doing it from the right, that might be the best place to learn.

Think of the gyro horizon as a game in a penny arcade; all you have to do is keep the toy airplane level and not lose interest. If you hold the wheel lightly, with only two fingers, you will be less prone to inadvertently climb or descend. When you have the hang of flying straight and level, start including the directional gyro in your game. If you keep the toy airplane in the gyro horizon scrupulously level, the directional gyro will not move either; but if you let the toy airplane tip a bit, the DG will start to rotate. Now you have to bring the pointer back to the number that represents the magnetic heading you are supposed to be flying; to do so, you tip the airplane in the artificial horizon the way you want the pointer in the DG to turn. Just tip it a little, and hold it tipped until the heading you want is nearing the pointer; then level out the airplane so that things end up where you want them.

It sounds like child's play—literally—and it is. Anyone can learn to do it in a few minutes; the reason, again, why pilots spend a long time getting their instrument rating is that they have to learn to do this and a dozen other things all at once. If all you have to do is keep the plane upright for a minute or two while the pilot fiddles with a computer or copies a clearance, you really need nothing more then eyes, hands, and a functioning brain.

Copying clearances is often difficult, especially for the IFR novice. Early in one's instrument-flying career, one is subject to a sort of stage fright when the clearance comes over the radio. The fear of appearing ridiculous before a lot of pros tuned to the same frequency, or of incurring a hint of irony in the voice of the clearance deliverer, makes the pilot so nervous that his memory goes to pot, his writing hand trembles, and he bungles the simplest clearance. The contrast with a more blasé pilot is striking; he may simply listen to the clearance, repeat it verbatim, and *then* write it down. The only difference between the two men is that one is confident and relaxed while the other is nervous. Put the nervous one in his car and have him ask directions at a gas station, and he will mentally record an involved series of left and right turns, stoplights, hamburger stands, gas stations, and road signs, and then navigate the route without writing it down.

You can assist in copying a clearance if you can remain relaxed enough to get it in memory, or even if you can scribble down parts of it. People with a knowledge of shorthand—an art that would be most useful to a pilot but is usually denied him in a society in which most pilots are men, most shorthand writers are women, and most women pilots never learned shorthand—have it made, not because shorthand is superior to any other code system but because they are used to copying down words at a normal speaking speed. The trick pilots use is to leave out all but the absolutely essential parts, reducing them to initials and symbols, and holding in memory everything they can't get down on paper. The stumbling blocks are intersections with long names that you have trouble understanding at all, and then have trouble abbreviating in a way that you feel you'll be able to figure out when you try to read the jumble back. One way around these might be for a pilot to copy the clearance in his normal manner while you follow the routing on a chart. If several charts were involved, this would be difficult or impossible; often, though, the routing would be clear to you, and obscure names could easily be deciphered. (All of this, of course, presupposes that you can understand radio talk at all. It used to be commonplace that people riding in light airplanes for the first time would exclaim in amazement at the pilot's ability to understand the garble coming over the radio. Better radios have made this rarer; but the trick, as with all foreign languages, is to know what the speaker is probably saying, and what key words to look for. Once you have a few guideposts established, everything else falls into place.)

Apart from the duties that are integral parts of IFR flying, you can do other little things to make the flight pleasant. Bring along a thermos of hot coffee, for instance, and a couple of sandwiches. (Keep the thermos open during the climb, though, or you can have a scorching surprise opening it at altitude.) Something to eat or drink, and a little conversation en route, help relieve the monotony for everybody—and the pilot, who is bearing whatever strains the flight might entail, is grateful for any diversion. My own feeling—and I don't know how widely it is shared—is that it's nicer for a passenger to stay awake than to go to sleep, though on a night IFR flight of three hours or so, staying awake with nothing to do is quite a trick. The thing is, it's sometimes hard for the pilot, too, and a general atmosphere of wakefulness and companionship helps a lot. In fact, just having an interested and comprehending passenger beside him can give a pilot a great lift; so if you want to break into the copiloting field as a nonprofessional, start out by paying close attention to what the pilot is doing and how he's doing it. He'll appreciate the interest, and most likely won't turn you down if you ask if there's anything you can do to help. You may feel like a fifth wheel for a while, but in the long run, you and the pilot will integrate yourselves into a team and make your IFR flying easier, pleasanter, and safer.

10 WORKING WITH THE ATC SYSTEM

by Richard L. Collins

WHEN A GENERAL-AVIATION instrument pilot winds up butting heads with the ATC system, it's often due to a lack of planning, logic, or understanding of how the system works. A high degree of uncertainty over the use of chart-in-lap and mike-in-hand to make peace with the mysterious man on the ground can be frustrating. Another frequent problem causer is that some pilots start with a chip on their shoulders and fight with every controller along the way.

The pilot with the chip usually gets the job done but makes few friends en route, and it certainly isn't necessary to do it that way. Any logical flight profile, pattern, or route can be flown IFR, and can be flown smoothly and calmly. All that's necessary is to know the answers—or know where to get the answers—and to approach the system as something that will serve us if we say the magic words and play ball.

General-aviation pilots most often become misfits in busy terminal areas. This usually happens when a stranger files to such an area without first giving some thought to how to best operate in the beehive. For instance, when a controller asks a Cherokee pilot what speed he'll be maintaining on final, he's almost always hoping for the highest possible number—one that will enable him to at least halfway fit the Cherokee into a flow of jets. If the Piper pilot is used to staggering in at 80 knots at North Cupcake Municipal and tells Washington National that's his speed, the poor controller can only mutter "sugar oscar bravo" before he keys the mike and says "roger." He'll then have to start figuring out how to make the slow bird fly a pretzel to be in the right place at the right time to fit into a gap that will have to be created in the line of jets. If you ever plan to go to the city to look at the big buildings, practice a few ILS finals at higher than normal airspeeds, so you can run with the pack.

Studying the charts beforehand can help when you're

headed for a busy spot. If they have standard terminal-arrival routes (STARs), use the appropriate one to file into the area and you'll avoid an inflight clearance change. Mentally stake it out well in advance. (Use the standard instrument departure—the SID—when leaving, too.)

All areas with heavy traffic have automatic terminal-information service (ATIS), and pilots are expected to listen and digest the facts before their initial call to ground control or approach control. This can create a problem when inbound, for it is not easy to listen for a call from a busy controller while also noting the ATIS information. There's no sure remedy for this, other than trying to get the ATIS as far in advance as possible.

If there's an equipment shortage (or outage) in the airplane, try to think of ways to ease the pain. There is no way to minimize the lack of a transponder if you're flying in a high-traffic area, but other shortages are easier to overcome: if you have only one operable VOR, for instance, and there is an intersection that must be identified as part of a STAR or SID, ask the controller if he'll sound off when you reach it. Don't take a lot of time asking, though; he doesn't care about all the money your wife spends, or the fact that your avionics technician is harder to see than a doctor, or whatever might be your reason for buzzing around in his airspace with but one VOR receiver. He wants to know only what you need: "Approach, Cherokee 15 Romeo, would you call Sassafras Intersection for me?"

At all times, brevity is a treasured virtue for pilots flying in busy areas. If a pilot will think before speaking, and decide on the quickest way to get the message across, everything works more smoothly. A really lousy initial call-up to an approach-control facility might be: "Ahhhhhhhhhhhm Dallas/Fort Worth Approach Control, this is Piper Cherokee November Four-Two-One-Five Romeo, IFR, flying level at 5,000 feet, squaking one five zero zero, listening one twenty-three point nine, and I have received Dallas Love Field Information Whiskey."

The announcement of One-Five Romeo's pending grand arrival at Love Field could just as well have been: "Approach, Four-Two-One-Five Romeo at 5,000 with Whiskey." A minute of a frequency will hold only 100 to 150 words (depending on whether the speaker is a mushmouth or a tobacco auctioneer), so every word is critical. If all pilots gave lengthy arrival speeches, many frequencies would become saturated, and controllers would all have ulcers. Say it clearly and say it briefly. Sound like a pro. If you want to know what pros sound like, take a transistor VHF with you and listen to the chatter whenever you have a few free minutes in your motel room.

Many pilots moan about the difficulty of flying single-pilot IFR in high-density areas, but their fears are unfounded. Terminal-area traffic is programmed to flow at a proper pace for jets, so when you come along in a slower piston plane, things don't happen so fast; a well-organized pilot will actually find the workload lighter than he expected. The controllers do talk a little faster in busy areas, but about all they can do is tell you which way to point the airplane, where to fly, and how high to fly—which is the same thing they tell you anywhere else. You may even end up preferring high-density areas to low-density airports: the controllers do better work, the traffic flows smoothly, and everybody's overall competence makes it a satisfying environment. When filing IFR to such a place, you have all the advantages of being expected and having procedures to fly, which actually makes it easier than showing up VFR, unannounced, and having to introduce yourself.

If a pilot has a bad hangup about big-city traffic, a little dual in such an area (with an instructor who flies actual IFR there) will usually sort things out. An autopilot can be a big help, too, but one isn't necessary if the pilot is worth his rating. Actually, the autopilot is at its best when you are flying IFR in a busy area in VFR conditions. Let the autopilot take care of precise flying while you look for other traffic. (Don't expect the controller to call all the traffic in busy areas. They often don't have time.)

One final word: do be aware of wake turbulence when flying to and from jetports, VFR or IFR. If you don't fly them frequently, review the avoidance principles before going. The primary avoidance principle is to stay above the flight path of a jet you are following on approach or after takeoff. The trailing vortex settles, so on an ILS, a preceding jet's wake should have settled below the glideslope when you come along behind. It's not a bad idea to stay a dot high on the glideslope, though, just in case that preceding jet did some swooping and dipping on his approach.

The city kid operating IFR from an uncontrolled airport without a flight service station can be just as anxious about IFR as Lil' Abner headed for La Guardia for the first time. The trouble in the boondocks is a lack of activity rather than a surfeit, and while anybody can get a clearance from a controller and make an approach into an out-of-the-way airport (if one is published), getting back out again is often not so simple.

Before trying that IFR departure from Dogpatch, let's say a few words about going there IFR if there is no approach procedure. One obvious way is to file to the nearest airport with an approach, let down, and then fly on to the destination VFR. You could also plan your flight along an airway that goes over or near the proposed destination, and you could then ask the controller for a descent to the minimum en-route altitude when you are in the vicinity of the airport. If you find yourself VFR at the MEA, cancel and land. Whatever you do, don't succumb—and the pun is intended—to the frequent urge to make up an approach using a convenient VOR or standard broadcast station. You could break out eyeball to eyeball with a TV tower or ridge.

When you are ready to leave, don't be discouraged if the airport manager thinks IFR is only for city slickers with wide ties and noisy jets, and that anybody who flies in clouds is crazy. There is a way to depart IFR from almost any airport—even those without approach procedures—and it's possible to fit smoothly into the system while doing so.

The flight service station is the key to the system in this situation. Call them, get the weather information you need, file an IFR flight plan, and discuss how you'll get your clearance. It is best to call at least 45 minutes in advance of such an operation, too, so the FSS will have time to coordinate it with the center or approach control into whose air you'll be flying.

If weather conditions permit, the best procedure might be to file, then take off VFR, and call the FSS for a clearance. If you're flying into air that belongs to an approach-control facility, they would be the ones to call; otherwise, if low-altitude center communications are possible in the area, the first call might be direct to them. At many big-city satellite airports and fields near remote center communications sites, you can even call for your IFR clearance while you're still on the ground.

Whatever the proper procedure is, the FSS will give you the word. If there's no way to get a clearance while seated in the airplane, the FSS will make a deal to give it to you on the telephone. After you file 45 minutes in advance of the proposed takeoff time, they'll probably tell you to call back 15 minutes before liftoff so they can read the clearance to you over the phone. Such clearances contain "effective" and "void" times, which simply means you can take off whenever you want during an agreed time span (usually from five to fifteen minutes), and center or approach control will be expecting you to blossom into the airspace over Dogpatch about then.

When you get a clearance on the phone, march right out to the airplane and get ready to go. It's a lot better to wait a couple of minutes at the end of the runway for the effective time, cocked and ready, then it is to have another Coke and then rush to get cranked up before the clearance is void. Telephone clearances will be specific about whom to call after takeoff.

If time is of the essence, and the weather is *good* VFR for your departure, there's a way to get airborne and get an IFR as soon as possible without filing 45 minutes in advance: file with the FSS on the phone, and tell them you'll pick up the IFR en route, whenever it is ready. Then take off and call center or the FSS on whatever frequency was agreed upon. Just be sure the weather is okay for the first 20 or 30 minutes of flight.

There is yet another way to leave Boondocks Municipal. If you happen to be inbound and you know you're going right back out, a center controller will sometimes give a departure clearance, with effective and void times, before you drop off his frequency to land.

The basic thing to remember is to communicate with the system in advance (through the FSS, by telephone) and give it time to perform its work. The pilot who jumps into his plan cold, takes off in very marginal VFR conditions, and then starts trying to make a quick deal for a clearance while dodging treetops and billboards is no better off than the man who tries to penetrate a high-density area with low-density habits.

If you ever find yourself making an IFR departure from an airport that does not have an approach procedure, make sure you have an idea of a place to shoot an approach if you should want to get back down soon after takeoff. Spend a few moments making certain that no obstructions—high terrain or tall towers—menace your proposed departure path, too. At IFR airports where special departure procedures are deemed necessary because of obstructions, these procedures are published on approach plates. You are on your own at airports without approaches.

Published preferred routes are important to any pilot who wants to fit gracefully into the IFR system. The computer age brought a degree of inflexibility to the system, and the computer rolls its eyes and belches ponderously whenever you file a route that is not a *preferred* route between two major cities. While the FAA usually says that the refiling required usually doesn't delay things, it does. You'll eventually be cleared to fly along the preferred route, so you might as well file that way in the first place. (They are listed in the *AIM* and the Jeppesen manuals.)

If you're en route and need a new altitude, and there's a reason other than sheer boredom, tell the controller. Again, don't recite a short story; be specific. "One-Five Romeo would like a higher altitude due to turbulence" is quite enough to get the message across. Plan ahead as much as possible in order to give the man time to do any coordinating necessary to approve an altitude change, and be calm when making your request. Nothing fouls up the efficient atmosphere of a center more than the squeaky voice of an excited instrument pilot.

Finally, one of the best ways to understand the system is to look at it from the other side. Every IFR pilot should visit a center and approach-control facility at least once. If you have a special question about some operation, that's the place to have it answered, and after watching them work, you can usually see why they do things the way they do.

The ATC system is designed to be *used*. Nobody ever intended it to be an obstacle course to convince lightplane pilots they are better off on the ground when the rains come—although that's unfortunately how many of us look at it. Though some of its components may seem to cater to the big boys, the most basically IFR-equipped general-aviation airplane can take advantage of the system—if its pilot plans ahead.

11 APPROACHES
by W. L. Traylor, Thomas H. Block, Robert Blodget and Archie Trammell

The Danger of the Duck-under   by W. L. Traylor

A SMILE CREPT across the pilot's face as he received clearance for the VOR approach. The thought of home base and the nearly completed flight lit him with secret self-pleasure. For three hours and almost 600 miles, he had been flying in solid instrument conditions. In spite of intermittent turbulence and a scum of rime ice on the wings, his Aztec tracked precisely along its course.

During the first awkward hours of instrument training, the pilot of Aztec 96 Yankee was intrigued with the detail and precision required in this type of flying. Now, after several hundred hours of experience, he found a great sense of accomplishment in IFR flight. He took well-deserved pride in his ability to plan and execute an instrument cross-country with accuracy and professionalism. Only the last phase of this flight remained before it could be logged: the VOR approach.

Outbound from the station to procedure turn, he got the current weather—measured ceiling 600 feet, visibility one mile in rain and fog, wind calm, ceiling ragged—exactly the minimums needed for the approach. Since departure, the weather at destination had slipped well below the forecast. The final hour of flight had been an undeclared race home before the weather went below minimums and forced the flight to an alternate airport more than an hour away.

Because of its coastal location, the airport was frequently closed during periods of poor weather, and the pilot often lamented the absence of an ILS system. Nonetheless, he had never failed to make it home, even though he had made the approach under less than the required 600-foot ceiling and one-mile visibility more than once.

The Aztec crossed the VOR inbound and the pilot began a 1,000-fpm descent. He reached the minimum descent altitude, came up on the power and held the aircraft in level flight. One minute and 10 seconds had passed since crossing the VOR inbound; two minutes and 45 seconds remained before reaching the missed-approach point. The aircraft was still enveloped in fog. One minute before his estimate to the runway, the pilot saw no improvement in visibility and eased the aircraft down another hundred feet. Making quick glances directly below, he caught sight of lights dimly glowing through the murk. With less than 30 seconds to the missed-approach point, the pilot descended another 100 feet, expecting to break out of the overcast any moment with the runway dead ahead. Finally, he realized the approach had been missed, and shoved the throttles forward to climb. No sooner had he done so than the aircraft suddenly flashed through a clear area, and the pilot saw the runway just off his left wing. A landing attempt was impossible; within three seconds, he was in the clouds again.

Seven minutes after declaring the missed approach, 96 Yankee proceeded inbound again on the final approach course. On his second attempt, the pilot was sure he could make a successful landing. The fleeting glimpse of the runway encouraged him, and he felt certain that if he had been only a few feet lower during the first approach, he would now be safely on the ground.

Dull sounds of impact as the Aztec struck a stand of tall trees one mile from the airport marked the flight's finish and the tragic finality of another duck-under approach.

The probable cause of this accident was inevitably found to be: "Pilot in command descended below minimum safe altitude while operating under instrument conditions." While this is an accurate description of why the aircraft crashed, it hardly explains why a competent pilot, experienced in IFR procedures, would commit such an obvious error. That the pilot was known to be a perfectionist in all aspects of flying only increases the mystery. It is axiomatic, however, that this pilot's high level of ability contributed to his last, fatal decision.

Although official records don't give exclusive license for duck-under accidents to highly experienced or professional pilots, statistics indicate that this type of accident more often than not involves pilots with above-average experience and proficiency. The urge to risk a below-MDA approach grows stronger as a pilot gains experience and confidence in his ability. While by far the greatest number of instrument approaches are made under VFR conditions, or become VFR approaches well above the MDA, there are times when he will find himself making approaches to actual minimums. Naturally, this happens most frequently to pilots who fly most often.

The first duck-under approach a pilot attempts will probably be more by chance than design. He finds himself nearing the airport at MDA and still in the clouds. But looking straight down, he discovers familiar landmarks. Obviously, the base of the ceiling is just a few feet below. To remain at the legal altitude means a sure missed approach; letting down another hundred feet seems like the only reasonable decision, and suddenly the runway is directly ahead.

The next attempt may be made under nearly the same conditions, only now the descent to a point 100 feet under the MDA is more easily justified. Soon, the pilot's personal minimums stray 200, then 300 below the MDA, and on each occasion, the lower descent is easier. Once the technique is mastered, the extent a pilot will descend below MDA reaches unbelievable proportions: an airliner that crashed during a VOR approach in Connecticut not long ago first impacted at a point *below* the elevation of the airport.

There is little doubt that in many instances, a pilot's decision to duck under is based largely on considerations totally unrelated to the physical problems of the approach. At least two of these factors influenced the pilot of Aztec 96 Yankee.

The first of these was his decision to attempt the second approach rather than proceed to his alternate. The unpleasant prospect of renting a car and driving 50 miles helped him to rationalize another try, even though he knew the weather was below legal minimums. The second consideration was the pilot's own record: he had never failed to make it home.

By his own standards, the pilot of 96 Yankee was in a position where the approach had to be not only attempted but completed. Not finishing the flight at his home airport, as planned, would be far more serious than mere inconvenience.

Pride is a very personal thing. Not all pilots are influenced by the opinion of others, but are judged by the most severe critic of all—themselves. Not long ago, the pilot of a Bonanza was established in a holding pattern waiting his turn for a VOR approach. The weather at the airport was reported at minimums. When clearance finally came, the pilot asked if the Piper Comanche making the approach ahead of him had landed. The reply was affirmative.

During the first attempt, the pilot declared a missed approach. When the tower requested his intentions, the pilot replied, "I'll go back to approach control for another try. If the Comanche got in, I guess I can." The second attempt followed immediately and was again terminated in a missed approach and a request for clearance to begin a third approach. The pilot became noticeably slower in his responses to radio contacts. After reporting the VOR inbound, an unusually long time passed without response from the Bonanza. Tower personnel noticed the aircraft some distance south of the approach course proceeding beyond the field at an extremely low altitude. They requested the Bonanza's position, and the pilot responded by declaring another missed approach. The aircraft then climbed into the overcast.

Perhaps the most fortunate moment in this pilot's life came when a special weather observation established that the ceiling was below minimums, and his request for a fourth approach was denied.

The stress on a pilot during an approach in poor weather is definitely a factor in his ability to perform. The well-trained instrument pilot should be able to cope with this stress; however, there are unknown variables. The amount of tension in the cockpit after the first or second missed approach cannot be exactly determined. Most second and third landing attempts do result in safe landings, but accident records show that crashes occur during the second or third attempt with disturbing frequency.

Of course, there are no regulations limiting the number of approaches a pilot may attempt, and such a limitation should never be imposed if the pilot is to remain the final authority. However, the pilot himself must be aware of the subconscious reaction that follows each missed approach. The first attempt can be considered routine; during the second try, a pilot will tend to rationalize his situation: "If I'd been only a little lower, I could have made it."

The second missed approach adds significantly to the pilot's stress and should be considered a danger signal. The decision for a third approach should be based on definite information of improving conditions, higher ceilings, or increased visibility. The third attempt under the same conditions becomes a challenge to the pilot, and when the situation deteriorates into a contest between man and machine or man and his pride, the pilot is least equipped to measure his own capabilities.

With continued expansion of ILS systems into smaller airports, the nonprecision approach is losing much of its popularity. Unfamiliarity with VOR approaches frightens some otherwise capable pilots who are shooting one down to minimums. Most weather-associated accidents occur in the approach and landing phase of flight, and these accidents happen predominantly to aircraft making nonprecision approaches. Despite these facts, the nonprecision approach itself is not at fault. In almost every case, accidents happened because the pilot descended below the MDA without visual reference to the runway.

Despite increasing use of precision approach systems, the nonprecision VOR approach will remain the only IFR approach at most airports for years to come. If used as intended, the nonprecision approach is simple and as safe as (if not safer than) its sophisticated counterparts. But one must accept and be aware of its limitations. Follow a few basic rules.

First, be certain you have the current altimeter setting; as you cross the VOR inbound to the field, be prepared to start the descent. Never continue to descend unless you are exactly on course. Note the time over the VOR inbound and the time you will be at the missed-approach point. After reaching the MDA, *never* allow the aircraft to dip below this altitude unless the runway is clearly in sight. If you reach the MDA while still in instrument conditions, make a missed approach. Start the missed approach exactly at your computed MDA time no matter what you think you see ahead.

Duck-under approaches are a lot like eating peanuts; once you taste one, you're going to want more. The chances are favorable that you'll make the first approach and the second, possibly even 10 or 20, but sooner or later, you're going to lose.

Nonprecision Approaches *by Thomas H. Block*

TAKE A PILOT who has been fed mostly ILS, then weaned on radar vectors, and brought to a simmer on autopilot plus flight director. Give him something like a VOR, ADF, or back-course approach to execute and, somewhere along the line, allow him to identify these maneuvers by their official title of "nonprecision approaches." What will happen? He'll commit almost every IFR error imaginable. As he fumbles his way through these stepchild instrument letdowns, two things will become readily apparent: that nonprecision approaches require more skill, cunning, and outright precision than any ILS; and that the term "nonprecision" refers to the manner in which most pilots will execute a VOR, ADF, or back-course approach.

In the context of instrument flying, the word "precision" refers to any instrument approach containing an electronic glideslope. Nonprecision is just a lack of the same electronic height guidance. A full ILS (localizer, glideslope, and related components) or precision approach radar (PAR—the procedure wherein the ground controller is constantly chattering about your being "on glideslope, on localizer," or "correcting nicely") are the only two letdowns that contain an electronic glideslope and hence are allowed the precision title. All other methods of approach, including such exotic modes as VOR-DME or surveillance radar (ASR—the procedure where in the controller only tells you if you're left or right of centerline, possibly with recommended altitudes each mile thrown in), fall in the nonprecision category. Obviously, nonprecision approaches outnumber their precise counterparts. In addition, any full ILS with an inoperative glideslope—an all-too-frequent occurrence—automatically transforms itself into a nonprecision approach.

Contributing to the second-rate reputation of nonprecision approaches is the fact that, in almost all cases, their minimum legal visibilities and descent altitudes (labeled MDA—"minimum descent altitude"—for nonprecision approaches) are considerably higher than the visibility and decision height would be for a full ILS or PAR. Although it might seem less honorable to descend to only 400 feet above the airport as opposed to 200 feet during the ILS, the fact is that *minimums on a nonprecision approach are higher because such letdowns are more difficult to perform.* Approaches are not legally terminated several hundred feet higher than otherwise just to allow a pilot to become sloppy; they remain higher because even a well-executed nonprecision approach requires higher parameters to assure safety. The airman who allows "nonprecision" to become synonymous with "sloppy" is no better off—and may be in a considerably worse predicament—than the pilot who descends a hundred feet below full-ILS minimums.

Knowing your position during a full ILS only requires a conscious pilot and the slightest amount of arithmetic. An aircraft "on glideslope" and halfway down to minimums is obviously nearly halfway between the outer marker and the airport. An "on glideslope, on localizer" condition at the middle marker would result in the runway being approximately half a mile ahead and 200 feet down. Due to terrain considerations, some airports have full ILSes with higher than the customary 200-and-a-half minimum (and airports with category II authorization have lower minimums, although you can't utilize them unless you're specifically authorized by the FAA), but regardless, all this appears on the instrument approach plate.

Knowing your position during a nonprecision approach requires a pure guess—hopefully an educated one—with the basis for that guess derived from certain performance levels produced by the airplane. If the instrument approach plate calls for the aircraft to cross the VOR inbound at 2,000 feet msl and descend to minimums of 400 feet msl in time to find the airport some four miles from the VOR, simple arithmetic tells the pilot he must lose 1,600 feet in four miles. If the aircraft's predicted groundspeed is 120 knots—i.e., two miles a minute—then two minutes will be required to get from the VOR to the airport. Two minutes—during which he must lose 1,600 feet of altitude—mean a rate of descent of 800 feet per minute, which will place the aircraft over the threshold at minimums. All this is made even easier by the inclusion of a speed/time/rate-of-descent chart on each nonprecision approach plate, so all the pilot needs to do is select the proper groundspeed and read off his time to missed approach and his minimum rate of descent. There is, however, one fly in the ointment.

All computations related to finding the airport revolve around aircraft *groundspeed,* and are, therefore, variable at the whim of wind direction and strength. This is the first area of guesswork for the pilot on a nonprecision approach. "How much," he must ask himself, "will the wind affect my groundspeed?" Most often, there will be a headwind and subsequent groundspeed reduction; but occasionally, the effects of a tailwind must be added. Regardless, a wind of any magnitude (or a VOR/beacon-to-airport distance of any length) will result in drastic time and rate-of-descent differences.

If the aircraft in the hypothetical 120-knot-airspeed and four-miles-to-airport situation encountered a wind that averaged 30 knots on the head, his new groundspeed of 90 knots would alter time to airport from two minutes to two minutes 40 seconds—and 40 seconds is a long time to be groping around in the clouds only 400 feet above the ground; his minimum rate of descent would, at the same time, be lowered from 800 to approximately 600 fpm. Put the aircraft in a 30-knot-tailwind situation and his time to airport becomes one minute 30 seconds, with minimum required rate of descent raised to slightly over 1,000 feet per minute. This is a far more dangerous situation, to the unsuspecting pilot, for a too-slow descent may ensure that he never sees the airport, and an overlong approach will concurrently bring him to his MDA well *past* the field, in an area where terrain- and obstruction-avoidance becomes a matter of sheer luck.

If all this appears to complicate what initially appeared to be a "nothing-to-it" nonprecision approach, remember also that in addition to wind effect being a guess, it is also an extremely variable factor—particularly within 3,000 feet of the surface, where theory alters the surface wind considerably in both direction and velocity.

The careless pilot complicates matters even further by taking the recommended crossing altitudes and airspeeds haphazardly. If a nonprecision approach requires guesswork even when flown perfectly, the man who allows his aircraft to cross the facility at any willy-nilly altitude and fly a fluctuating airspeed/rate of descent to some wildly arrived-at time to airport had better hope his destination weather is above both minimums and obstructions in the area. Unless a man stringently adheres to the approach chart's plan of action for attacking a nonprecision approach, he will be adding so many variables that the word "educated" will have to be dropped from his guesswork.

So many pilots *do* take their performance indications for granted during a nonprecision approach that an explanation of their actions seems required. While no one with a functioning mind would attempt to fly a full ILS with the glideslope needle showing a substantial "aircraft low" position, the dangers of the same aircraft position during a nonprecision approach are not nearly so graphic; while the glideslope needle on a full ILS literally shouts to the pilot that something is askew, a too-low (or too-high) position without a glideslope is a more subtle thing. Without a needle to shout in his ear, many a pilot will happily fall asleep at the switch from VOR/beacon inbound, waking up only occasionally to look out the window in hopes of finding the airport.

Naturally, there are tricks for making a nonprecision approach easier— little techniques such as slightly exceeding the computed rate of descent so as to arrive at the MDA somewhat before arrival at the missed-approach point, thus giving the pilot a few seconds in level, stabilized flight while looking for the airport. The catch is that this trick—or any other trick, for that matter—is no substitute for precise, thoughtful flying.

A man who tries to fly a nonprecision approach precisely will build a foundation upon which any future knowledge will find a ready footing. The man who looks upon the nonprecision approach as one of the last areas of instrument flying requiring constant attention, superior insight, and implicit understanding will find himself happily executing many safe and precise letdowns. After all, the pilot is the precision factor in any nonprecision approach.

On the ILS *by Robert Blodget*

IF YOU LOOK at the approach plate for an instrument-landing-system (ILS) approach, it's easy to come to the conclusion that the approach begins at the outer marker. True, terminal routes are shown, but they are no more than transitions; the approach itself obviously doesn't begin until we have lined ourselves up with the localizer and have passed the outer marker.

The truth is that ILS approaches, like all other IFR and VFR approaches, begin long before the flight has reached the final approach fix, inbound. A careful study of the approach plate is an important part of flight planning, even if the approach will be made to a familiar airport; and from the time the pilot begins maneuvering in the letdown, the approach plate should be in full view. But before we start crossing the needles, let's take a brief look at the ILS system.

The ILS is quite old: the first experimental system was tested in the late 1920s. It is still the most accurate landing aid we have. The heart of the ILS is two directional radio signals, one of which—the localizer—provides lateral guidance to line the airplane up with the runway, and other—called the glideslope—does exactly what it says: it allows the pilot to put the airplane on the correct vertical path to touch down at the right spot on the runway.

Both transmitted guidance signals produce a pattern shaped like a fan, with the point of the fan at the touchdown point. The two signals combined provide a guidance coverage that can be thought of as a flattened funnel, the point of which touches the runway at the touchdown zone.

The in-cockpit display is one of the best in terms of ease of interpretation. The localizer signal is shown on a pointer that moves left and right, and the pilot simply flies toward the pointer. In the same way, the glideslope signal moves another pointer up and down; once again, the pilot flies to the needle. If the localizer needle is left, the pilot flies left; if the glideslope needle is above center, the pilot flies up. All modern indicators combine the two pointers in one case. The left/right needle is the same instrument that displays course information from the en-route VORs, and the same receiver drives it. The ILS localizer course is much sharper than a VOR course, so the receiver and indicator are automatically provided with means to adjust for this. When the system is operating in the localizer mode, the left/right needle also automatically finds its own center, without the need of setting the inbound course on the omni-bearing selector. (Some pilots do this anyway, simply as a reminder, but the system doesn't require it.)

The glideslope signal is transmitted on a different band than the localizer signal (UHF for glideslope, VHF for localizer), so it needs a separate receiver. All localizer and glideslope signals are "paired," meaning that the glideslope signal for each localizer frequency is always the same. To simplify the pilot's job, modern airborne equipment is arranged so that the glideslope receiver is automatically turned on and "channeled" to the correct frequency whenever the navigation receiver is set to any localizer frequency. Warning flags are provided for each signal to show when the system (either ground or air) isn't working. In most indicators, both needles come to rest in the center when no signals are being received. This can lead to some problems, as we will see. In more expensive and complex airborne systems, the needles are removed from view, in addition to showing warning flags, when the systems aren't working.

Although the two directional signals are the heart of an ILS, there are several other elements needed for a complete system. These include marker beacons, transmitting on the frequency 75 mHz: an outer marker (OM) in all cases, a middle marker in many (MM), and an inner marker (IM) in Category II installations. The purpose of the markers is to tell the pilot how far he is from touchdown. In addition, the glideslope signal is generally intercepted at the outer marker, which is usually the point at which the final-approach vertical path, or "glide," is established.

In many cases, nondirectional beacons (NDBs) are also installed at the marker locations. These are known as "compass locators," the compass referred to being the radio compass, or automatic direction finder (ADF). In these cases, the outer marker is shown on the charts as "LOM," and the middle marker "LMM," standing for "locator outer marker" and "locator middle marker." The advantage of the nondirectional beacon is that it allows aircraft equipped with ADFs to fly directly to the localizer just by homing on the NDB. In the absence of an NDB, other transitions must be used, the most common of which now is guidance by ground radar.

Other integral elements of a complete ILS include an approach-light system (ALS) and, in a few places, distance-measuring equipment working in conjunction with the glideslope. This is called ILS/DME. Absence of any of these elements requires higher landing minimums. For example, the decision height (DH) for Runway 25 at Los Angeles International is 200 feet above ground level, and visibility one-half mile, or 2,400 feet runway visual range (RVR) when the full ILS is in operation. If the approach-light system is out, DH is 250 feet, but visibility goes up to three-quarters of a mile (4,000 feet RVR). Without the glideslope, but with the ILS/DME working, decision height is replaced by minimum descent altitude (MDA), and various visibilities are specified, depending on what other ILS elements aren't working. The most severe penalty occurs when everything but the localizer is out of service, when the MDA becomes 520 feet and the visibility one mile.

All these factors must be considered when the flight is being planned; and part of the planning must be to find out if any of the elements of the ILS at the destination airport are out of service.

There is another, more subtle, planning element that deserves attention. Except for professional crews and a limited number of other highly qualified pilots, prudence suggests that weather standards above the absolute minimums be established, particularly for unfamiliar airports. This provides a safety factor of considerable importance, and simplifies the problem of becoming fully familiar with all the different possible conditions.

In the current air-traffic-control system the transition from cruise to approach is made by the pilot simply following the controller's instructions; and this will increasingly become the case. This inevitably tempts pilots to rely entirely on the controllers, and pilots often have no firm idea in mind as to how to perform the transition in case anything interferes with the radio-communications link. The transitions (terminal routes) are shown on the approach plates, so it's not a big chore. The important thing is to keep surprises out of the cockpit, especially during the final approach.

Now for the final approach itself. Thanks to either radar vectors or our own navigation, let us assume we are tracking inbound on the localizer, at the proper altitude to intercept the glideslope. The vertical-guidance needle is now active, and is above the center position, indicating "fly up." As we approach the outer marker, the vertical needle moves down, and will reach center as we pass the outer marker if we are at the right height. Before reaching the outer marker, we should have extended the landing gear and whatever approach flaps are called for in our particular airplane. (In some places, the approach controller may try to con us into maintaining a higher speed during final than we would like. It's best to ignore the request, and stay firmly on whatever speed is best for our airplane.)

From the outer marker to visual sighting of the runway, our task is simple: keep the left/right and up/down needles somewhere near center. If we happen to find that we are able to keep them exactly centered, all the way down, think of it as an accident. Why? Because about the only way to keep both needles precisely centered for any length of time is to turn the radios off. It's safe to say that all glideslopes and localizers have small kinks in them; and the kinks, since they are within tolerance, aren't published. At several airports, you can position yourself below the final approach path and watch big jet airplanes faithfully following the localizer kinks. This is a good indication that the approach is being flown by the autopilot coupler, since airplane captains seldom bother with such small deviations, especially once they know they are present.

In addition to kinks, there is also usually some sort of turbulence and wind shear; here again, the needles will move about a bit. The temptation is strong to go after them and get them nailed down, just as your friendly instrument instructor insisted you must.

This is not the best way. The on-course path keeps getting sharper and sharper the closer we get to touchdown; and if we keep trying to center the needles, there is an excellent chance of missing the runway entirely, since the rate of change gets faster and faster as we approach the end of the runway, but the airplane cannot respond faster and faster.

It is much better to fly a "stop-needle" technique. If the localizer goes off a bit to the right, start a gentle turn after it, and stop the turn as soon as the needle stops moving. Even if it begins to move back toward center, don't wait for it to bisect the bull's eye, but instead start another turn and stop it when the needle stops.

Do the same thing with the glideslope needle. If it says "fly up," add a bit of power until it stops moving up. If it says "fly down," ease power off until it starts to center, then restore the power, or most of it. Be more concerned with flyup than with flydown, unless you are flying a turbojet that really needs all the concrete in front of you.

In addition to making your job much easier, this technique will be valuable in another way. Remember that most ILS indicators, even when they show failure flags, have their needles return to center in case of failure of either airborne or ground equipment. It's uncomfortably easy to fail to see the flags; and the spectacle of the two needles beautifully centered can cause the pilot to smile to himself and think that the world is all roses.

On the other hand, if you are willing to accept needles that are half a dot off center, you will not only find the runway at the right altitude, but you will also have the best possible indication if anything goes wrong with the signals: you will automatically associate precise centering with something unusual.

From the ground, nobody will know. In the cockpit, nobody will care, unless you have asked your complaining old instrument instructor to fly copilot for you. If you have done that, ask him to make the approach while you watch. Chances are that even he will not be able to keep those needles exactly on the bull's eye.

The ADF Approach *by Thomas H. Block*

ADF, THE OLD workhorse. It went hand in hand with the DC-3, cones of silence, and flashing on-course lights. Unlike most other relics, however, the ADF keeps rearing its ugly head, like an uninvited relative who shows up at a picnic. Let an instrument pilot try to fly without a good foundation in ADF needlework and he'll very quickly run smack into an indigestible hash of headings and bearings. The ability to fly the ADF correctly is still very much a necessity in modern instrument work—particularly at the many airports where a beacon approach may be the only IFR method of getting down. The ADF approach may not be quite as simple as some, but with proper pilot technique, it can usually be as effective as any.

The radio signal utilized during an ADF approach is radiated omnidirectionally from a nondirectional beacon. Unlike VOR or ILS, there is no definite beam or radial laid down for the approach; the pilot must figure out his course by converting a combination of aircraft magnetic heading and relative direction to the radio beacon as indicated by the ADF needle. The potentially dangerous aspect of the whole affair is that there are no warning flags or full-scale needle deviations to alert the pilot that he has fouled up. A man incorrectly executing an ADF approach will receive nothing more than the subtle but potent indications of an ADF needle and magnetic compass to show that he is grossly off course.

The combination of aircraft magnetic heading and relative bearing between aircraft and radio beacon will result in a line of position denoting aircraft location. In other words, if you can determine aircraft magnetic heading (from the compass and directional gyro) and relative bearing to the radio beacon (tune in the appropriate station and watch the ADF needle), the resulting number will allow you to draw a line on the chart—and the aircraft will be located somewhere along the line. It is neither a mystical nor a difficult procedure, but it does require the pilot to possess one important ability: *the ability to visualize aircraft position after determining aircraft-to-station bearing.*

If an aircraft heading east is pointed directly toward a nondirectional beacon, that aircraft is obviously located somewhere along a line of position lying due west of the station. If the pilot had intended to be inbound toward the station from the northwest, he would alter his aricraft's heading to the left until crossing the desired bearing, then turn again to a direct-to-station heading (southeast). Flying a given course is nothing more than making constant corrections back toward the desired line of position when the computations of magnetic heading and relative beacon position show you've drifted to one side or the other of the intended bearing to station.

Adding aircraft magnetic heading (compass reading) to relative bearing to the radio beacon (ADF reading) will result in the magnetic bearing through the station, somewhere along which the aircraft is traveling. To make this admittedly confusing relationship easier, we have the electronic wizard of ADF approaches—the radio magnetic indicator, or RMI.

If an aircraft has an RMI, it has a mechanical/pictorial method of adding and subtracting. RMI consists simply of an ADF needle mounted on the face of a direction-sensing compass card. Since the card under the needle rotates to indicate aircraft heading, just like a vertical-card gyrocompass, the pilot need do nothing more than read "bearing to station" under the front end of the ADF needle and bearing from station" at the needle's tail end; the arithmetic is done by the mechanical relationship between needle and card. There remains, however, the very important requirement that the pilot be able to visualize aircraft position after establishing what the bearing to station is. Instant access to the fact that bearing to station is 030 (the number under the front end of the ADF needle) and bearing from station is 210 (the reciprocal; also the reading from the needle's tail end) is of little use to the man who cannot visualize his position as southwest of the radio beacon. If the instrument-approach plate required flight along the 190 bearing from station, the pilot who had a clear picture of his position southwest would know that flying an east heading until the ADF needle said 010 (tail-end reading of 190) would result in the most expeditious intercept of the desired bearing from station. The ability to accurately visualize position is the most valuable asset an ADF pilot can have.

Despite its talents, the RMI does not take the place of a sharp pilot—but it does take the place of cumbersome arithmetic. A 245 heading on the DG when the ADF reads 160 makes for no joy for the calculating airmen—and it often produces embarrassing results caused by faulty arithmetic. There is also the problem of having to subtract the unworkable number "360" whenever the sum of magnetic heading and relative bearing exceeds the dimensions of the compass rose—an all-too-frequent occurrence. Although the mental calculations required might not seem very weighty, remember that all this arithmetic takes place simultaneously with aviating and communicating at two or more miles per minute.

Salvation for the RMI-less pilot is simple. Even when a man must fly an ADF approach with what is known as "fixed card"—in which the ADF card does not rotate to sense direction but an ADF needle pointing toward the right wing always reads 090, the left wing 270, the airplane's tail 180—he can beat the math painlessly by using an easy gimmick.

Naturally, it is simpler to work with small numbers. Also, there is a distinct advantage in utilizing the facial characteristics of a gauge, because it helps prevent errors by at least starting the pilot in the proper direction. If the man who faced the fixed-card ADF problem had synthetically divided his ADF card into four appropriate labeled quadrants, he would reap both the above benefits.

Relative bearings from the upper-right quadrant are *added* to the aircraft magnetic heading to produce aircraft bearing to station; bearings from the upper left quadrant are *subtracted*. The resulting number is an almost instant readout of direction to the radio beacon—but again, it is nothing more than a useless number to the pilot who cannot picture where a 320 bearing to station actually places him.

After passing the radio beacon inbound, the ADF needle will reverse itself and point toward the tail of the aircraft. By then utilizing the two lower quadrants of his synthetic compass rose, the pilot will be solving for bearing from station—which is all the information he needs to get from beacon to airport.

The subtraction and addition of the lower quadrants are the reverse of what takes place in the upper hemisphere: a bearing from the lower right is *subtracted,* while bearings from the lower left are *added* to aircraft heading to arrive at bearing from station.

Guesses for wind and resulting crab angle can be thrown in any time—but regardless of their correctness, any error or drift will show very quickly because the pilot will discover his latest "line of position" to one side or the other of the sought-after course. Once the approximate wind direction and velocity have been discovered, the pilot should try to produce the proper amount of crab angle, so he will not be repeatedly blown back off the desired line of position.

Practice is the key to this gimmick, yet the practice itself can be nothing more elaborate than an hour or two spent mentally flying armchair ADF approaches in the quiet of a living room. The "synthetic-bearing" trick does no more than allow the pilot to play with smaller, handier, facially oriented numbers. But the pilot who needs to spend only a minimal time figuring numbers and also has the ability to locate instantly any line of position on the approach chart will never feel that horrible sinking sensation whenever ATC unexpectedly clears him for an ADF approach. As one pilot was heard to say as he nervously meandered over the radio beacon and then outbound: "It sure is a shame to be so awfully near the airport—and yet still so very, very far."

Pull Up!   *by Archie Trammell*

THE MISSED APPROACH is by far the most difficult maneuver an IFR pilot is ever called upon to perform. It always comes as the unhoped-for climax of a tight approach, and it requires a total and instantaneous reversal of the pilot's objectives just when he is at the most critical altitude and in the most critical configuration possible. One moment he's descending quietly, taking quick looks out the windshield for a runway he knows is there and fully expects to see with each glance; in the next breath, he's surrounded by the roar of a full-throttle climb, an out-of-trim airplane, and the guilty sensations of one who has just tried and failed.

The consequences of not performing the maneuver properly are recorded in accident statistics, and surprisingly often, the guilty pilot is an ATR professional with a right-seat backup. This means that cockpit discipline and mental attitude are more often the weak links than skill and technique, and the most unnecessary missed-approach accidents are those resulting from a breakdown in discipline. After an air-carrier accident a year ago, in which the aircraft flew into some beach cottages off the approach end of the runway, cockpit tapes were played back; passing through decision height, the copilot was heard to call out to the captain that he was "sinking five" with the airport not in sight. In a moment, the copilot said, "This is too low!" to which the captain replied, "I can see the water . . . I got straight down." Copilot: "Ah, yeah, I can see the water . . . Man, we ain't 20 feet off the water." The next sound on the tape was of crumpling metal. Another crew last year hit a hill three miles short of the runway. Another hit wires on the airport boundary. Another hit the approach lights.

These are usually called approach accidents, but they are actually missed-approach accidents. In each instance, the pilot was in a missed-approach situation, but at his decision height, discipline went out the window and he failed to switch his thinking from get-down to get-up, which is the most critical and most often committed mistake in a missed approach.

Maintain strict cockpit discipline and that kind of accident will never happen to you. All you have to do is resolve never to go below minimums unless the runway is in sight and the aircraft is in a position to land on it. This can be a difficult resolve to follow, because often a hole will open up directly below, through which you may see a familiar landmark or perhaps a piece of the airport itself, and you'll have an overpowering urge to duck under and complete the approach visually. In instances where you actually see the airport, instinct will make you want to close the throttle, put on full flaps, and dive at the runway. Don't.

The insidious thing about the duck-under or dive-for-the-runway approach is that you may get away with it for years before you dive under one day and discover that it's only the hole that goes all the way to the ground. Or you may discover that you can see down through a thin fog but can't see through it horizontally. That is what happens to those trained professionals: their years of experience are terminated by one last discovery that what works most of the time doesn't work every time.

Those of us who fly little airplanes have a particular cockpit-discipline problem. Sometimes our passengers, in their urgent wish to be helpful, can talk us into a bad situation. On a tight ADF approach one night, a lady in the back seat suddenly shouted, "Here's the runway back here." I instinctively looked back, and sure enough, there were the runway endlights just behind the left wing. A missed approach begun on a foggy night with you looking back over your left shoulder is no fun. So, before beginning a tight approach with passengers, tell them to keep their mouths shut until they hear the squeak of the tires on the runway.

If you have a pilot whom you can trust in the right seat, you might ask him to watch first for the approach lights and then the runway through the windshield, but he shouldn't utter a peep until they are firmly in sight. By no means should he, or you, pay any attention to what might be seen out the side windows. Until you can see the runway through the windshield, over the cowling while in a normal glide, you can't possibly land on it.

An important corollary to cockpit discipline is choosing your own personal minimums, below which you will not wander. The minimums printed on approach plates are not for every pilot. They presume a certain skill at executing a missed approach, and a certain level of imperturbability, which the new IFR pilot may not have. He needs a little extra cushion. Even airline captains moving into a new type, or to the left seat after years on the right side, have higher minimums for the first 50 to 100 hours.

Many new instrument pilots choose arbitrary numbers—such as 500 and one, or 200 feet and a half mile above published minimums—for their personal minimums. That can lead into a trap. Terrain and proximity to antennas or buildings can make 500 and one, or plus 200, too thin a cushion. A better plan is to use circling minimums for the first few dozen actual approaches. These do allow for obstructions in the vicinity of the airport, so the missed approach must be badly mangled before serious trouble is encountered. (On some runways, however, circling minimums are actually lower than straight in. In that case, add 20 percent to the straight-in minimums).

The new man should use circling minimums even on a full ILS approach— *especially* on a full ILS approach, in fact. The reason is that more altitude will probably be lost in a missed approach when transitioning from a glideslope descent to a climb than from a nonprecision descent. Analyzing the pilot's workload on the two types of approaches will reveal why this is so. (Actually, the new man should be wary of shooting glideslope approaches at all during his first few approaches in actual conditions. The chances of making a serious blunder are much less when descending to an indicated altitude than when chasing a glideslope needle.)

Having the proper mental attitude goes hand-in-glove with cockpit discipline. It's safe to say that every duck-under accident has been the result of a pilot being more willing to accept the consequences of busting minimums that the consequences of a missed approach. This might be simply the pressure of pride. If it is, it's the worst kind of false pride. Many veteran IFR pilots will tell you unashamedly that as time goes on, they make more and more missed approaches.

Sometimes the pressures are even more compelling than pride, though: low fuel, fear that the alternate will be as bad as the airport being approached, a load of ice or lack of preparation for a missed approach. Developing the proper mental attitude, then, is as simple as beginning every IFR flight with the expectation that the destination approach will be a missed one. Have plenty of reserve fuel; have not one but two alternates that you are confident will be good; don't continue letting down into ice until the airplane is carrying so much that it can't get up again; study the missed-approach procedure and having the routing, altitude, and first navigation fix to your alternate written down. If you have an extra navigation radio, set it up for the missed-approach procedure before beginning the approach itself. In short, don't let yourself get squeezed into a corner from which the only escape is a landing. This is *the* coffin corner in IFR flying. Following this rule will take the pressure off and make the actual technique of a missed approach as simple as rice pudding.

As with every other IFR procedure, preparation for the execution of a missed approach must begin on a bright, sunny day with the visibility stretching into tomorrow. Go up to about a thousand feet (higher than that and the airplane will respond differently to go-around power); set up a standard instrument-approach configuration and rate of descent; then go to missed-approach power and begin taking careful note of precisely what happens as you transition to climb. How many swipes at the trim wheel are required? How many "counts" does it take to bring the flaps from approach to climb position? Which sequence yields the most immediate and certain rate of climb in *your* airplane with *you* flying it: flaps and then gear; or gear, then flaps? (Each aircraft model is different, and each airplane may be different in the hands of different pilots.) What is the *precise* pitch attitude shown on your artificial horizon at the beginning of a missed approach? As the flaps and gear come in? Lightly loaded? With all seats full? How much roll error does your horizon have in going from the approach to climb configuration? Which direction is it? (All artificial horizons have acceleration errors, which vary with the rate and magnitude of the acceleration. It shows up as roll error in straight-ahead flight; both roll and pitch, in turns.)

Go through this exercise at least a dozen times, both visually and with a hood and check pilot in place, until you've got it all down pat. Now you're ready for some practice actuals.

That may shake you up; practice actual missed approaches? Hardly anyone flies into an airport where the chances are less than 50/50 of getting in, but until you've actually missed a few approaches, you're really not a complete IFR aviator. Because we tend to shy away from marginal IFR, many of us fly instruments for years before we have an actual. As a consequence, we make even the easy approaches with sweaty palms because the possibility of a missed one hangs menacingly in the regions of the unknown. This will pass after a few actual missed approaches, but it's better to experience them at your convenience, under carefully selected weather conditions, than at the whim of Dame Fortune.

The perfect weather conditions for missed-approach practice are low ceilings (about 100 feet below your personal minimums but no lower than 300 feet agl) over flat terrain with good visibility underneath (two miles or more), low tops, and excellent VFR within 50 or 100 miles.

Good visibility underneath will give you confidence that you aren't going to fly into the ground without ever seeing it, and will make the approaches legal. FAR 91.116 (b) allows us to shoot all the approaches we want when the ceilings are below minimums, provided the visibility is above published minimums. On those kinds of days, the traffic is light, so you probably won't be inconveniencing other pilots, particularly at the smaller airports. Also, although it's embarrassing to shuttle back and forth between good weather and bad, don't hesitate to shoot a couple of approaches, then go home for fuel and a cup of coffee. My copilot and I made three trips in and out of below-IFR weather one day. The center controller for the sector guessed what we were up to on the second trip, and issued us a clearance out while we were still going in. He was saying, in effect, "Have fun, and let us know when you're ready to give up." You can take an instructor with you if you like, but it's best if you do it yourself. A trusted copilot won't hurt, but since the object is to gain confidence, you should be the undisputed pilot in command.

You'll find that actual missed approaches are relatively simple after your VFR practice. If you follow the cardinal IFR rule—aviate, navigate, communicate, in that order—you won't go far astray. When commencing the go-around, first go up on the power and simultaneously plant your eyes on the artificial horizon. Level the wings, set the proper pitch angle, then with a glance at the turn and bank, center the ball.

Failure to follow this sequence is the most common, and most hazardous, mistake of low-time IFR pilots. They are too concerned about airspeed, altitude, and heading at this critical moment, and their eyes dart from the altimeter to the DG to the rate of climb to the airspeed as they pump the wheel trying to make everything move in the desired direction. If nothing else, the VFR practice should convince you that if you go to the artificial horizon first, level the wings, and establish the proper pitch angle, everything else will fall into place without the necessity of ever looking at the other instruments. You can then devote your attention to getting the airplane cleaned up, the trim set, and the cowl flaps open.

Getting the wings level is first on the list of things to do with your eyes. The reason is that the most critical missed approaches usually begin with the pilot looking for (or at) the runway. Additionally, the airplane has often been placed in a bank in a futile attempt to get lined up on a fuzzily seen runway. Therefore, you want to give yourself something simple and useful to hang onto for an instant while you rearrange your thoughts for the climb. Leveling the wings will do you the most good at this instant. This stops any turn that may have developed as you were looking outside the cockpit or making one last stab at the localizer, and puts the airplane in the best attitude for climbing. As the wings come level, you must, without delay, set a pitch attitude that experience has taught you will result in immediate climb, then glance at the turn-and-bank indicator. The ball should be centered, because airplanes climb best when the controls are coordinated, but also check the turn needle to be certain it agrees with the artificial horizon. This is no time for a delay in discovering that your vacuum system has failed. If it is a partial panel go-around (which you will have previously practiced, of course), knowledge of your airplane and precisely how to trim it for a smooth and certain transition to climb will make all the difference. Chasing a lagging airspeed and rate of climb on an IFR go-around can be exciting.

After the airplane is leveled and a positive climb is established, begin the navigating phase. Turn to the first heading called for in the missed-approach procedure, and if you didn't set your VOR or ADF for a missed approach earlier, do it now. The primary thing to remember, though, is to aviate; in other words, climb. Navigation is secondary to that.

Communication is last on the list of things to do. Don't wait until you've gone to the fix and established a holding pattern before you tell ATC you've missed the approach, but do give yourself a minute to get the airplane firmly under control and the new navigation situation firmly in mind. Then decide what you're going to do next, and finally, pick up the mike.

If you saw a piece of the airport but weren't in a position to land on it, or if the missed approach was due to straying from the final approach course, you may elect to shoot another approach. You needn't be embarrassed about this. Grizzled airline captains sometimes shoot several before getting in or giving up. If another approach would be a waste of time and fuel, go to your alternate. In either case, it's more professional to make a decision before communicating with ATC, so you can tell them on the first call what you're going to do next. Otherwise, they'll come right back after you've called the missed approach and ask, "What are your intentions?" This tends to put pressure on you to make a hasty decision, and hasty decisions have no place in IFR flying.

If you break the missed approach into its three natural parts—planning, technique, discipline—and tackle them one at a time before the approach commences, the missed approach will never be a problem.

The Solid-gold Alternate   *by Thomas H. Block*

ALTERNATE AIRPORTS are the single most neglected area of instrument flying. They are a mere technicality on the face of the flight plan to which most

IFR pilots allot a slight, grudging slice of attention. Even the words "alternate airport" have a repugnant ring to them—a reminder that the best-laid schemes of men sometimes go agley. Yet if the phrase "alternate airport" does hide darkly in the back of the pilot's mind, it takes only an occasional whim of weather or a machine's irregularity to force it to the forefront. A solid alternate, like an understanding wife and a working flashlight, is something that seems unnecessary until suddenly, it becomes indispensable.

The selection of an alternate airport can, without exaggeration, be the important difference between life and death. Yet despite the gamble that a hastily chosen alternate represents, the odds are overwhelmingly stacked in favor of the most lackadaisical pilot. Advances in weather forecasting and the ability of well-equipped pilots to execute the lowest instrument approaches are among the reasons that the art of alternate-airport selection is becoming an arcane talent. After all, how often does a pilot on an IFR flight plan actually need an airport other than his destination? Very seldom indeed.

Unfortunately, this has resulted in many pilots having an unwarranted sense of security over their selections. They continue to choose alternates in a haphazard manner—and are reinforced by the observation that every flight seems to terminate at destination. The number of instances in which a pilot merrily tools toward his destination while unbeknownst to him his alternate lies flat on its back would make a startling statistic. The pilots themselves are seldom aware that they are included in this group, since they consider alternate selection purely a preflight activity anyway, and never bother to investigate the results of their methods.

Methods of picking which airport will attain the official yet dubious status of "alternate" range anywhere from a detailed analysis of weather to simply running a hand down teletype sequences until the tip of the forefinger reaches a station with legal weather. In almost every instance, the pilot who winds up with a useless alternate has committed the all-too-common error of failing to regard the weather as a combination of dynamic factors in a changing environment. To put it more simply, the assumption that the alternate weather will remain what the sequence report says it is (or that the forecast and the reality will always coincide) is an assumption that may leave you with no place left to land on some cloudy afternoon.

The proper process of alternate selection starts well before one considers the weather reports. The first area of evaluation surrounds the pilot/airplane combination. An inexperienced pilot in a sparsely radioed aircraft should obviously have higher alternate requirements than the grizzled airline veteran with Category II authorization. This obvious factor, however, seems to escape those who regard the FAA alternate minimums (usually listed on the instrument approach chart as 800-foot ceiling and two miles of visibility, although it does vary at different airports) as the only requirement. The object of an alternate is to provide the pilot in command with an easy out should landing at destination become impossible. There is nothing "easy" about executing an eight-and-two shot under pressure of quickly sinking fuel gauges when you have never actually done one before; nor is a moderately low approach to an alternate airport "easy" when the only radio on board is acting sick.

If the pilot/aircraft combination does seem competent to handle the minimum required alternate criteria, the next analysis regards the weather itself. When a pilot inspects his alternate with an eye to only the sequence reports and terminal forecasts, he's in trouble. Certainly there is no advantage in trying to outforecast the weatherman, but a pilot who possesses a general knowledge of the day's weather will be measurably ahead of one who just scans a list of station symbols. A general outlook is readily available from the weather map and a few minutes spent with the appropriate area forecasts—an expenditure of time that will give the pilot a valid basis for his preflight and en-route decisions.

If the most convenient alternates are legal but within an area of worsening weather (fronts moving in, increasing rainshowers, narrowing temperature-dewpoint gap or whatever), the prudent thing to do is list two alternates, with one away from the area containing the downward weather trend. FAA regulations require that flights contain enough fuel to reach "the farthest alternate listed," not both. The advantage of having two (or more, if you believe in wearing a belt with your suspenders) is that if destination weather collapses, the pilot has additional planned diversion points within his precomputed range (plus the necessary 45-minute reserve of fuel, of course). A simple weather check will then tell him if legal, convenient, but marginal alternate A is better or worse than forecast, and the decision to divert there or to alternate B, far from the worsening trend, becomes a painless selection based on previously thought-out criteria.

Of all the errors involving alternates, the one that produces the most devastating results is the technique of completely forgetting about the alternate between the time of flight-plan filing and time the throttle goes forward for the missed approach. Unless a pilot watches both destination and alternate weather while en route, he is apt to be in for quite a surprise during the latter portion of the flight. Obviously, the science of weather forecasting is somewhat less than perfect, and burying your head in en-route work will do little to improve the accuracy of the terminal forecast upon which you are now so dependent.

The test of the good instrument pilot lies in his ability to think rationally, and a pilot who turns to a heading with some idea of the weather trends ahead will be secure in his basis for en-route evaluations. After checking weather during the trip, the pilot who can say: "It sounds like the front is moving faster than predicted, making my westbound alternate a little shaky. If I miss the approach at destination, I'll go to my eastward alternate instead'' is a pilot who sounds like he knows what he is doing.

Although the first few attempts at alternate analysis may provide unnecessarily prudent results, you'll be better off for your efforts—efforts that will be substituting for the en-route blindness so prevalent in so many pilots. Practicing the ability to think is energy that is never wasted.

12 NO WAY DOWN
by Peter Garrison

MOST AIRBORNE EMERGENCIES share a common ancestor: the pilot. There seems, in hindsight, to have been a point in almost every accident or emergency history at which the pilot had it in his power to avoid the eventual crisis and failed to do so—sometimes by a simple omission and sometimes by a conscious decision to take a risk. This generalization does not particularly apply to situations in which a mechanical or radio failure jeopardizes a flight—though even here, some negligence by the pilot may have played a role. I am talking, mainly, about the classic weather emergency in which a pilot suddenly finds himself in over his head and is either saved by ATC or by his own actions, or simply isn't saved at all.

It is partly because most pilots have had a few brushes with disaster from which they escaped safely that they are willing to take the little risks that one day may pile up to trap them. It is not uncommon for a novice instrument pilot to, say, become disoriented while holding at an approach fix without radar assistance; or to descend, inadvertently or not, below MDA; or to run uncomfortably short of fuel on a night IFR flight when every semi-imaginary roughness in the engine seemed to be a signal to enrich the mixture. Most of the time, one gets away with these indiscretions, and a feeling develops that one is "lucky," or that cautionary advice is generally exaggerated or meant for someone else, or that one is a sufficiently "good" pilot to be able to live on the ragged edge with aplomb. None of these comforting conclusions about oneself has much meaning; many a "good" pilot didn't find out the difference between good and bad until he made one mistake too many and paid for it. "Good," after all, doesn't just mean that you can always be counted on to get there, and the only way to be sure of that is to keep getting there. There is no room for the feeling that if you foul up, the airplane will take over where you left off.

Embryonic emergencies that mercifully go away are not only responsible for the excessive confidence and self-indulgence of many pilots; they can also be credited with creating the fascination that imaginary emergencies have for most of us. The mere words "total electrical" or "both engines out" or "losing altitude with a load of ice" have a mesmeric quality. Who has not fantasized himself into some crisis, and then fantasized his way back out of it? Pilots are the biggest Walter Mittys of all.

Sad to say, those awful hangar emergencies sometimes unfold in the sky, and to the unlucky fellow who has to ride them out, the Walter Mitty aspect of the case is altogether lost. There are a lot of amazing tales: one that comes to mind right away concerns a Baron running low on fuel in zero-zero weather somewhere in the Midwest, and after several attempted landings in fog, finally running out of gas and bellying safely in on a flat field without the occupants ever seeing the ground until they climbed out of the airplane. Another concerns a fellow who had the tip fuel cell on his Cherokee Six explode during a snowstorm. Such stories are always repeated with gusto, like the endless parachutists' tales of people who jumped without chutes and somehow survived. It would take a planeload of Zen masters to actually experience such an event as anything but the most awful moment of their lives. Panic is so far from a living-room armchair that it is practically impossible for the idle imaginer of an airborne emergency to feel the sensation of looking death in the face.

One classic IFR bugaboo is some sort of partial equipment failure—either vacuum, which is not unusual (some vacuum pumps have surprisingly short lives), pitot-static, or communications. A pitot failure sounds pretty serious, but really it isn't too bad; if you know the approximate power settings that you use for cruise, climb, and descent, you can control airspeed with the power setting and the rate-of-climb indicator and/or altimeter. If the static ports ice up or clog, the altimeter, rate of climb, and airspeed begin to work sluggishly or not at all. Some airplanes have alternate static provisions, controlled by a valve in the cockpit. Sometimes, the valve simply opens the static manifold to the cockpit; in other cases, the alternate static inlet is inside the wing.

Where there is no alternate static inlet, the glass face of an instrument can be broken to open the static system to cabin air. The instrument to break is the VSI, since it is the least vital of the three, and breaking the glass may also damage the instrument. In connection with pitot ice, it should be mentioned that while we don't usually have the problem—pitot heat is an inexpensive, commonplace accessory—an airplane with an automatic gear-extension system and an unheated airspeed sensor to operate it can be a monster: if the pitot freezes, the gear goes down at cruising speed, and there is no way to get it back up unless the automatic system can be bypassed or disabled.

Gyro failure due to loss of suction is a problem, but not a huge one; the panel should be set up for full reliance on *either* suction or electrically operated instruments. Imagine, though, losing an altimeter. It does happen. If you haven't been so extravagantly provident as to buy a backup altimeter (the small, ruinously expensive kind or simply a surplus standard instrument), you can firewall the throttle and note the indicated manifold pressure. By comparing the full-throttle manifold pressure with values given on the engine power charts, which are usually in the operator's handbook on the airplane, you can guess your altitude within a thousand feet or so—not much accuracy, but better than none, and enough to get you down under a fairly high overcast in the worst pinch, or to stay clear of mountains. An alternative is to subtract the MP indicated at full throttle from 30, and read the result as altitude in thousands of feet, give or take 500.

Then there is the matter of radios. The FAA publishes a set of emergency procedures for various types of radio failures, including special transponder squawks; it is all in the *Airman's Information Manual,* and it is a good idea to carry the *AIM,* or at least the relevant pages, in the airplane.

Loss of communication is one thing; loss of nav radio is quite another. A flight, including an approach to minimums, could be completed with no nav but with at least one working com radio; but what do you do if all the electrical equipment goes out at once? There is no choice but to proceed by dead reckoning to a place—if you aren't in one already—where the ground is as far as possible below the overcast, and begin to descend gradually until you break out.

Descending gradually into unknown country is, needless to say, extremely unsafe. In mountainous or even hilly terrain, it can be practically hopeless; in flat desert, farm country, or even woodland, there is a fair chance of seeing the ground before running into it; or, in fog, of more or less landing rather than simply crashing into an obstacle. Such a descent should be made at the minimum controllable speed, with gear up (if possible) and flaps down—for the last thousand feet, at least—with sufficient power to keep the rate of descent to a minimum. Finally, shut off the master switch to reduce the chance of fire due to sparking from whatever portions of the electrical system may still be working.

The first choice, if you have any choices, is to turn toward a place where there is better weather; and for this reason, one part of the preflight weather briefing should be a determination of where within the range of the airplane fair weather exists—if anywhere. The usable area roughly consists of a circle drawn about the departure point with a radius equal to the plane's range at the power setting normally used. As you approach the destination, the available area gets smaller and smaller, but by throttling back to the best-range speed, you may be able to keep the outside perimeter of your operating area close to the perimeter of the original circle within 90 degrees or so of your course. If there is no good weather anywhere within range of your airplane, it is still advisable, during the preflight, to be aware of the lay of the land within your radius of action; it takes no extra time and very little effort to form a mental overview of the terrain and weather through which you will be flying.

If an uncontrolled emergency descent must be made, it should be made with some fuel remaining. To wait until the fuel runs out and your hand is forced would be foolhardy; what if you came out of the bottom of the overcast and there was an airport in sight, but out of gliding range?

The ultimate equipment failure is failure of the engine itself in a single. An engine-out descent through overcast should be made at minimum forward speed (unless it is certain that you will break out before reaching the ground) with one notch of flap—10 to 15 degrees—gear up, fuel and switches off, and all safety harnesses cinched up tight. If there are pillows, blankets, extra clothing, small suitcases, briefcases, or the like in reach, they should be placed in the pilot's and front-seat passenger's laps or against the instrument panel, as the case warrants, leaving the airspeed indicator and altimeter in view. A small suitcase between your chest and the control yoke will prevent chest injury in a survivable crash in the absence of shoulder harness—which, we must sadly recognize, is still absent in the great majority of airplanes. The seats should be moved all the way back, even depriving the pilot of rudder control if necessary, and all harnesses tightened with the seats in the rearmost position. The propeller should be rotated with the starter to a horizontal position to minimize damage should it be possible to make a controlled landing; and keep in mind that if you break out over a landable area, you have to turn on the master switch again before electric landing gear will work.

Another grim possibility to mull over during the empty moments of your instrument flights is that of your destination and all alternates going below minimums or zero-zero. This is not a likely development unless you flight-plan beyond the edge of safety, but you could conceivably have a fuel-system failure, say, that narrowed your range and limited your choice of alternates. Assuming that everything in reach is below minimums, your first choice of an approach facility ought to be a military base, notwithstanding the fact that you might be greeted on the runway with a jeepload of machine guns or, even worse, a handful of papers to fill out. Military bases usually have extremely long, wide runways, good GCA or PAR facilities, and practice in using them. Usually, their aircraft are parked well clear of the runway, whereas general-aviation parking areas are often close enough to the runway for an airplane slightly off course in fog to take out a row or two of them.

An ILS approach to 100 and one-quarter or less is not much different from any other ILS if you have glideslope. The glideslope usually intersects the runway more than a thousand feet past the threshold, and so long as you keep the glideslope bar away from full-up and full-down indications, you're sure to land on concrete. The main chore is to keep the localizer needle centered and stable. The last part of the approach should be made at minimum speed, with gear down and flaps *up*. The reason for keeping the flaps up is that the nose of the plane stays higher that way, and if you are going to fly into the runway without flaring, it's best to hit on the main wheels first. Flying at 70 knots, which most single-engine airplanes will do without difficulty even on instruments, vertical speed on the glideslope is less than 400 fpm, a rate of descent at which any airplane can strike the runway without damage. Thus, even in a dense ground fog where there is no chance of seeing the runway before touching down, you can make a safe landing by flying a good ILS approach.

Such an approach without glideslope is more difficult, but possible with the help of GCA, or even without any help at all. A little preliminary math will give the rate of descent on the glideslope from the outer marker inbound (if it is not already on the plate), and the time from the outer marker to the threshold. Experimentation during the transition from en-route altitude to the approach fix will give the power setting for the calculated rate of descent. My own inclination would then be to descend a little more rapidly than necessary from the outer marker to the middle marker, reach the middle marker in level flight at the precise speed and altitude planned in advance, and reduce the power to the preestablished value on entering the MM cone. I would then hold my breath.

The published distance between the MM and the end of the runway is just that, to the nearest 250 feet; it is *not* the distance from the MM to the touchdown point, which should be another 1,000 feet or so down the runway.

Allowance must be made for this difference. If you are using the glideslope, allow for the fact that the glideslope indication is commonly unreliable below about 250 feet; the last 40 seconds or so of the approach must be flown in a steady descent state, without reference to the glideslope bar.

Normally, zero-zero conditions occur in still weather, and a low approach speed is safe; when visibility is limited by blowing snow, for instance, or in any other condition where there might be gusty winds during the approach, normal safety factors have to be added to the approach speed. Landing gears will withstand touchdowns at more than 1,000 fpm of descent without flare; the problem of high approach speeds, then, is not so much the impact velocity (since a higher approach speed calls for a higher rate of descent) as the relatively nose-low attitude such a speed requires.

Each type of emergency is somewhat different from all others. Some arise from equipment failures that are more or less beyond the pilot's control, and that he can cope with, at least some of the time, if he has a basic familiarity with how the instrumentation of the airplane works and what it is meant to tell him. Others arise from chance events—freak weather conditions, for instance, or completely inaccurate forecasts. These it is more in the pilot's power to forestall, if he keeps himself abreast of weather conditions at his destination and alternates during the entire flight. There is no excuse for cruising along for four hours only to arrive and find your destination below minimums, when periodic weather checks would have revealed the change in time to permit an early landing, a deviation, or a return to the starting point.

Just as it is important to check weather periodically, it is important to check fuel consumption. Particularly in rented airplanes, book values are useless. Because of the taxi, takeoff and climb segments of the flight, excess fuel consumption due to the use of carb heat, incomplete filling of the tanks, and so on, it is rare to find an airplane with an advertised five-hour range that actually has half its fuel left after two and a half hours. In fact, it is a great mistake to think of range in terms of distance; range is, really, a duration, and time—not space—sets the limits to a flight. Fuel gauges are generally lousy, and there's no reliable way, short of long experimentation, to find out exactly what their indications mean. By making periodic checks of remaining fuel and fuel consumption, you can keep abreast of time remaining and probably avoid last-minute fuel emergencies. At the same time, you can use spare moments en route to plan time and fuel to several alternates.

Considering how many accidents arise from absurd causes such as fuel mismanagement, it is perhaps optimstic to think that most pilots caught in an IFR emergency would do better than simply to blunder their way out of it, or else into the ground. The pressures of an emergency can be so great that thought falters and knowledge evaporates; the only solution to the human problem is to do your thinking in advance. Armchair flying may not be quite like the real thing, but you are much more likely to think of using your manifold-pressure gauge for an altimeter when pondering the riddle at leisure than when making a blind descent with a stuck altimeter, and the preferability of a broken VSI to a broken ASI or altimeter comes more easily to the mind of a relaxing pilot than a panicking one. By thinking over the ramifications of an IFR problem in advance, discussing them with other pilots, and keeping them in mind in the course of every flight, you stand the best chance of avoiding emergencies altogether, and of coping successfully with them if they come.

13 STAYING IN PRACTICE

by Peter Garrison

WITH ITS characteristic bureaucratic lust for shoveling smoke, the FAA sets certain minimum requirements for currency for instrument pilots. The present requirements are for at least six hours of actual or simulated instrument time within the last six months; new rules about to come into effect add the requirement of six instrument approaches. It would be hard to argue that there should be no currency requirements at all, but it is equally hard to argue that these have any particular validity. For instance, a pilot who has just completed a three-hour instrument flight—illegally—after not flying IFR for eight months is probably more "current" than one who flew eight hours five months ago and has not punctured a cloud since. I doubt that there is any scientific basis for the FAA's requirements; they probably just sounded reasonable and acceptable to whoever drafted the rule, and tolerable to the respondents to the NPRMs.

Real currency is something else again. Amazingly enough, you can get rusty in a week out of the saddle. Like most new rust, it wears away quickly and leaves you fit and shiny as ever; but what is remarkable is how completely strange you can sometimes feel while trying to reintegrate your instrument-flying techniques after you've been idle for a while. It makes you wonder if there isn't something you can be doing to keep in practice during good weather. Doubtless this is more of a problem in the Southwest than elsewhere; there, months may go by without a cloud in the sky, and since the only instrument time that counts toward recency requirements is *actual* (in cloud) or under-the-hood time, staying current can get rather inconvenient. A lot of people don't have the time or the will to hire an instructor to ride with them, but the alternative—getting a friend to ride along and watch for traffic—is not very satisfying to anyone seriously wanting to improve his instrument technique, since he has to be his own critic.

If the recency regulations are to be borne like a cross, the cross is at least an ethereal one, weighing more upon the conscience than upon the back. The only person concerned about your recency might be an accident investigator, and then only because lack of recent experience provides an easy way for him to fill in some blanks if you ball it up. Rarely does anyone ask to see your logbook or, for that matter, your license. Anyway, falsifying logs is such a simple matter that there is hardly a pilot who is ignorant of what "Parker Pen time" is. To those who are worried about recency for some technical reason such as insurance, the conventional solution is to fight paper with paper and jot down a few fictitious but current flights.

If you worry about recency because you seriously believe that your safety or that of your passengers is at stake, the solution is not so easy. How do you stay current if you can't find enough bad weather to stay current in? I'm talking about *real* currency now, not legal currency; different pilots with various levels of ability will argue on just how much IFR flying is required for them to stay in practice; but they all know that when they start feeling the rust, it's been too long.

There are two different functions of IFR flying that require practice: the bookkeeping and the piloting. By bookkeeping I mean flight planning, filing, clearance copying, navigating, and approach comprehension; by piloting I mean holding and changing altitude, heading, and airspeed without effort and without error, smoothly, for hours at a time.

The bookkeeping side is the easier to practice. You can file an IFR flight plan any time you want to; sometimes it may mean a slight additional delay or a few extra miles of routing, but 15 or 20 minutes are not a high price to pay for keeping fit. Many pilots file IFR at night routinely for several reasons: the chance of quick rescue is better if they make a forced landing; the delays are slight or nonexistent; the periodic conversations with controllers and the rhythm of navigational tasks fight fatigue; and you don't have to remember to close an IFR flight plan. The point is to practice clearance copying, radio communication, chart handling, cockpit organization—nothing very difficult, really, but a lot of little things that can get out of hand if you forget how to deal with them.

Another important part of the bookkeeping is chart reading. Again, it is nothing difficult, but if you forget, for instance, that to turn a Jepp chart over you switch it end for end rather than top for bottom, you double your chart-turning-over time and effort—time taken away from instrument scan, time in which to drop your pencil on the floor, time wasted on irritation and confusion. This may seem an odd thing to forget, but since sectional charts work the opposite way, confusion is easy.

It is also easy to become confused about parts of chart symbolism, to forget where certain information is found, to get out of the habit of using the Jepp "zigdex"—a city index across the bias-cut top of the en-route charts. You can also manage to forget how your approach plates are organized and where certain minima or departure information is to be found.

To promote facility with en-route charts, it is helpful to use them in VFR flying. I would not use them to the exclusion of sectionals, though this may be stodgy of me; I know that many people do, and it has not been demonstrated that the practice is unsafe. It is *theoretically* a little risky, however. The rate of information change on the sectionals is much lower than on the IFR charts, but there are a lot of little airports that would be very hard to find without the tips on highways, railroads, drive-in theaters, and storage tanks supplied by the sectionals. The Jeppesen low-altitude planning chart for the whole country should, at any rate, be part of every flight kit, and if you use Jepp charts and keep the low-altitude chart with the others, you might as well bring along the whole binder. (If you don't have the whole Jepp set or don't want to carry the whole binder, an outdated planning chart scrounged here or there is as good as a new one.)

It may go without saying, but you should also use the full complement of radio equipment that you have aboard—even on VFR flights. Most pilots do this anyway. It might be argued that constant use of all the radios wears them out faster, but standing idle is probably no better for them, especially in polluted-air areas where contact corrosion can be a problem. By using them all the time, you are always fully aware of their condition and performance, and you stay in the habit of digesting doses of information at a certain rate. Using DME and VOR in connection with en-route charts, for instance, you can keep a continuous check on the accuracy of the equipment by tuning in neighboring stations when you are passing over an omni and checking their distances and bearings against the charted figures. This is an activity for which you don't always have time in actual IFR conditions, so do it when you're VFR.

If you want to take the time, you can request practice instrument approaches at your destination airports. This will keep you somewhat attuned to the nuances of approach plates and to the habit patterns of the approach. Practice approaches have another value, too, which is rarely mentioned. One of the difficult aspects of an instrument approach, especially for a relatively inexperienced pilot, is spatial orientation and time sense. Approaches usually involve a lot of turning and sometimes some timing, and they require that certain chores be completed within time limits that are set by the distances between fixes. Now, if you are nervous while flying an approach, you tend to distort your time sense, usually by compression, and to become anxious by anticipating fixes that are still a minute or two off. A minute may seem like nothing, but a pressure situation can make it quite long. Concern about when a fix will arrive or whether you will have reached an assigned altitude in time blocks other awareness.

There is an analogy to cooking, curiously enough. When you first learn to cook, you feel that you have to watch everything very carefully because it is going to burn, boil, curdle, or sink if you don't. After practice, however, you become rather calm and indifferent about things because you have developed a sense about how long they should take and how quickly you can deal with them when you decide to do so. A practiced cook, therefore, appears to be ignoring everything; he is, in fact, immersed in the rhythms of his work and quite abreast of everything. Ideally, an instrument approach should be flown with the same subliminal sense of overall timing and the willingness to ignore certain tasks while attending to others.

What do practice approaches in VFR conditions have to do with all this? The reason time anxiety does not appear when the weather is good is that you have space confidence; time and space, like altitude and airspeed, are readily translated into each other. As long as you can see the airport, you can easily assess how long it will take to reach it, how high you are above it, and so on. There is value, therefore, in developing a habitual sense of the average proportions of instrument approaches. For instance, all outer markers are five or six miles from the end of the runway, all glideslopes are at a three-degree angle or so, all rates of descent for a 100-knot approach are 500 or 600 feet per minute, and so on. All ILS approaches have about the same shape. There is more variety in nonprecision approaches, but they still have an obvious spatial analogy to their depiction on charts; you can fly a procedure turn to taste, using the same timing everywhere. If you can develop a sense of spatial orientation that will spare you the nagging feeling that you are about to run outside the limits of the approach and into a mountain, time anxiety will go away. Flying approaches in VFR conditions helps you develop that spatial sense.

Once you are in the habit of flying approaches, you can imagine your way through an approach while sitting at home. A convenient time to do it is when updating your approach plates. Pick a few plates at random; review the key fixes and the crossing altitudes for those fixes and try to imagine how long it would take you to get from one to the other, what rates of descent you would select, where radio-frequency changes would be involved, what frequencies would be used, where they are noted on the plate, and so on. If you have a clock handy, note the time lapses between fixes. You will be surprised at the length of time allowed for each part of the approach, and puzzled that you ever felt anxious or rushed at being required to do so few things in so much time.

While updating plates, it is a good idea to take note of the reasons for reissue on plates that you are likely to use. Sometimes changes can be major—an entire missed-approach procedure redesigned, for instance.

Practicing IFR *flying* techniques—as opposed to bookkeeping techniques—without IFR weather is more difficult. The bookkeeping side is intellectual, and imagination can to some extent take the place of real environment in developing intellectual habits. The actual flying is less intellectual than physical; it is done with no thought—contrary to what many flight instructors seem to believe—and any attempt to think merely hinders it. I suppose one might practice one's scan, though this is rather like practicing one's breathing: after a minute you forget that you're doing it. Various theories of how to scan instruments are presented by various authorities; my own system might be called a random scan. I just try to keep track of what's going on, and I have no idea of how I do it. I have tried to use geometrically logical scanning patterns, and I find that they are useless; the idea that one first looks at one instrument and then at another presupposes a pilot with tunnel vision. In fact, you see three or four instruments at a time, and if the altimeter is doing something funny while you're looking at the gyro horizon, you'll probably notice it even though it isn't that instrument's turn to be noticed yet. In fact, the only information requiring scanning is digital information, in the sense that there are several discrete pieces of digital information that must frequently be reviewed, and none should be omitted while another is fixated upon. This is something that can perhaps be practiced, under the heading of good pilot technique in general: keeping your heading, altitude, and airspeed under control without fixating on those items. The new DGs with movable heading bugs practically eliminate the need for conscious interpretation of the DG, and airspeed can usually be left up to the trim; so if you have a DG bug, you can confine your practice to a regular check of the altimeter—the one instrument for which simplification does not yet seem in sight. (A vertical-tape altimeter with a bug that could be placed on a selected point on the tape would do the trick, but I've never seen one. Jeppesen came up with a moving bug to use with a conventional pointer altimeter, but it met with massive indifference.)

Anyway, a good awareness of the instruments is something to be practiced in any weather. Another technique that can be constantly worked on is that of using predetermined power and trim settings to obtain certain rates of descent. If you feel like going to the trouble, you can note the relation between manifold pressure and rate of descent for a couple of typical weights on a simple graph, and by studying it, perhaps develop an ability to guess quite closely the required MP for a certain approach. Speed must be held constant, of course, though small adjustments in sink rate can be made by adjusting speed as well as by adjusting power.

It is commonplace, on nonprecision approaches, to begin the descent on passing the final approach fix. Since there is some lag in establishing a steady rate of descent after reducing power, you often end up above the approach path. One way to avoid this is to reduce power momentarily well below the approach setting just as you arrive over the FAF; then, as the VSI needle passes through the value selected for the approach, bring the power back up to what you want for the descent. The effect is to bring you out more or less on the approach slope. To get the rhythm of the movement, try it on the glideslope; on passage of the glideslope bar through center, cut the power as you would on passing over the FAF, and bring it back on cue from the VSI; see if you end up on the glideslope. If not, adjust the VSI reading at which you restore power.

By doing all these things in VFR conditions, you establish a mental picture of what is happening when you do them in IFR conditions—what I have referred to as a spatial orientation. The more habitual your landing routines become, the less attention you have to give them during an actual approach to minimums, and the easier it is to execute the approach. On an ILS approach, for instance, the primary approach cue is the cross-pointer display; but if you have trouble with trim, airspeed, and power, your attention is continually being taken away from this primary cue. If, therefore, you can get into the habit of flying at all times as though you were making an instrument approach, using the speeds, power settings, and configurations that are typical of instrument approaches, then the difference of being or not being in cloud becomes unimportant; the only additional burden of the instrument approach is that of having to navigate.

Much of the apparent difficulty of instrument flying arises from the emphasis on how different it is from visual flying; if you emphasize—and by your techniques try to maximize—what is common to both, you find that even if you do not get to practice the real thing as often as you would like, you can keep the rust in your joints down to a bare minimum.

14 HOW GOOD AN INSTRUMENT PILOT ARE YOU?
by Richard L. Collins

WHO KNOWS how much IFR talent lurks in the mind of a pilot? The pilot knows. Any pilot can make an accurate and forthright assessment of his talents, weak points, and strengths simply by evaluating each flight—wih no kidding allowed. The airman who swoops and dips on his approach, sneaks below MDA for a peek, finally sees the runway at the eleventh hour, and pounces on it like a dive-bomber pilot has obvious problems, but the secret lies in recognizing them. The *truly* dangerous pilot is the one who dismisses a sloppy approach, or any other sign of poor technique, with the thought that it couldn't have been too bad because at least he made it. (Wait until *next* time.)

How can a pilot grade his own flights? It does take a conscious effort to be successful at self-appraisal, but by dividing an instrument flight into sections and putting a measure reflecting importance on each part of the flight, we can expose areas in need of attention. We've done it on a 100-point basis, with the idea being to score your own flights just as you score your own golf game or measure your fish (without cheating, of course). The areas to study and the points per flight scored in each are:

Preflight and weather briefing: 10
Takeoff and departure: 10
En route: 15
Descent/transition into terminal area: 10
Approach and landing: 25
Detail work: 10
Communication procedures: 5
Overall planning: 15

You get a perfect grade to start with, and points are taken away for transgressions; remember, we are evaluating actual instrument flights, in the real world, as opposed to check rides—which are often hasty, unrealistic trained-seal acts.

Before going into each area and the basis for grading it, I'll point out that the approach and landing get the highest score because this is where most pilots break their airplanes—which is what we are trying to avoid. Overall planning gets a lot of points, too, for a pilot who starts early in the game to plot against the forces that work against him, and methodically plots against them until the airplane is parked, is the one who flies smoothly and safely.

Weather briefing and preflight planning are the place to start. The goal of a weather briefing is to get the big picture, the effect the big picture will have on weather conditions along the proposed route of flight, and a verification that the information obtained appears to be accurate. Start by subtracting a point from your score if an honest effort is not made to learn something about weather *before* calling the FSS. Digesting the transcribed weather broadcast (TWEB) or pilot's automatic telephone weather service (Patwas) before calling the FSS is an excellent habit. If neither of those is available, the TV weather map—especially the one on the *Today* show—can help start your day's store of weather wisdom. The newspaper map doesn't hurt, but the information it contains is pretty old.

Was your overview of the weather—including the blank spaces to be filled in by the FSS— well organized before you picked up the telephone? Subtract a point if a list of desired information was not made before calling, and subtract another if the briefing information was not neatly recorded and taken along on the flight.

It is difficult to grade the briefing itself. What you need depends on the type of day it is. But if you remembered some necessary bit of wisdom you needed after hanging up—something like the freezing level, the winds and temperatures aloft, the radar summary, the forecast for your alternate, or the temperatures and dew points—subtract a point for each item forgotten.

Score your preflight on the basis of IFR-related items. (After all, checking the airplane's soundness and the fuel and oil levels is something you assumedly do before *every* flight.) If a thorough checklist tailored for IFR operation is not used, subtract a couple of points. IFR flying is *procedural* flying, with the settings and readings of everything in the airplane playing an almost absolute role in getting the airplane where it is going. A checklist that demands that the altimeter, DG, and clock be set, and that all radio frequencies and bearing selectors are set in advance for the first phase of flight, is a good hedge against confusion.

The checklist should even concern itself with charts, for arranging the paperwork properly is an important confusion fighter. The approach-plate book should be opened to the page showing the approach in use at the departure point—in case something makes you want to return and land soon after takeoff. The standard instrument departure, area, and/or en-route chart applicable to the first part of the route should also be at hand.

In essence, the goal of an IFR preflight is to make every possible setting in advance so that when you are ready to taxi, you are ready to *fly*. Many things, such as the DG and altimeter, are checked again before takeoff, but if the IFR checklist was properly used in the beginning, everything should be found in the proper place when double-checked. Subtract a point for anything missed in the initial checklist used.

While you're taxiing, check the operation of the flight instruments. On the runup pad, double-check that the electrical and vacuum systems are working properly, since they are such strongly IFR-related items.

Copying the clearance is a ground item, too. This used to be a complicated process, but today, most clearances read something like: "Cleared as filed, maintain 7,000." Anybody can copy one of those. Where you might lose a point or two on a clearance would be in not managing to copy a complicated one correctly. Don't feel too bad about that, though, and don't grade yourself harshly, because you are not going to fly unless you get it right.

If the pilot managed to keep all 10 preflight points, he should be cocked and ready as the airplane moves onto the runway. Everything is set, everything is checked, and the pilot is confident that the weather contains nothing that can't be handled with a good margin of safety. If points were lost, there might be a nagging apprehension about the weather, the DG might not be set right, the chart might be in the back seat, or the radios might still be set for yesterday's approach.

The takeoff and terminal-area departure are an important part of an instrument flight. The pilot thrusts himself into the IFR environment, and his performance as he does so sets the stage for the rest of the flight. Some might argue with my first takeoff item, and with the fact that I'd subtract from my grade if it wasn't done on an IFR takeoff. As soon as the power is fully forward, I glance at several things: the manifold pressure, rpm, fuel pressure, alternator output, and vacuum. The check is not to verify a specific numerical reading. It is just a glance to see that the needles are in roughly the right places and that they are steady. If everything spins up and works smoothly, it is likely to keep working for a while.

If the takeoff is into real IFR conditions, it is best to go on instruments at the time of liftoff. (If it was an instrument takeoff, the roll might have been by the gauges, but I don't think many pilots actually take off in zero-zero conditions. They just say they do when swapping lies in bars and around the hangar.) A transition from visual to the gauges has to be made, and the sooner it is, the better. If you didn't look at the gauges until punching into the bottom of the ceiling, subtract a couple of points.

The pilot often encounters diversions right after takeoff, almost all of which are manufactured on the ground. Just as you are settling down to the gauge flying, the tower tells you to switch to departure control. The next message might be to squawk ident, turn right to a heading of 270 degrees, maintain 3,000, and expect 7,000 in 10 miles. The pilot who overreacts loses points. It is best to pay almost absolute attention to flying the airplane during the first stage of the climbout. The wise pilot would say "Roger" to such a barrage of instructions. Then, when ready, he'd start a *gentle* turn to 270; that first turn of the day on the gauges can be the toughest one, and a gentle transition into a standard-rate turn is the way to go. Don't try to immediately record the clearance to maintain 3,000 and expect 7,000 later, either. At least don't record it until the airplane is around on that 270 heading and has been steady on it for a moment. In this phase of flight, as in all the rest, subtract two points anytime something is written down, or a chart is consulted, when the airplane is turning.

The pilot can provide confusing distractions for the man on the ground, too, and should lose points for doing so. When a controller gives a departure instruction to fly a heading of 270 degrees, he expects the pilot to do just that, and as he follows the airplane's blip across the scope, he considers that its track is the product of a 270-degree heading. If the pilot is flying 260 instead of 270, it becomes very confusing to everyone when the controller tells the pilot to turn right to 280. He expects the pilot to turn 10 degrees right, but if the message wakes the pilot up and he decides to really fly the assigned heading, he turns 20 degrees to the right. Subtract a couple of points every time you find your heading in substantial disagreement with an assigned vector.

If the departure is a nonradar one, grade strictly on flying the clearance with precision. (If the controller has radar, he'll grade it if you stray.) Also on nonradar departures, subtract three points if the approach plate was not checked for special departure and climb instructions for terrain or obstruction avoidance. Most places don't have such procedures, but you'd only have to forget once at a place that does.

En-route IFR is often dull, but there are things to do to add interest, or at least to make work. Even if in radar contact and DME-equipped, estimates should always be calculated. If the controller calls with word that "radar contact is lost" and asks for an estimate for the next reporting point, it's nice to have one waiting for him. Subtract a point for each estimate not figured while en route. Also subtract one for each navigational aid utilized and not positively identified aurally.

En route is a time to check weather, too. It is foolish to check the weather before takeoff and never check it again—especially when you spend a long time sitting there flying straight and level with nothing else to do. You can check either by asking to leave the frequency to call the FSS or by listening to the TWEB playing in the background—not loud enough to obliterate any call from center.

Everyone should strive for perfection, but few attain it. If the altitude varies a hundred feet and the heading five degrees, or if the nav needle wanders a couple of dots from center while en route, it's no major tragedy. It's not sharp flying, either, and should call for the subtraction of a point. If at any point in the flight the altitude varies as much as 300 feet or the heading 15 degrees, or if the nav needle reaches more than a half-scale deflection, take away two or more points, depending on the severity of the infraction. (It's interesting that those who say IFR is dull are often the sloppiest pilots. If they would put out the effort to keep all the needles right where they belong, they would find it less than dull.)

The importance of looking for other traffic when you're on an IFR flight plan in VFR conditions can't be overemphasized, and if you catch yourself flying for long stretches in VFR weather staring at nothing but the instrument panel, or with a fixed gaze into space, that must be judged poor IFR practice. The radar controller will not be able to call all traffic for you, and it's your responsibility to see and avoid VFR airplanes. Grade very critically on this point as a part of the departure, en-route, and arrival phases of the flight.

The transmitting of pilot reports is an en-route function that is immensely helpful to your fellow aviator. If you encounter en-route conditions that are different from those forecast, give the nearest FSS a call. A pilot who plunges through unforecast turbulent rain or collects a moderate amount of ice without passing on a pirep should have his en-route performance graded down. Likewise, a pilot who leaves on a morning when rain, turbulence, and high cloud tops are forecast but breaks out at 2,500 feet and sees a beautiful sunrise is a pure jackass if he doesn't make a fast pilot report to relieve the sweating of his fellow aviators.

Some items of aircraft operation are more critical to IFR than VFR flight. One of these is the selection of power and the leaning of the mixture. An IFR flight is always fuel-critical if there is real weather about. One's ability to cope with unexpected weather changes or marginal conditions is in almost direct proportion to the amount of fuel on board. The pilot who doesn't select a power setting that will take maximum advantage of fuel, and who doesn't lean to best economy, might someday wish mightily that he had done so. If a flight is flown with high power settings leaned to best power instead of best economy, subtract en-route points unless the weather was lead-pipe cinch at the destination and alternate.

The descent and transition into the terminal area is where many pilots become disorganized. This is especially true if the flight is into one of the higher-density terminal areas. First, deduct a couple of points if you didn't listen to the automatic terminal information service in advance. About 40 miles from the average terminal is a good time to tune it in, for handoff from center to approach control will come at around 30 miles. Much of the information on the ATIS is often useless, but it does give the altimeter setting and word of what kind of approach to expect.

Another thing to work at on an arrival is the descent from cruising altitude. ATC's usual procedure is to run the slower airplanes into the terminal area at a fairly low altitude—around 5,000 to 6,000 feet, in most instances. This is fine, and usually means that they pretty well plan the descent for you. However, if the air is turbulent and you are going to have to restrict airspeed in the descent and also keep the engine warm, necessitating a relatively slow rate of descent, subtract a couple of points for not anticipating this and speaking to the controller about it in advance.

Many pilots get sloppy as they move into the terminal area, because they feel the press of many duties other than that of controlling the aircraft. Even the autopilot user can find himself behind things if he spends his arrival minutes in confusion over communications and arrival routing. This is no place for sloppy flying. Remember the rule about not consulting a chart or writing anything down while turning, and also the fact that terminal-area flying is really simpler than it sounds. The controller will call with instructions, but they will be simple enough. He can only tell you to do one thing at a time. And even though he'll be talking to a lot of airplanes, yours is the only one that really counts. If the activity and increased communications pace in a terminal area cause heading excursions of 15 degrees or altitude variations of 200 feet, that's reason for about five points to come off the score. Why so many points? Because you are leading up to the finale—the instrument approach—and that's a hell of a place to start forgetting the notes to the music.

Another place to be severe with grading is on position awareness in the terminal area. Many pilots emit a sigh of relief when the approach controller says "Radar contact" and gives a heading and altitude to fly. It's as if the navigational load shifted from the pilot's shoulders to the controller's. That shouldn't be, though. The pilot must always fly as though the controller's radar—or his own transponder—might fail, which would mean the navigational chores would shift back to the cockpit.

An awareness of position also tells a pilot what to expect next. If the airplane is a fast one with collapsible wheels, speed can be kept up as long as practical, and the airplane can be slowed and changed to the approach configuration at the appropriate time, for example. If a pilot does not use every tool in the airplane to keep up with position as he moves around the terminal area, he introduces the possibility of confusion, and loses points.

There's one paperwork chore—the preparation for the approach—that must be handled during the descent and transition into the terminal area. It would be nice if this could be done while droning along en route, but the pilot often doesn't know for sure what approach he'll be shooting. What preparation is necessary? A lot more than opening the approach-plate book to the correct page. The key points to remember on an approach are the final approach course, the minimum descent altitude or decision height, and what to do if you miss the approach. These are things you must have in mind before turning final. If I have time, I write the MDA or DH on a little piece of paper, along with an arrow showing whether the missed approach is straight ahead, right, or left, and tape this small piece of paper under the altimeter. My alternative to this is to write the MDA or DH very prominently on the kneepad, along with the missed-approach direction.

Regardless of how you do it, the goal is to have the necessary information for the final approach extracted from the approach plate *before* hitting the final approach fix and starting down. If this isn't done, subtract five points from your score for the descent and transition to the terminal area. That's a lot of points, but not being ready for the approach is a bad mistake. You're flying the airplane into a funnel that will disgorge it on the end of the runway, if it's all done properly, and the farther into the funnel you fly, the more serious any error will be.

Altitude and position awareness is an extremely important part of the actual approach. It's obvious how critical this can be during the last mile of an approach, but it can also be pretty important farther out. In hilly terrain, some approach segments before reaching the final approach fix might provide only 500-foot clearance from, say, a sheer cliff. (I put a plate in my book today that calls for a descent to 2,000 feet in the approach segment preceding final; the descent starts as soon as a 2,200-foot TV tower is passed.) One must thus be precise, for while some approach procedures tolerate sloppy flying and allow a pilot to rescue himself because the terrain is flat, others reward sloppiness with a resounding bang. If altitude strays more than 18 inches below the minimum on an approach segment before final, subtract 10 points.

As we reach the final approach fix, where we are going to start a descent and fly into the neck of the funnel, there are a lot of factors to bear in mind. One is wind. What is the surface wind, and what actual wind aloft has been encountered? This is often of little actual importance, for the wind change is usually gradual as you descend. But if there is a strong southwesterly flow aloft, for example, and a southeasterly (or even northeasterly) surface wind, you *know* something is going to change pretty drastically somewhere during the final descent. Being able to anticipate the shift helps, and anytime you don't put together a little mental cross section of the final approach course before starting down the slide, subtract five points.

Position awareness is important as you fly toward the last part of the approach, too, even if you are being radar-vectored. Do you know about how far outside the marker you'll intercept the final approach course? Do you know the angle of intercept? If you have DME, it can often be used to keep track of approximate distances to the marker (and exact distances, if the vortac is on the airport or there is DME in conjunction with the ILS system). If you have an ADF, the needle pointing toward the locator at the outer marker is handy in telling you about angles. You can also ask the controller how far out you'll intercept, and if it appears questionable, what the intercept angle will be. Some controllers vector airplanes around for extreme intercept angles, and you must anticipate this to be able to handle it gracefully. If at any time toward the last of the approach you find you aren't sure of position, or aren't using every available aid in the airplane and on the ground to maintain an awareness of how far you have to go and how you'll intercept the final approach course, subtract three points.

When you reach the final approach fix, it is time to start down. What is the power setting for a normal rate of descent in your airplane? If you don't know, subtract 3 points. It's foolish to have to search for a proper power setting at that time. Airspeed excursions of more than five knots cost you 5 points, and you lose 3 points for any nav-needle excursion exceeding two dots after passing the final approach fix inbound. The same goes for the glideslope if it's an ILS approach. Exceeding three dots costs 5 points. Four dots cost 10. What about full-scale excursions? If a missed approach is executed as a result of the full-scale deflection, there is no penalty. Otherwise, you lose 20 points but get our wish for lots of luck; you'll need it. The key to an approach is keeping the airplane on the track. If it jumps the rails, those final moments provide a very poor environment for an instant salvage operation.

Toward the end of the approach, we get to the most critical point. A great many IFR accidents are caused by the pilot descending below the minimum descent altitude or decision height before the runway lights or the runway is in sight. Anytime you go below the MDA or DH before the landing is assured by visual contact, deduct 25 points and consider yourself lucky the penalty was not more severe. The visual illusions that can bug a pilot at this critical point in an approach necessitate absolute discipline on not shaving the minimum altitude. Pilots have been proving this for years, by littering the final approach courses to airports with airplanes of all sizes—large and small.

There should be no hesitation about going into the missed approach, either. When you've gone as far as you can go and the runway is not in sight, take positive action. Judge your missed-approach performance by the timeliness of the maneuver as well as by how well the attitude you selected for the pullup worked out. Was the airspeed on the correct value and did the airplane's heading remain relatively constant until any required turn was started?

One very experienced pilot told me that he never tried a second approach (unless the weather improved after the first one) because he felt that was asking for trouble; if he couldn't do it the first time, he'd go somewhere else.

Other pilots will keep missing until they hit the runway—or something else—or until they play out their fuel reserve. If the needles were in the right place at the MDA or DH and a look up for the runway revealed nothing but the inside of a cloud, it does seem logical to give up.

When your instrument flight is over, there are still three areas to grade when you review the flight after the airplane is parked: detail work, radio procedures, and overall planning.

Detailwork is bookkeeping. Everyone has his own system on this, and it is foolish to say that a pilot should use one system or another. The goal is to have everything written down that needs to be written down. The clearance was graded during the preflight, but other items that should find their way on to the clipboard in the course of the flight will actually take up more paper than the clearance. Altitude assignments should be recorded, as well as requests to report leaving or passing altitudes. Every new frequency assignment should be noted; in case you are unable to reach the man they told you to call, you'll have the number of the previous controller in front of you so you can call him back. Every en-route clearance change should be noted. Anytime one of these is not recorded, or recorded so poorly that you can't read it later, subtract a point. But don't be a slave to the pencil. If you wrote something while in a turn, or if recording something caused a distraction that led to a noticeable deterioration of heading, attitude or altitude, subtract all five detailwork points for the flight. Recordkeeping should be done during slack moments.

Communications procedures are important to instrument flight because you are sharing a party line. The object is to speak clearly, say no more than is necessary, and be considerate of other pilots by listening for a moment before calling on a new frequency to make sure the frequency is clear.

In grading your communications, be very critical of times when you were not clear in your request or response. Did you know what you were going to say before you keyed the mike? Were you as brief as possible? Nothing is more aggravating than a long and uncertain speech on a busy frequency. A pilot who goes into detail far beyond what is necessary to get the message to the man on the ground is only hastening the day when the FAA will say they have to have more frequencies and controllers, and we have to buy new radios and pay more taxes. Were you courteous? If you smile while you talk, you'll find the service from the controller a lot better. Also, when you run into the occasional grumpy controller, just be agreeable and pleasant when you talk to him. If he came to work unhappy and pilots make him unhappier, that means we'll all be hurtling though space being vectored and separated by a man who is angry. If you have a complaint about the service, don't bitch on the air. Instead, call the watch supervisor on the telephone as soon as you land. Be as nasty and long-winded as you wish then; it's your dime.

Finally, overall planning: this is what keeps the pilot confident by eliminating surprises and tedious situations. It is the talent that handles the things that could be encountered on any flight but aren't encountered on every flight. Planning is something to grade based on total mental performance, but there are a lot of individual things that can be singled out as examples.

Think of it as long-term or short-term planning. Long-term planning is something we do to identify and minimize possible risks on a systematic basis. It would include things like plans made in case the destination *and* alternate go sour. It can happen. The last time it happened to me, I took off on a 300-nm trip with the destination forecast to be excellent when I got there, and with an alternate that was forecast to be barely adequate. My approach to the destination was to minimums; my alternate was zero-zero. I had a hunch that might happen, because I was familiar with the area plus the fact that actual weather was running worse than forecast when I checked before takeoff. In this case, I had enough fuel to fly several hundred more miles, which would have been necessary if I had missed the approach at the destination. If I hadn't had the extra fuel, my long-term planning would have been lousy.

Long-term planning also bears on route selection. If one route goes over the rocks for 200 miles and an alternate route remains over a broad valley, the good planner would go a bit out of the way to enjoy better terrain.

Long-term planning covers the ideas a pilot formulates and keeps simmering to use in case equipment starts going sour. What is the plan if the vacuum pump goes out? What if the alternator goes out? The pilot who thinks through all such things and develops firm ideas of what he will do can meet almost any problem with a minimum of hesitation and sweat. The guy who doesn't plan ahead seems to fly from one barely manageable crisis to another.

Short-term planning comes more under the heading of how we handle the minute-to-minute events and keep the airplane moving through the friendliest sky possible. For example, a very poor short-term planner is the lad who sits at 9,000, with the temperature just below freezing, and watches the cloud tops slowly rise below him until they finally envelop the airplane and start coating it with ice. (Ice is often heavier in the cloud tops than anywhere else.) Then, a sweaty hand reaches for the mike and a squeaky voice announces to the controller that a new altitude is needed *now*. Another poor short-term planner is the pilot who started picking up ice and didn't tell the controller about it until he was up to his pitot tube in icicles.

Thunder and lightning probably place more demand on short-term planning ability than ice because a thunderstorm is a more sudden event. Ice generally puts the screws to us slowly; a thunderstorm envelops the airplane all at once and is as bad as it is going to get in a matter of seconds. With ice, we tend to react after the fact—after a little of the frosty begins collecting. With thunderstorms, the object must be to keep the airplane clear of the beast.

The good short-term planner flies the airplane so as to always have an out in a thunderstorm situation. When the pilot is forced into a position of thinking "I hope this works" and has to scoot down in the seat and tighten the belt until it hurts, he is a very poor short-term planner.

It might appear a little juvenile to think in terms of pilots flying along filling out their own report cards, but it works—for pilots who are honest in their self-appraisal. No pilot is likely to ground himself, but when areas of weakness are identified, anybody with sense is going to quietly move to correct them. All the wise pilot must do is keep a running tab on his performance during each flight, and conduct a critical mental review after the flight is over. You might even derive a lot of satisfaction out of keeping up with each flight on a grade basis, and working toward the high 90s—or even a perfect flight.

PART III

1 WAY-OUT WONDERS
by George C. Larson

YOU ARE STANDING in an utterly dark room. It is either necessary or desirable that you get across the room, and you need to know how you are progressing while you make the trip. You are equipped with a scooter and a flash camera. Luckily, there is someone in the room with you. This person has one end of a long clothesline that stretches from one side of the room to the other. You ask this person to produce a standing wave in the clothesline by whipping the end up and down. When the wave has stabilized, you set off on your scooter. Every two seconds, for you are traveling very slowly, you take a flash picture of the clothesline and get it developed instantly on a special film that reveals to you a section of the clothesline. By determining where you stand along the waves in the clothesline, you are able to determine your position and progress so as to cross the room without banging into the opposite wall.

That complex analogy describes the fundamental principle by which VLF navigation systems work. Such advanced systems offer benefits that should make these advances attractive to all of us, even if the current prices make them appear to be out of anyone's class but the jets. Microcomputers are already performing small miracles for tiny prices in things like postal meters and microwave ovens. With the Bendix BX2000 avionics family, they made their debut in the aviation world and what is particularly significant, they made that debut in general aviation, not in airline equipment, as used to be the case. As one Bendix engineer put it at the unveiling: "From now on, the innovative thinking will be found in products for general aviation. The volume is there, and the price is right. Furthermore, the general-aviation market has a flexibility you can't expect of an airline. The airlines buy for 20 years' worth of service, and then they live with what they bought."

If you plan to cash in on all this and enjoy the spotlight, you're going to have to know what you're talking about. It may help you to take a look at where we've been and how we've come to where we are now, so let's start at the beginning.

1 Dead Reckoning

It has been said that this method took its name from the fact that if you used it, you'd end up dead; however, it is startlingly accurate when done properly and when it is based on good data.

Dead reckoning makes use of two principal instruments: a compass and a clock. The compass is a highly reliable instrument that does not have to use electrical power and that has a very low rate of failure. It makes use of the earth's own magnetic field to derive navigation information. It is limited to providing orientation only to magnetic North, which "broadcasts" from an area near the true geometric North Pole. The pilot must calculate a course relative to the compass reading. The compass' principal weakness is the variation of the earth's lines of magnetism; this variation is so severe near the Pole that the instrument is useless there; additionally, because the field being detected is very weak, the compass is subject to local errors from spurious magnetic sources, either from within the airframe or from outside forces.

The clock is one of humankind's oldest instruments. It is used to divide the period of the earth's rotation into 86,400 equal increments called seconds, which in turn comprise 1,440 minutes. Since most aircraft travel at a rate that has been well established by test experience, a pilot who is able to keep an accurate record of time is able to calculate the distance the aircraft has flown (through the air) by multiplying the time it has flown by the known rate of the airplane.

Dead reckoning has some serious drawbacks that can make it very inaccurate. It provides no intrinsic source of wind data. If the pilot is fortunate enough to obtain some observations based on meteorological measurements or on the time between observable landmarks the positions of which are accurately known, it is relatively easy for him to minimize this disadvantage. The advantages of dead reckoning lie in its simplicity and modest cost.

2 Radio Ranges

The magnetic field generated by the North Pole is very weak, but that might be overcome by supplementing its information which some relatively strong electromagnetic radio signals that would be detectable by a radio receiver aboard the aircraft. Radio ranges were set up so that they would broadcast an on-course signal—a steady audio tone—only along the airway. A deviation off course altered the audio the pilot heard to the Morse-code letter A or N. Actually, the on-course signal was merely a blending of those A or N signals so that the resulting tone sounded steady.

The ranges offered an immense improvement in the accuracy of air navigation over what had been provided by the compass and clock; for the first time, there was literally a "roadway" defined by an electronic signal that was affixed at a known ground reference and was detectable through most weather and other obstructions to vision. The principle of a beam to guide the aircraft, with only left-of-course, right-of-course information, is still at work, though in a different electronic form, in the localizer and similar directional aids.

The disadvantage of ranges was that it helped if the pilot already knew where the airplane was to begin with. If you had a reasonable idea of your position, the range would help you to confirm it and would do so with much greater accuracy than you'd obtain with a compass alone. However, as the beam usually produced only a single electronic line and two left/right areas, a pilot with no prior information would have a difficult time orienting the airplane. The use of audio commands was another disadvantage, since it depended largely upon the subjective impression of the pilot's hearing (in a very loud cockpit) for decision making. If the signal seemed to fade, you were supposed to turn the other way, for you were flying away from the beam. Trouble was, the type of radio frequency used for the ranges was easily disturbed. Electrical storms and terrain affected the ranges and altered significantly the information that reached the pilot.

3 Direction Finding

This represented the direct substitution of a radio beacon for the magnetic pole. Of course, one could locate the beacon anywhere one chose, which was nice. Another improvement was that guidance was provided electronically instead of by an audio signal. Loops and goniometers were able to resolve very accurately the bearings at which signals were at peak and null strengths. This, coupled with a reference antenna to resolve 180-degree ambiguity, provided heretofore unmatched accuracy. Once it was discovered how to make this function automatic, to make an automatic direction finder out of simple coffee-grinder DF, the pilot was in hog heaven, with an arrow on the panel that pointed to the relative bearing of any nondirectional beacon that was tuned on the receiver.

Direction finding produced navigational information that was accurate enough for approaches in low visibility, and it eliminated the gnawing doubt that had plagued the captain who had to interpret ranges. Its disadvantage was that it was foolproof only when you were flying inbound to the beacon. You could always "home" to the beacon, even though "tracking" was the desired technique, and know you would reach the destination if you were inbound, for all courses led to the station and all errors and drift would be irrelevant. Flying away from the station was another matter, however, for all courses would put the arrow on the center of the tail—but with no guarantee that the desired destination was off the nose. You could fly with perfect precision and, without the use of an adjunct (usually the old-reliable compass), you would fly yourself to Nowhere.

Nondirectional beacons broadcast in the part of the frequency spectrum that included all the atmospheric disturbances from electrical storms, which meant you would home toward the lightning if some were around. They were also affected by something called the "shoreline effect," a phenomenon attributable to the different propagation characteristics of land and water. Along the shoreline, the beacon's apparent location would be refracted, as if by a lens. A canny pilot could minimize that effect, but again, only if the aircraft's position were known within a reasonable degree of doubt.

4 Omnis, Vortacs, and DME

Omni, or omnirange navigation, took care of most of the disadvantages of direction finding at a very small sacrifice. It introduced VHF radio transmission, which eliminated at a stroke much of the noise interference from electrical activity in the atmosphere but with a slight concession: VHF propagation is strictly line-of-sight. If your receiver can't "see" the station, you won't receive a signal. Prior families of radio-navigation equipment generated signals that had longer wavelengths that were refracted by the layer of ionized particles in the upper atmosphere as if by a huge mirror; they would return to earth at great distances from the transmitter. The loss of this advantage wasn't particularly felt at first, since the obvious substitute was a network of omni stations so numerous that an aircraft need be out of range of one only rarely. It was expensive, but the net was built and is being added to and maintained by the Federal Aviation Administration, which has found the burden so imposing that it has gone to the supreme effort of busting into the airport trust fund to find more money.

Omni transmitters are really two transmitters that operate on a timed cycle. One is an omnidirectional type that is simply pulsed every one-thirtieth of a second. A second transmitter/antenna is of the highly directional type, and it rotates 360 degrees every one-thirtieth of a second, like a light in a lighthouse. Because this second signal is very directional, the airplane's receiver sees the signal only as it sweeps past the airplane. Here is where the timing becomes important: the first transmitter is calibrated to pulse at the exact moment when the "lighthouse" beam lines up with magnetic north. Obviously, if the airplane is some angle away from magnetic north—east of the station at 090 degrees, say—the lighthouse beam will take a fourth of a thirtieth of a second to sweep around and be seen. The time difference between the two is detectable by the receiver electronics as a phase difference. Detecting very short intervals of time is relatively easy for electronic instruments. A phase difference as small as one degree is translated into a needle deflection, giving the pilot a fly-left or fly-right command.

The geometric advantage of omnirange navigation is that it produces 360 lines of reference from the station that are, relatively speaking, fixed in space. Simply center the needle, and the system will tell you which spoke you are positioned upon. As with ADF, it is easy to tell whether you are tuning a course toward or away from the station. Unlike the ADF, omni is not heading-conscious; put the airplane into a tight turn at a generous distance from the station and nothing on the omni indicator will even budge. Thus, a pilot could fly around in circles forever and receive no indication that the airplane wasn't on course to the planned destination. Unlike ADF, omni is a positioning system, not a homing system, although an airplane flying on a course that matches the one in the OBS will fly either toward or away from the station and always on the correct spoke of the wheel. Therefore, you can rely on VOR guidance *away* from a station as long as you fly relatively straight over a long period of time; that's one big operational advantage over direction finding.

Although VOR eliminates angular ambiguity, the other dimension—the one that places the airplane at a particular spot along its selected spoke—is not available until we add a pulse system called DME. Distance-measuring equipment, although quite expensive, is extremely simple in principle. A very brief pair of pulses goes out to the DME ground transceiver. The DME ground station replies on another frequency, and the airborne receiver merely counts how much time passes between its first transmission and the receipt of the return. A radio pulse requires 12.37 microseconds to travel one mile and return, so you can divide the round-trip electronically into the total time counted to obtain the airplane's distance from the DME station. DME stations are usually colocated with VORs, and their frequencies are now paired systematically.

Thus, the pilot within line-of-sight range of a vortac can determine the radial from the station and the airplane's slant-range distance from it. This is called *rho-theta* navigation, and it uses polar-coordinated geometry in order to solve, for the first time in the history of air navigation, the question of precise position from a single source. (It's fudging to say that, since the source really has two kinds of radios, VOR and DME, but the important point is that the site is common and the measurement is based on a single origin.) Even DME is not without disadvantages, however. There is just so much time for each DME transceiver to answer every aircraft that interrogates it, and the DME can become overcrowded, since it is an "active reply" navaid. When that happens, it simply replies to the strongest signals and ignores the others. (That's one reason why airlines like to buy very powerful DMEs, and it's the biggest reason why your DME may occasionally lose its lock and show intermittent flags.)

There is also a geometric fallacy in DMEs due to the altitude of the aircraft. The answer you get on the indicator is really the slant distance through the air between the aircraft and the DME transmitter, and that is always slightly greater than the planar distance of the aircraft from the station. That error is usually insignificant enough to be ignored except at those times when the aircraft is very close to the station. The percentage of error changes with distance, and a pilot must bear that in mind.

It may seem to be stretching a point, but a lot of the underlying principle of dead reckoning is still at work here. Counting time, even though it is done internally, electronically, and using very short intervals, is still the basis for distance measurement, and angular displacement from a known reference is still the principal source of position information. The ordinary mechanical clock is, needless to say, no longer much help, but the distance equation, the product of rate times measured time, has become an absolute essential.

The compass, too, is no longer the prominent angular-reference instrument, but the use of a radio station to establish a heading based on angular displacement is merely the substitution of a radio transmitter for the magnetic North Pole. A VOR station was, in fact, purposely designed to be a new "pole," with 360 radials being broadcast from it, which, if they were able to stretch beyond line-of-sight range, would have produced a whole new set of longitundinal lines. VOR and DME were equally important to the way navaid design was developing because they established the usefulness of time to work out the mathematics of position. VOR used the phase shift (which is nothing more than a time delay detected in an oscillating signal) to place the airplane at the proper direction from the station. DME used the speed of light (and radio signals) as a constant for measuring straight-line distance based on a different kind of delay measurement.

5 Loran

This long-range navigation system is like DME in its use of time to establish a straight-line distance from several stations. Loran uses station pairs that transmit, while the airplane carries a receiver only. This is a very different system from that of DME, which is based on two transceivers, one on the ground and one in the airplane, talking to each other. The airplane becomes entirely passive in loran, which eliminates the need to put a powerful transmitter on the aircraft. It is easier to build ground-based transmitters of very high power and simply equip the airplane with a decent receiver, and for long-distance navigation, such an arrangement is practically mandatory.

Loran station pairs transmit pulses. A master station pulses once; this pulse is received by a slave station, which waits for a set interval, and then transmits its own pulse. If the airplane is somewhere between the two stations, it can determine its position by counting electronically the difference in time each pulse took to reach the receiver. There is only one line, which happens to be a hyperbola, along which the airplane could be located, given the time difference measured. Having obtained that line, the same thing is done again with a different station pair. Thus, two hyperbolas can be drawn; the airplane's position is the point where they intersect. Obviously, the technique will work well only if there are two station pairs available to the loran receiver. Also, to be accurate, the plotting of position is dependent upon those two hyperbolas intersecting at a fairly generous angle—90 degrees is best. Loran is fairly demanding because the only readout from the loran itself is the measured time difference between stations; the pilot has to take that information and perform the computation in order to determine his position.

6 INS and Doppler

What do you use for navigation when there are no radio aids around—not even loran? One approach to an entirely self-contained system—one that needed no outside source of signals—was Doppler. It got its name from the Doppler effect, which is the *apparent* shift in frequency due to movement. The simplest example of Doppler effect is the train whistle that, on approaching, sounds higher (an upward frequency shift) than it does going away (a shift downward). In astronomy, the discovery that many stars generated light that seemed to be shifted toward the red end of the visible spectrum led to the first recognition of the fact that the universe is expanding at a great rate, a fact that in turn led to the origin of the "big bang" theory of creation. The Doppler effect could be put to work on an airplane to keep track of how fast the terrain below seemed to be moving past the airplane. A pair of radar transceivers focused in a V from the nose of the airplane sent signals to the ground. The return would be shifted in frequency by an amount that would indicate how fast the airplane was moving forward. Why two beams? If the airplane were drifting, cross-track, the two readings would be slightly dissimilar, and the computer that did the counting would take that into consideration. The disadvantage of this was that radar reflected reliably only from solid targets with good relative facets.

One of the earliest applications of inertial navigation systems was for military submarines. Subs had unique navigation problems, since they had to determine their location while deeply submerged and could not count on radio navaids. Inertial navigation works backward through the mathematics of motion to arrive at a way of figuring present position and velocity. Some very sensitive accelerometers in the INS system are the key to its operation.

The mathematics works in steps: the rate of change of position is called velocity. The rate of change of velocity is called acceleration. If you can measure acceleration, it's not difficult to integrate acceleration to obtain velocity at any instant and then to integrate again to obtain instantaneous position. The way to measure acceleration is simply to use a mass and its inertia—its resistance to change of position. If you move a mass, the mass will oppose the motive force with its own force of inertia, which is easy to measure. This is not to belittle the incredible complexity and accuracy of the stable platforms in the INS sensors; they are marvels of precision engineering.

INS systems stand alone, utterly unique. They are totally self-contained and unaffected by any outside disturbance. The present generation of INS, such as the Litton or the Carousel systems, is the king of dead reckoning. The INS needs only the eternal force of inertia to provide its data, a force that is part of the immutable physical world. It is law. No shoreline effect. No static. None of the gripes that subtly render the other modes of navigation less than completely accurate.

Inertial systems have their drawbacks, though. Left to themselves, they are utterly unable to tell where they are. They measure only acceleration, and it is therefore up to the pilot to tell them exactly what point to begin measuring from. Once airborne, they cannot "find" themselves as a radio navaid can, unless, of course, the aircraft stumbles luckily into a known reference. Thus, although INS systems are built so well that they produce only the tiniest error, that error grows with time. Without periodic updating information (most are set up to accommodate some source of updating), they will be truthful only at the start and then gradually become liars.

Inertial navigation was most important for its contribution to the mathematics of navigation and for its use of a continual, real-time assessment of position. By always knowing how fast the position was changing, it always knew the present position, and that led naturally to the total automation of the navigation function. The final step was at hand.

7 Area Navigation

But first, a parenthetical aside. Once the principle of *rho-theta* position determination was established, it became obvious that a single vortac could describe any point on a plane, since it was its own little self-contained set of polar coordinates. A straight-line course can be described very easily using polar coordinates, and it was a very easy step, therefore, to build a computer that would handle exactly that computation. Suddenly, that omni-DME combo on the panel, when coupled to such a computer, could derive courses that did more than just connect vortac to vortac; indeed, it could virtually ignore where the vortac was located and simply use the signals to generate a geometric system across which *any line* could be computed and drawn. The first ones were called course-line computers, and the principle of direct navigation from point to point using vortacs as an interlocking network of polar coordinates became known as area navigation. For a time, any navigation system that could compute a direct course was called an area-navigation computer, but the avionics industry is beginning to eschew that term, at least when it is applied to a device that does more than simply draw the direct course from offset waypoints and give left/right deviation indications to the CDI. The simple course-line computer "area navigator" is still a boon to the pilot, but it requires that prior to a flight, waypoints be calculated and drawn on a chart. If ATC alters even one part of the carefully constructed house of cards, you've had it. Obviously, something more was needed.

8 Advanced Navigation Systems

What was needed has come. Instead of separate boxes advising us of raw radio data from which we derive our position and try to compute our velocity, we now have the computer box that constantly updates position from a dead-reckoning computation; the DR box is the heart of every advanced nav system because it contains within it the capability to build a mathematical model of our changing position trend. A system of inertial sensors is only one way to feed such a DR box the direction and rate information it needs. The vast web of vortacs and their associated polar-coordinate geometric systems can feed an omni-DME system of sensors, which, when translated into the DR-box language, yields accurate corrections to the basic DR computation. These corrections, no matter what sensors are used, eventually translate into the average long-term error in the DR computation; and if you think about that, you will realize that the average long-term error can be only one thing—the *wind*.

Another family of automatic navigation systems uses very low-frequency stations combined with a new navigation net called omega; of late, the direction seems to be toward new systems using omega alone. INS systems taught us the worth of thinking about position in terms of latitude and longitude, for such a definition readily plugs into a digital-computer language and allows the computer to do its internal thinking in rectangular coordinates, which computers handle easily. There may be a software package that can retranslate back into polar coordinates if you need to think in terms of vortacs, but the DR box does its little constant-update number in lat/long only.

Thus, the format of INS systems has come to influence the design of most of the advanced navigation systems on the market now. Their primary language is lat/long, and the control units on most of them are very similar to those of INS. The *only* significant difference between INS and the other two families of advanced nav systems, in fact, is that INS uses accelerometers as sensors, whereas the radio-based units process an electromagnetic signal to accomplish the same thing. It is no secret that INS was limited to a certain class of airplane—very heavy. INS systems are large and expensive. The new VLF and omega families are "poor man's INS," to use the term the trade uses. They are designed to plug into the airplane and perform as a substitute for INS.

9 Automatic Nav Systems Based on Vortacs

Examples of these systems that now seem to be attracting attention are the Garrett AiRNAV-100 and -200 (the latter with VLF/INS interface), the Sperry TERN 100, and the New Communications Components Corporation Vortrac.

The Garrett AiRNAV systems are total nav-management units that select the best geometry from the array of vortacs, VORs, and DMEs presented. They tune the selected stations automatically and resolve vortac geometry into lat/long language (though you can speak vortac-oriented *rho-theta* to the box as well) and compute direct great-circle courses to selected waypoints. The complete network of omni-DME stations is stored, along with pertinent data for each station, such as magnetic variation and elevation, on a magnetic-tape memory. The memory also stores the current RNAV SIDs and STARs and area-nav route structure, though Garrett is already making noises about how few people actually ever use those routes now that direct clearances from present position are becoming available from an enlightened ATC. A pilot talks to the AiRNAV in terms of waypoints or routes, and the system takes over complete management of navigation from there. It handles omni-DME navigation using DME-DME measurement as its primary mode, but it is capable of recognizing unsatisfactory signals in that mode and switching to combinations of VOR-DME or VOR-VOR tuning in a logical priority to obtain a position. If no valid signal is received, the DR box will continue to manage navigation based on the last known position and velocity.

The Sperry TERN 100 differs from the Garrett in some significant features. It uses a disc memory of very high capacity, which enables it to store up to 50 flight plans and 204 waypoints defined by the pilot. Access to the memory is extremely fast, taking 10 milliseconds at maximum. The TERN 100 uses *rho-theta* navigation (VOR-DME) for its geometric base, which differs from the Garrett units' *rho-rho* (DME-DME) geometry. There are also differences in the programming of the control-display units and the sequence of pilot action to obtain a response from the computer. The TERN 100 is also a total management system in that it automatically tunes the best station to obtain the most accurate geometry for the DR box to use in its computations and derives great-circle direct courses.

The Communications Components Corporation Vortrac is representative of a marketing strategy. CCC has, in this system, a relatively inexpensive DR box that handles most of the nav chores but without the route-storage capacity of the other units' memories and without the automatic tuning capability. Instead of auto-tune, the Vortrac system advises the pilot which station it wishes to have tuned and the pilot turns the knobs or pushes the buttons, depending. In all other respects, its mathematics is similar to that of the other units, and it can direct-route the aircraft along a great-circle course. In a way, the Vortrac represents a middle ground. It eliminates the difficulty of entering bearing and distance offset, as the most primitive area-nav units require, thereby entitling it to proper claim to being an automatic nav system. The automatic tuning function is gone, but you have to sacrifice something for economy, and the Vortrac is meant to sell for less than half the price of the fully automatic systems. It has a memory chip that stores every omni-DME station in the country; unlike the more expensive systems, which use magnetic tapes to update their station data, if anything were to change in the omni-DME network, you'd have to change the chip in the Vortrac. The Vortrac also lacks a vertical-nav capability, which both the AiRNAV and the TERN have.

10 VLF Omega and "Pure" Omega

The U.S. Navy found out a long time ago that very low-frequency radio waves carry for extreme distances, so it built several transmitters that operate in the VLF spectrum (10 to 30 kHz), a frequency range so low that part of it corresponds to the range within human hearing. In other words, you could simply wire a speaker to a VLF antenna, and if the signal were strong enough to drive the speaker, you'd actually hear the carrier itself! Last year, after navigation systems for aircraft had reached the market that uses these very stable VLF transmissions for position determination, the Department of Defense announced that it could not be held responsible for maintaining those stations on the air constantly, and since they had to be closed down occasionally for maintenance, it could not handle the chore of announcing when one was out of service. It also stated that it would be closing down these stations for good at some future date, but it didn't say when. After a period of some heated discussion, this situation has now resolved itself. VLF users have been assured that the stations won't disappear tomorrow and that the system is a viable one for a good while. There is a long story filled with political skullduggery behind much of this, but the important point is that the VLF system has become semipermanent.

The principle by which VLF navigation systems derive their position and velocity information was described earlier in this article. VLF systems have no innate ability to determine their position, at least not from the radio data. Therefore, they must be "initialized," just like INS systems, with the lat/long of their starting position (or updated by plugging in a known position while in flight). In addition, they must be told the date and time of day in order to compute the effect of atmospheric bending of the transmissions, an effect called "diurnal shift."

Using an extremely accurate atomic clock to calibrate themselves, VLF sensors begin by locking onto the phase of stations they select for tracking; if they can receive all the stations that are up, they will use them, for there are receiver boards for all the VLF frequencies in the VLF box. The older Global GNS 200s used to require that stations be selected manually, but such pedestrian activity has long since become automatic. The VLF transmitters are extremely strong, some as high as a million watts.

Once the station or stations is phase-locked, the sensor detects aircraft motion by noting that it is seeing a different portion of the wave as the aircraft moves (like the person on the scooter moving past the clothesline). That's all the DR box needs. With a pair of stations to the right and left as well as some arranged nearly along the track, the geometry is ideal for producing a ground track and velocity vector with wind drift already considered. Since there are no angular determinations or measurements, we can consider this a *rho-rho* mode of navigation, similar in geometry to a DME-DME navigation system.

Most combined VLF-omega systems, like the Ontrac III and the Global GNS-500A, use omega as they use the VLF stations; they merely keep track of the phase shift of the signal and derive vectored motion computations from that. Although the principles upon which both these competitive systems operate are the same, they offer different features. CCC says its offset-waypoint capability is an exclusive that its customers have found desirable. It enables you to key in the direction "10 north of waypoint three," for example, should you determine at some point that you might be vectored to hold there. (Most features are designed into these boxes to cope with actual ATC situations and to ease the workload of responding to them.) Global is quite proud of its "pseudovortac" feature, which enables you to treat any waypoint as if it were a VOR-DME instead of a phantom position determined by counting VLF waves. This enables the pilot to select a particular bearing to the waypoint and to derive left/right guidance onto that course. Since that is essentially the way we use vortacs, the term "pseudovortac" seems a natural choice. Global also has one rather witty piece of software. Having initialized the system and being aware that it is locked and ready, you are supposed to engage the VLF mode from the takeoff position at the end of the departure runway. The usual ATC directive from that situation is "November 15R, position and hold." Global engages its VLF mode by having the pilot select *position* on the mode selector and punching *hold* on the keyboard. Hard to forget.

Pure-omega systems are following hard on the heels of the VLF-based ones. Omega transmissions are considerably weaker than VLF; therefore, processing the signal is a considerably more sophisticated task. In order to solve that problem, computers are being put to work as signal processors, being moved closer to the antenna, if you will. Omega differs from VLF in that it uses an art known as "propagation prediction" to determine position. You must program your initial position, just as you do with all INS-related systems, but the omega navigator has incorporated within its software a model of the atmosphere and other factors that affect the signal. The effects of conductivity over land and water, the height of the ionosphere, and the shape of the earth's magnetic field are a changing regime. The omega navigator's computer applies itself to the task of determining how many cycles it should be receiving from a given station. If the prediction says 361.5 cycles and you're getting 361.4, it will do an averaging problem and see that your position is changing.

The omega transmitters are set up to operate in a known sequence every 10 seconds on three frequencies: 10.2 kHz, 11.3 kHz, and 13.6 kHz. These three frequencies produce wavelengths that have a common null only once every 72 miles. Thus, if you think of it as a matrix, the receiver can "find" itself in a circle 72 miles in diameter. Very recently, it was proposed that a fourth frequency—11.05 kHz—be added to the signal format. This would be no disadvantage to present omega users in the marine application, and it offers aviation users one very great advantage; now the common null would occur only once in 288 miles, so a high-performance aircraft could lose the system for a considerable period of time and still find itself after it came back on if it hadn't yet traveled 288 miles.

There are, at present, eight omega transmitters located around the world and operated under an international agreement. There are still some problems to be worked out with some of the stations' signals, but it is hoped that eventually, they will all produce a usable signal during most of the day. Each station is identifiable to the receiver because each has a unique set of times in the 10-second sequence when it transmits on each one of the three (and soon four) frequencies. The omega sensors and the computer's logic are able to make sense of the sequence to the broadcasts and identify which signal comes from which station. Once that's done, the phase measurement of each is accomplished and compared to the computer's predicted results. The error is sent to the DR computer for further processing into nav information.

Some of the omega systems that are coming to general aviation are the Litton LTN-201, the Tracor, the Dynel ONS VII, The Bendix ONS-25, and the Hoffman HON-360. Because they are the newest units, they are expected to offer more features and fanciness than some of the recent VLF-omegas. While the present VLF systems offer warning before change of leg and turns, some of these omegas may actually compute turns and lead them for the pilot, just as some of the more costly, more complex units do.

The VLF-omega and pure-omega systems are intended for en-route navigation only, although some manufacturers add the words "and terminal acquisition," which, one supposes, means they'll get you within the approach navaid gate to the airport. They are required by Advisory Circular 90-45A to be accurate within 2.5 nm cross-track and within 1.5 nm along the track with 90-percent confidence. Most of the manufacturers have been required to demonstrate during flight testing that their units are able to navigate successfully with several critical handicaps, such as loss of all but three stations; if the aircraft is equipped with an atomic frequency standard, two stations are enough. A TSO that will cover omega systems is also being published, and all new and current systems are expected to comply. The possible addition of the fourth frequency and of other unique frequencies would not render obsolete those units already operating—including both pure-omega and VLF systems.

Because of the frequency range at which they operate, they are very sensitive to precipitation static, the accumulation of an electrostatic charge in any portion of the airframe, and its consequent discharge and corona, which generates a very noisy racket electronically. Installation of this equipment is normally preceded by a thorough check of the aircraft's bonding, the static wick system—buy some good, heavy-duty ones—and any conductive coatings that might be present on all nonmetallic components (which means mostly fiberglass parts). Newer omegas have shielded E-field magnetic-loop antennas that pick up the magnetic portion of the radio transmission rather than the changing electrical field. These are less sensitive to corona discharges but more sensitive to 400-cycle noise from inverters. They must be located very precisely on the exterior skin.

In all cases, advanced navigation systems have a backup, should all the sensors fail. All of them will fall back on dead reckoning. The dead-reckoning capability is not, as one might suppose, a feature that has been added on, like HSI coupling or airspeed input from a data box; instead, the DR function is something that is going on all the time anyway, whether the radio data is coming in or not. It is there always, except that you seldom pay attention to it unless something goes wrong. In that sense, it is very like a heartbeat; like a heartbeat or like the ticking of a clock . . . which is where we came in.

2 YOURS FAITHFULLY, ADF
by Richard L. Collins

THE ADF (automatic direction finder) is the grand old transistor of avionics. ADF led the way in radio navigation before the youngest of today's airline captains were born, and except for technological advances that have made them smaller, lighter and more reliable, ADFs have remained functionally about the same over the years.

The ADF automatically points to stations in the low-frequency (200 to 400 kHz) and broadcast bands, telling the pilot where a tuned-to station is located in relation to the nose or tail of the airplane. Perhaps that simplicity is responsible for the long history of the ADF. Or perhaps the fact that it makes music, news, and sports available in the cockpit has helped it to endure. True, there are drawbacks to it: the ADF's vulnerability to static, its demand for thinking by the pilot, and its lack of precision compared to that of a VOR have reduced it to a secondary role on many panels. You can still go anywhere you wish using the ADF, however, and there are many airports in the U.S. where the only available IFR access is an ADF approach. The ADF is also useful on many instrument landing systems, where a beacon located at the outer marker site (or the middle marker at a few) gives a good means of navigation to the marker plus a redundant indication of passage.

There is more to the ADF than immediately meets the eye. If a pilot really learns to use ADF, things other than the obvious can be added to its bag of tricks. To name a few, the ADF can almost become a makeshift area-navigation set and an airport finder in VFR conditions; it can be a valuable position reference when you are being radar-vectored for an approach or are maneuvering on your own; it can be used to determine distance to the station; and it can even be a backup for a failing directional gyro.

The ADF's area-navigation and airport-finding possibilities stem from the fact that it works on a lot of ground stations—far more than does VOR equipment—so there's a greater likelihood of finding a station close to a direct line from here to there. On an Iowa aeronautical chart, issued by that state, I counted 20 low-frequency beacons (excluding those located on instrument-landing systems) and 18 VORs. That's two up for the ADF and it's just a beginning. Jeppesen's *J-Aid* lists airports and their relationships (bearing and distance) to standard broadcast stations; there is a total of 126 listings in Iowa—which is quite handy. If you want to go to Elkader Municipal Airport, Iowa, for example, tune in WPRE (in Prairie du Chien, Wisconsin) on 980 kHz; the airport is 16 miles from WPRE on a magnetic bearing of 212 degrees.

The mention of magnetic bearing brings us to the thinking that is required to fly the ADF precisely. To track out on 212 by VOR, you would have only to set the course selection on 212 and fly (with a "from" indication) as necessary to keep the needle centered. The VOR needle will tell you all about the plane's position in relation to the desired track. With the ADF, you have to fly a magnetic heading of 212 with the ADF needle on the 180, or tail, position to track from WPRE to Elkader. In other words, two things—the heading and the ADF indication—must be combined to get a position in relation to a desired track.

Even with two things to consider, flying across a station and tracking outbound are not difficult when there is no crosswind component. Throw in a stout northwest wind on the southwestbound track to Elkader, though, and the challenge begins. If you fly away from the station with the compass on 212, the ADF needle won't stay on the 180 position. It will move to the right of the tail position, an indication that the northwest wind is causing the airplane to drift leftward, south of the desired track, and that the track is off to the right. That is logical—the needle is to the right of the tail, so the desired track is to the right—but mental gymnastics become necessary when a correction is made.

Let's assume the needle was 10 degrees off to the right of the tail before you definitely established the need for a correction and that the correction amount selected is 30 degrees. Anytime the airplane is turned, the ADF needle moves, and when this 30-degree correction to the right is made, the ADF needle will move even farther from the tail position. The total deflection will be 40 degrees to the right of the tail, 10 for the off-track, and 30 for the correction. How do you know when you are back on track? Logic returns: when the correction angle, 30 degrees, is matched by the deflection of the ADF from the tail position, you are back on the track. In this case, the correction heading would be 242. When the ADF needle reaches a point 30 degrees to the right of the tail position, simple arithmetic tells you that the airplane is back on the 212 track from the station. Turn to 222, leaving in 10 degrees off to the right of the tail, which is indicative of the correction for a crosswind from the right, and the track being made good will be 212.

If the correction is excessive, as would be indicated if the needle moved from 10 degrees to the right of the tail back toward the tail, showing that the airplane is moving to the right of the 212 track, a turn back to the left and an adjustment in the correction after interception are necessary. If the needle drifts to the right, the 10-degree wind correction would have to be judged inadequate, and more correction would be necessary after another interception of the desired track. If confusion takes over at any time, the simplest way to defog the brain is to turn to the heading of the desired track, 212, and look at the ADF needle. Its position to the right or left of the tail will indicate where the desired track is located. (Just for comparison, a full deflection of a VOR needle is 10 degrees, so if the ADF needle is 10 degrees off the tail when you are flying a heading corresponding to a desired track, it's the same as having the VOR needle a full deflection to one side.)

If you've decided that tracking away from a station with an ADF doesn't sound precise, you are right. Precision seems even more elusive when you consider that any compass error will also induce track error. But, if the compass is right and you can hold headings with some precision, the tracking will be close—at least a lot closer than you might get by guess and by golly.

Flying to a station, one can take the easy way out and the long way around (if there is a crosswind component) by just homing to the station. Fly with the ADF needle on zero, and you'll get there. If there is a crosswind, the heading will have to change as you fly toward the station, and the track will be curved.

Making good a desired track to a station is preferable and works much the same as tracking from the station. For example, say you are approaching Elkader from the Southwest and wish to track inbound and over the airport on an inbound bearing of 032 degrees (the reciprocal of 212). The on-track situation with *no* crosswind would find the heading on 032 and the ADF needle on the nose of the airplane. If the strong northwest wind starts blowing the airplane south off course, the needle would begin pointing to the desired track as it moved to the left. When it moves 10 degrees off and a 30-degree correction is made, though, the difference between inbound and outbound tracking shows up. As the airplane is turned to 002 for the correction, the needle would move to the right, through the nose position, and would be 20 degrees off to the right of the nose when flying the correction heading.

Similarity returns on interception; the 032 track would be intercepted when the ADF moved to a position 30 degrees off the nose position. Then, if a 10-degree correction is kept for drift, the compass would be on 022, and the ADF would be 10 degrees off to the right, where it would have to stay for the track to be made. Remember that on an inbound track with correction for crosswind, the ADF needle is on the side opposite the wind; on an outbound track, the ADF needle is on the upwind side.

In those two examples, the exercise started with the airplane flying either over the station or beginning on the desired track. There is also a simple way to intercept a track, employing the same principles used in correcting back to a track after drifting away. The first step in interception is to turn to a heading corresponding to the desired track, or 032 degrees, if the plan is to track in over Elkader. Note the ADF needle's position. If it is, for example, 30 degrees off to the left of the nose, you are 30 degrees from the desired track. To get there, turn 60 degrees left (the number of degrees off track plus 30) to a heading of 332. At that time, the needle will swing to a position 30 degrees to the right of the nose, and it will move farther from the nose position as you move toward the desired track. When it is 60 off to the right (the amount of the correction), you are there. In the case of intercepting an outbound track, the same procedure would apply except that when you turned to the intercept heading, the needle would move to a position 90 degrees from the tail position and would move back toward the tail as the track was approached. When it reaches 60 degrees off the tail position, you are there. The difference between inbound and outbound interception is this: when you are moving toward a desired inbound track, the ADF needle moves away from the nose position; when you are moving toward a desired outbound track, the needle moves toward the tail position.

Spending just a little time developing a sense of the relationship between the heading, the desired track, and the ADF deflection and movement from the nose or tail position will make it all seem very logical.

Once the basics of intercepting desired tracks and tracking are mastered, the ADF becomes a real cross-country friend. Jeppesen's *J-Aid* or other listings of standard broadcast stations can be used to determine which stations are along a route to supplement nondirectional beacons so that more direct flights can be planned and wind drift can be handled with ease. One thing about broadcast stations: they are required to identify themselves by call letters and location once an hour, but not at a precise time. Normally, stations do it within two minutes of either the hour or half hour, according to their choice. If your ADF is not digitally tuned and is off calibration, using broadcast stations can be a bit more difficult. But when you are VFR, which is when an ADF would be used for this type of navigation, you can catch the identifcation by listening to the station as you fly (if you can stand the music).

One other thing about standard broadcast stations: the IFR pilot should never take the listing of bearing and distance to airports as an invitation to develop an impromptu instrument approach based on a standard broadcast station. There are valid reasons for not making up your own approach, some of which are trees, hills, and man-made obstacles. If one of these should intervene, your precise tracking would be for naught.

The principles of interception and tracking are of great value when you are maneuvering or being vectored for an ILS approach. When tuned to the compass locator at the outer marker (which most ILS systems have), the ADF gives a constant reference to the outer marker and can help the pilot anticipate a lot of things. If the ILS course is aligned on a 045 bearing, for example, and the transition to final is from the South on an assigned heading of 345, the ADF would be 60 degrees off to the right of the nose position if the final approach course was intercepted on that heading. Before interception, the needle will be less than 60 degrees to the right of the nose position and moving away from the nose. The controller would normally shallow the intercept angle to 30 degrees before actual interception of the final approach course; in this case, the heading would be 015, and the ADF would be 30 degrees to the right of the nose at the time of intercept.

The ADF is also quite useful when shooting an ILS without vectors. It can be used after a procedure turn to anticipate an intercept on the final-approach course: at the completion of the turn, the ADF should be less than 45 degrees to the right of the nose position. It will move away from the nose, toward the 45 position, which it will reach as the final approach course is intercepted. If the ADF needle happened already to be past the 45-degree point after you roll out of the procedure turn and start inbound to intercept the final approach course, something would be amiss—the airplane would already be through the final approach course. That would also show on the localizer needle, but the ADF would give a verification.

In case of an extremely strong crosswind, the ADF can save the day completely in the initial phase of an ILS approach. For example, if the airplane wound up left of the final approach course before reaching the outer marker inbound and the pilot corrected 30 degrees to the right, the ADF would tell very quickly whether or not this correction would do the job. If the 30-degree correction put the ADF needle on the left side of the nose position, and if it remained there or increased its left deflection, the localizer would be intercepted before the airplane reached the marker. But if the needle moved back toward the nose position, the ADF would be telling the pilot that the localizer would not be intercepted before he reached the marker.

On an ADF approach (correctly called an NDB approach, after the non-directional beacon on the ground instead of the equipment used in the airplane), the ADF and the aircraft heading tell all. The principles outlined for interception and tracking are the keys to success. If the beacon is located on the airport, there's not too much of a chance to miss. With the beacon away from the airport, finding the airport is a matter of accurate tracking away from the station.

Using the ADF to tell distance to a station is another simple geometry problem. Turn the airplane until the ADF needle is 5 degrees ahead of either the left or right wing-tip position. Note the time and fly until the needle moves 10 degrees and is 5 degrees behind the chosen wing-tip position. Again, note the time and multiply the true airspeed (in knots) by the number of minutes it took for the ADF needle to move 10 degrees. Drop the last zero from the answer, and, presto, you have the approximate distance to the station in nautical miles. The answer is approximate, because it's based on true airspeed. If you knew the groundspeed, the number would be closer but still not exact, because nobody could really get the exact time for a 10-degree change on an ADF.

Using the ADF as a standby directional gyro is simple enough. When tuned to a station, the ADF needle turns whenever the airplane turns, so it can be used as a reference that is as reliable as a DG. In fact, you can fly an airplane on the gauges with nothing but an ADF plus one instrument that would indicate pitch attitude. This is something that should be practiced under the hood, for it could be a useful life preserver on a day when everything seems to go wrong at once.

A final benefit of the ADF is related to one of its shortcomings—static. Enough static makes it completely useless, but its ability to receive the static makes it a valuable indicator of thunderstorm activity in the area. Conversely, when storms are forecast but there's no static on the ADF, the pilot can fly with a bit more peace of mind.

Many pilots use the ADF for nothing more than homing on an occasional beacon and for listening to music, news, and sports, but they are missing a lot of available and valuable information and guidance.

③ RMI: EASY GUIDER
by John W. Olcott

ORIENTATION, the principle of knowing the relative position of an aircraft to a fixed location, is a fundamental part of every navigation problem. Particularly for instrument flying, knowing where the aircraft is, relative to the outer compass locator while being vectored for an ILS approach, relative to the VOR while entering a holding pattern, or relative to the appropriate radio facility during a missed approach, is the difference between positional awareness and potential confusion.

Unlike tracking—the technique of flying a straight-line course directly to or from a nondirectional beacon or an omni station—orientation involves knowing where a particular radio facility is relative to the aircraft's position and knowing what course to fly in order to reach that facility. Orientation procedures are not overly complicated, especially where an NDB is involved. An ADF automatically provides the relative bearing between the aircraft's nose and a direct line to a low-frequency radio facility. Adding the relative bearing to the aircraft's heading provides the magnetic bearing, of course, to that fixed location. The procedures for determining the bearing to or the radial from a VOR require a mere turning of the omni-bearing selector until the flight-path-deviation needle centers and then noting the OBS reading and the to/from flag.

When a pilot is bogged down with instrument chores, however, even these simple orientation procedures will add to his overall workload. The petty arithmetic required to determine a bearing to an NDB becomes an annoying exercise, particularly when a heading greater than 360 degrees results from adding the ADF's relative-bearing indication to the aircraft's heading. The constant twirling of the OBS knob and recentering of the needle makes omni orientation a time-and-attention absorber. Furthermore, the relative position between the aircraft and the VOR station is not as easily visualized as it is with an ADF type of presentation.

Presto, the RMI! The radio-magnetic indicator provides orientation information from both low-frequency and VOR stations, and it does so without imposing the mental gymnastics of an ADF or the knob twiddling and visualization problems of an omni.

An RMI is the superposition of a rotating azimuth, which is slaved to the aircraft's compass system, and an ADF-like needle, which points directly toward any selected low-frequency or omni facility. The relative bearing between the aircraft's nose and a direct line to the NDB or VOR is presented in a manner similar to that of an ADF indicator. The arithmetic, which is the *bête noire* of the ADF, is eliminated, because the azimuth card behind the needle coincides with the aircraft's heading. (The azimuth card is usually tied to the aircraft's compass system by an electrical servo and just repeats the heading information presented elsewhere on the instrument panel.) The azimuth reading to which the needle points is the sum of the relative bearing and the aircraft's heading; it automatically is the bearing to fly in order to reach the selected radio facility. Therefore, the head of the RMI needle not only gives the relative position of the airplane to the NDB or VOR but it also shows the bearing to the selected radio facility, directly and without any computation by the pilot. The tail of the needle is, naturally, 180 degrees opposed to its head; thus, the tail shows the radial from the station automatically.

RMIs provide an added dimension to normal VOR presentations by providing a clear picture of where the omni station is, relative to the aircraft. Regardless of the OBS setting, the VOR needle position or the to/from indication, the RMI needle head gives the bearing that will lead to the station, and its tail gives a continuous readout of what radial the plane is transversing. The automatic readout of omni bearings and radials is available on the new Collins Micro Line, but the ability to see where the station lies relative to the aircraft's nose is found only with an RMI.

Because RMIs work on both low-frequency and omni signals, they usually are outfitted with dual needles. Needle one has a point shaft similar to the familiar ADF pointer, and needle two has a dual shaft but only a single point at its head and tail positions. A four-pole switch associated with the RMI indicator allows the pilot to select which needle represents which radio. If the aircraft has dual ADFs and dual VORs, most RMI installations will allow matching any needle to any VOR or ADF at any time and according to the pilot's wishes. For example, during an ILS approach, the number-one needle could be selected for the outer marker while number two is linked to the VOR to be used for a missed approach; or perhaps for an omni approach, needle one would be coupled to the number-one VOR, while needle two was set to the number-one ADF. The correspondence between the needle and radio can be changed at will to suit the situation at hand.

The orientation capability of an RMI can be seen best in several IFR tasks. For example, a teardrop or parallel entry into a holding pattern can be confusing unless the pilot maintains his orientation relative to the radio facility. It helps to visualize the familiar racetrack on the face of the RMI, with the inbound leg of the track on the RMI face coinciding with the radial that defines the holding pattern. If ATC's instructions were to hold on the 060-degree radial or bearing outbound from the XYZ facility, mentally place the racetrack on the RMI dial face with the inbound leg lying between the center of the indicator and the 060-degree reading on the azimuth. If left-hand turns were specified, shape the ends of the racetrack accordingly. Then, with an RMI needle tuned to the holding facility, the orientation the aircraft will have when it enters the holding pattern is presented on the RMI face with the needle intersecting the pattern just as the aircraft will, regardless of the aircraft's heading.

Upon entering the hold, the RMI needle continues to present the relative position of the aircraft within the pattern. When it is initially flying outbound on either the teardrop or parallel entry, the plane is flying away from the holding fix, so it is convenient to think in terms of what radial the plane is on. The tail of the RMI needle gives the desired information, since it shows numerically what radial the plane is on and visually where that radial is with respect to the holding pattern. Once established in the pattern, it is still important and comforting for the pilot to use the tail of the RMI needle to monitor the plane's position in terms of radials, and to use the nose of the RMI needle to maintain the orientation of the aircraft relative to the holding fix.

On a missed approach or in any situation where the pilot must fly directly to a particular radio facility, an RMI provides exactly the heading information required and it does so automatically, with the relative position of the aircraft and radio facility clearly visible.

The navigational principles that give the RMI its usefulness—automatically adding the aircraft's heading to the relative bearing between the aircraft and radio facility, plus maintaining proper orientation by considering either bearings of radials—apply to garden-variety ADF indicators, although with somewhat less flexibility. Most ADFs made today have manually rotatable azimuths, so determining the magnetic bearing by superimposing the ADF needle on a duplicate of the directional-gyro face can be done manually. Provided a lot of heading changes are not required, the manually rotatable ADF face works well and can be quite useful for determining the proper procedures for entering holding patterns.

There are many times when it not possible or convenient to use a rotatable ADF azimuth, either because of frequent heading changes, because of pilot workload, or because the ADF indicator does not have that feature. However, it still is possible and most helpful to maintain a sense of orientation by determining what bearing or radial the airplane is on. The procedures are relatively simple, provided some basic rules are adhered to.

Whether the pilot works with bearings or radials will be strictly a matter of convenience. To keep the arithmetic simple, divide the ADF azimuth into four equal quadrants; zero to 90 degrees, 90 to 180 degrees, 180 to 270, and 270 to 360. Label mentally the first and third quadrants + and the second and fourth quadrants −. All ADF readings will be taken relative to either the zero-degree azimuth (12 o'clock) or the 180-degree azimuth (six o'clock) positions, so, in essence, they might be called "modified relative bearings." According to this scheme, a modified relative bearing will never be more than 90 degrees and will be considered a + MRB if the needle falls within the first or third quadrants and a − MRB if it falls within the second or fourth quadrants. If the head of the ADF needle lies in the top hemisphere (first and fourth quadrants), consider the orientation in terms of bearings; if it lies in the bottom hemisphere (second and third quadrants), consider the orientation in terms of radials.

Using this technique of quartering the face of the ADF indicator, it is quite easy to determine the aircraft's orientation in terms of what bearing or radial it is on, relative to the LF radio facility. First, note in which hemisphere the head of the needle lies. If, for example, it points to 150 degrees, it relates to a radial and lies in the minus quadrant, 30 degrees off the 180-degree or six o'clock position. The radial from the station, therefore, is the magnetic heading of the aircraft minus 30 degrees. Assuming an aircraft heading of 300 degrees, the radial from the facility is 270 degrees. That calculation is far easier than using the conventional method of adding the relative bearing of 150 to the heading of 300 degrees and subtracting 360 to obtain a usable magnetic bearing, particularly when it is considerably more effective to think in terms of radials when a station is behind the airplane. If the station lies ahead, the head of the ADF needle will lie in the first or fourth quadrants, it will relate to a bearing, and its modified relative bearing added to or subtracted from the airplane's magnetic heading will yield the bearing to the station.

Without an RMI, it is most important to do the simple arithmetic required for determining what bearing or radial the airplane is on, and then visualize the orientation for that bearing or radial with respect to the bearing or radial that defines the task at hand. On an ADF approach, for example, visualize all outbound radials as spokes of a wheel, with the approach radial being the most prominent. Radials that are larger (in degrees) than the approach radial lie to its right; smaller radials lie to its left. Determine what radial the aircraft is on and visualize where that radial is, relative to the approach. When flying inbound to the station, visualize where bearings lie with respect to each other and then relate the aircraft's present bearing to the desired bearing.

Particularly in strong wind conditions where a crab angle is required to maintain a desired track, the visualization procedure helps the pilot remain properly oriented. While it is always possible to determine whether a particular bearing or radial lies off the right or left wingtip by turning to that particular heading and noting where the ADF needle points (it points toward the desired bearing or radial), it is not always convenient to do so, and it does not allow a continuous monitoring of position, which is the essence of being oriented.

If the airplane has a full-face directional gyro, such as the type found on most newer instrument panels, there is a simple way to determine whether the pilot has calculated his bearings or radials correctly. Mentally superimpose the position of the ADF needle over the face of the directional gyro, thus creating, in effect, an imaginary RMI. The procedure may not be exact, but it has the character of an RMI, and it will protect the pilot against making a blunder in his ADF calculations.

While it is possible to mentally construct an RMI through the superposition of the directional gyro and the ADF, there is no complete substitute for the real thing. The ease with which an RMI displays the aircraft's position relative to LF and VOR stations encourages planning and reduces workload, two keys in the principle of orientation.

4 FLIGHT DIRECTORS
by Robert Blodget

THE PURPOSE OF a flight-director system is to make things easier for the pilot under instrument-flight conditions. It does this in two ways, with two instrument displays: it collects information from several instruments and presents it all on a single instrument, which reduces the pilot's job of scanning; and it is equipped to compute, as the pilot directs, the required flight path to make a chosen maneuver come out right. The installation consists of three elements installed on the panel: the computed display, which is part of the attitude gyro; the control panel; and a collected display of uncomputed data, which has as many names as there are companies making it. Of them all, perhaps "navigation-situation display" comes closest to describing what it does.

The flight director can do simple problems—the kind human pilots do easily but with some workload penalty, especially during stress periods—and it can also do complex computations, some of which are too hard for cockpit solutions.

On the simpler side, if a pilot is holding a heading of 270 degrees and wants to turn to 030 degrees, he must first compute which way to turn, which is easy. Then, depending on what sort of turn-indication equipment he has, he must do another computation to find which heading he should have when he begins to roll out so that he will end up precisely on 030 degrees.

With the flight director, he sets in the heading to which he wants to turn on the navigation-situation display, and then engages the heading function. The computer not only starts turning the right way, but will also do all the necessary figuring, so that if the pilot simply keeps the command indicator satisfied by matching the movement of the little "airplane" on his attitude gyro to that of the command indicator—he will roll out precisely on the heading selected.

The problem of intercepting and tracking a particular course to or from a VOR station is somewhat more difficult. Precise solutions are usually incompatible with cockpit facilities and workload. The flight director's computer takes such things in stride. The pilot sets in the radial (or course) he wants, tunes up the proper receiver, and feeds its signals into the flight director. The command indicator promptly tells him what to do. As long as the command is satisfied, the airplane will intercept the course and roll out on it precisely, even including whatever crosswind compensation is needed. The same goes for interception and tracking of localizers and glideslopes, plus other useful functions such as executing missed approaches.

Different makes of flight directors use different kinds of command displays, each of which takes some getting used to. Collins, Bendix, and Edo-Aire (formerly Mitchell) use what is called "single cue." In these, some sort of symbol moves both in pitch and roll, and the pilot maneuvers to keep the attitude-gyro reference symbol aligned with it.

The Collins system uses V-bars colored yellow to show command, and the pilot satisfies the commands by keeping an orange delta symbol tightly snugged into the inverted V of the bars. Edo-Aire's display is quite similar. Bendix uses two small yellow paddles at the circumference of the attitude display. The reference symbol for the airplane is very much like those in conventional artificial horizons, except that there are red dots at each "wingtip." The pilot simply strives to keep the red dots centered in the yellow paddles.

Sperry uses two cues—two crossed pointers, which look exactly like the needles of a conventional ILS indicator. The horizontal needle moves up and down, and thus commands pitch, and the vertical needle moves from side to side to command turns. The airplane reference symbol in the Sperry is indistinguishable from the one in a standard attitude gyro. In the Sperry system, the pilot's task is simply to center both needles.

Color is vitally important in flight-director displays. A lot of information is being presented in a small space, and without bright colors to separate the needles and their functions, a pilot can easily get confused. This is what has made black-and-white TV-tube displays so unsatisfactory, despite much experimentation with concept. When picture tubes do replace the present electromechanical displays, they will have to be in color. (Sperry, for one, has such a device in development.)

The computer-command system is a big help to the pilot, but it can't give him the whole story. He has no way of knowing, just by looking at the command display, how far the maneuver has progressed, because once the command is satisfied, it will show nothing different from the beginning of the maneuver to the time when the airplane is actually on the selected track, whatever it may be.

Here is where the navigation-situation display comes in. The heart of this is a directional gyro, slaved to magnetic North. Around its circumference are two bugs—one for heading, the other for selecting course. In the center is an airplane symbol, a to/from indicator, and a bar representing the selected track. When the bar is in line with the "fuselage" of the airplane symbol, the airplane is on track. Until then, the bar will show, by its displacement to one side or the other, how close the flight is to the desired track. So a glance at the navigation-situation display tells the pilot how he is doing.

This display is usually mounted directly below the horizon/attitude indicator. The information it displays—except for heading—is also available elsewhere on the panel, but not in a form that can be so quickly understood.

The control panel must also be placed close to the two displays, and this is more important than sometimes realized. The explanation involves the underlying philosophy of both pilots and flight directors. To understand this, let's look briefly at autopilots.

Modern full-service autopilots and flight directors have a great deal in common. In fact, it is fairly close to the truth to say that a flight director is the same as an autopilot without control servos. In flying with an autopilot, the human pilot sets the problem he wants done, activates the autopilot to do it, and then becomes the flight supervisor to watch and see that the autopilot does what it was told. When using a flight director, the pilot has a different role to play: he becomes both supervisor and servant. First, the pilot sets in the task he wants done, then he executes the commands given him by the flight director. His ability to function as a supervisor is limited, however, since his muscles are busy replacing the servos of an autopilot.

If the flight-director control panel has been put somewhere well outside the pilot's normal panel scan—and this is unfortunately quite common—he can set in a task and get so busy helping to execute it that he forgets what it was he told himself to do. There's a good chance that this is responsible for much of the trouble pilots have in learning to adapt to flight directors.

Flight directors are coming down in price, and improving in reliability, flexibility, and human-factors engineering. There are already several combinations of autopilots and flight directors; and since they both have many points and functions in common, a combination costs less than one of each separate system.

The autopilot has the advantage that it can release the human pilot from direct participation in any phase of the flight so that he can do the other chores connected with it. It has the companion disadvantage of taking the human pilot out of the chain of control, and if anything goes wrong, it will take some time and considerable skill to resume personal control. This is especially important during instrument approaches, or in rare cases of catastrophic failure of the autopilot system.

With a flight director, the pilot is always an integral link in the control chain. If failures occur, he at least can take over at once.

The ideal equipment choice is a combination autopilot and flight director, so that one can use the autopilot during noncritical phases of the flight, and switch to the flight director during approaches and other maneuvers in which it is desirable to have the pilot exerting not only command but control. For general-aviation operations, since a choice usually has to be made between one or the other, the autopilot is preferred for single-pilot work, because it allows the pilot to be relieved of airplane management during much of every flight. It can also be used to set up the approach, with the pilot taking over manually at the final approach fix inbound. Properly understood and used, however, flight directors make easy jobs out of hard work. They are still expensive, but if you do a lot of instrument flying, they soon pay for themselves.

5 DECODING ENCODING ALTIMETERS

by John W. Olcott

EVERYONE WHO flies IFR and many who fly only VFR should consider encoding altimeters. The reasons go beyond the FAA regulations that require encoders for all flights (VFR as well as IFR) that enter Group I terminal-control areas or fly in controlled air-space above 12,500 feet msl, except when 12,500 feet msl is less than 2,500 feet from the surface. The reasons certainly go beyond an appreciation of the emerging technology that enables flight data to be transmitted via transponder energy pulses to ground-based, narrow-band radar receivers used by air traffic control. The rationale for a serious look at altimeters stems from the fundamental importance of accurate altitude information.

Altitude is one of the factors upon which air safety is based. An accurate knowledge of altitude is necessary for obstacle clearance, naturally. A consistent measurement of altitude is the basis for air traffic separation and control. Precise determination of altitude is needed for efficient utilization of the airspace.

Yet pilots accept altitude information with a nonchalance that borders on contempt. The altimeter was one of the first instruments we saw as student pilots. It was always there, even on the crudest instrument panel, and it always worked. An altimeter could never be as exciting as a horizontal-situation indicator or a flight director. Isn't it only a barometer whose aneroid sensing element expands or contracts in relation to the pressure of the ambient air? The pointers on its face just revolve slowly and display altitude, as the hands of a kitchen clock indicate time, or so we thought.

The altimeter often is taken for granted, although it is truly a unique instrument, far more precise in certain areas than the finest Swiss watch. Manufacturing an altimeter requires a dedication to perfection that is awesome. The linear movement of an aneroid capsule, which is the brain of an altimeter, is translated into rotational motion by means of a rocking shaft arrangement and then multiplied nearly a million times by means of a perfectly shaped and matched gear train to achieve accurate movement of the altimeter's altitude display. The gear train also is coupled to the mechanism that senses altitude in encoding altimeters. The smallest error in shaping the teeth of the first gears used to convert aneroid motion into display motion can cause a gross distortion in the hundreds-foot pointer and produce incorrect altitude data.

The aneroid itself is made from two thin, corrugated disks, about 2½ inches in diameter, which look like miniature cereal dishes. These are joined with an airtight seal and then evacuated. Some manufacturers join the disks by electron-beam welding, thus creating an aneroid capsule automatically, since the welding process requires that the work be done in a vacuum. Older construction techniques involve soldering the disks together and then creating the internal vacuum. With either method, the result is the same: the thickness of the aneroid capsule depends upon the atmospheric pressure that is trying to force the two disks together. Temperature-compensating bimetal strips are applied either directly to the aneroid capsule or to the linkages between the capsule and the gear train to assure that the aneroid expands according to a prescribed relationship between the aneroid's thickness and the ambient pressure.

In altimeters that do not use servo motors to help the movement of the gear train and indicator, the aneroid is the instrument's muscle as well as its brain. Most mechanical altimeters use two aneroid capsules linked together, because their individual force outputs are limited. They also use very lightweight moving parts to reduce inertia effects that would inhibit a quick response to altitude changes. For example, the rotating drum on Bendix's −1,000-to-50,000-foot-range Model 3252010 altimeter is made from metal four-thousandths of an inch thick, and Bendix engineers are considering using material one-thousandth of an inch thick on future models. Like many other altimeters, this instrument incorporates an internal vibrator to minimize the possibility of friction or inertia causing anything to stick. Servo-type altimeters, such as Cessna's 400 Series opti-servoed encoding altimeter, incorporate a servo motor that senses the movement of the aneroid and provides the torque to rotate the drums, pointers, and encoder components that are driven by the altimeter's gear train. Thus, in a servo-type instrument, the need for low inertia and minimum friction of internal components is less critical, but precision manfacture and assembly techniques still must be maintained. Also, loss of power to the servo constitutes a failure of the altimeter.

An encoding altimeter converts the mechanical movement of a conventional altimeter into an electrical signal that can be transmitted, in 100-foot increments, as part of the normal signals pulsed out by a 4,096-code transponder with Mode C capability. The process of sensing the rotation of the altimeter's gear train requires the use of electrical pickoffs, and additions of any devices to the gear-train mechanism add unwanted inertia and possibly friction to the altimeter. With servo-altimeters, brush-type electrical sensors that determine gear-train rotation can be used because the friction caused by the brushes can be canceled by the torque of the servo motor. Problems associated with brush wear and mechanical contacts, plus the drag of the brushes, have caused most manufacturers to use optical means to convert the mechanical motion of an altimeter into an electrical signal.

Optical encoders use a low-inertia risk that is attached to the main shaft of the altimeter in such a position that, as the altitude changes, the disk rotates between several light-emitting diodes and a set of light-sensitive transducers. Cutouts in the disk allow the light from the LEDs either to contact or not to contact the transducers, thereby triggering them either to send or not to send an electrical signal to the transponder. In the same way that on/off electrical signals form the basis for all digital-computer functions, the encoder mechanism converts mechanical altitude information into electrical altitude information, using a binary representation of the numbers zero to nine known as the "gray code" in computer circles and the Gilaham Code among encoder manufacturers.

There are no mechanical contacts between the altimeter's gear train and the optical encoder's electrical signal-conditioning elements, thus there is no friction to interfere with the normal functioning of the altimeter. If the encoder mechanism fails, the altimeter's operation is unaffected.

Blind encoders work on the same principle as encoders that are an integral part of conventional altimeters. However, they are separate units connected neither physically nor electrically to the aircraft's altimeter. A blind encoder has its own aneroid and gear train, and it converts mechanical representation of altitude into electrical signals by means of contact brushes or optics; the usual method of conversion is optical. Because they do not have the pointer or drum presentation of an altimeter, blind encoders are usually less expensive than encoding altimeters. As self-contained, separate units, they can be installed anywhere in the aircraft, although they must be connected to the aircraft's static source, preferably the same static source the altimeter uses. Blind encoders, as well as the encoder portions of encoding altimeters, must meet the specifications of TSO C88.

The FAA requires that the altitude transmitted by an encoder be within 125 feet of the pilot's altimeter referenced to 29.92 inches Hg (that is, pressure altitude). The criterion is achieved automatically with an encoding altimeter since the encoder and the altimeter are combined. With a blind encoder, the correspondence must be accomplished by adjusting the calibration of the blind encoder to match the calibration of the cockpit altimeter. Blind encoders are built with a means for such matching. Some units can do so using only a few bias points, while others use more adjustments and are able to match more closely the desired calibration characteristics of an altimeter. The matching process must be done at the time of installation, and it should be checked after any servicing of either the cockpit altimeter or the blind encoder. Leigh Systems satisfies the requirement of the FAA's 125-foot-agreement rule by offering a digital altitude-reporting module, which is a blind encoder with a cockpit-mounted LED readout for displaying the same 100-foot altitude increments that are sent to the transponder.

Regardless of what altimeter setting appears in the altimeter's baroset window, all encoding altimeters and blind encoders provide altitude data referenced to pressure altitude. Thus, the LED display of the Leigh altitude-reporting module would differ from the cockpit altimeter by 100 feet for each tenth-of-an-inch difference between the actual altimeter setting and 29.92 inches Hg; the cockpit altimeter would read higher than the Leigh display for an altimeter setting higher than 29.92, and lower for a setting lower than standard.

Provided an aircraft's transponder is functioning properly on Mode C or A/C, air traffic control can send a radar signal to that aircraft, and, without any action by the pilot, the transponder will send back to the ATC radar receiver a pulse of energy that contains the code describing the aircraft's altitude, computed to the nearest 100-foot increment. The transponder does not transmit any altitude data until stimulated to so by a Mode C pulse from a ground station. (Modes, incidentally, denote the spacing between electronic pulses. Mode C has a pulse pair spacing of 21 microseconds, while Mode A's spacing is eight microseconds. An airborne transponder differentiates among modes by their pulse spacing.) When the transponder equipped with an encoding altimeter receives a Mode C interrogation it replies by sending a series of one-half-microsecond pulses spaced in a pulse group that can be repeated at a rate of about 450 per second. The ground-based radar receiver used by ATC accepts these bursts of data and feeds them into a computer, where an altitude correction based upon the local altimeter setting is applied. The data are then converted to an alphanumeric (letter/number) display that includes the aircraft's assigned and actual altitudes, N or flight number, and an arrow that indicates whether the aircraft is climbing or descending. This data is stored in the computer, from which it can be called up for display on a controller's radar screen. Computer groundspeed also can be displayed. Because of this store-and-call-up feature, controllers can effect handoffs between ATC facilities more rapidly.

The altitude shown on ATC's radarscope should be exactly the same as the altitude presented on the pilot's altimeter, but the 125-foot allowable difference between encoder and observed altitude, plus small differences in the altimeter settings used by the ATC computer and the pilot, probably will produce some discrepancies. A total difference of 250 feet is considered tolerable for the entire system. If the difference between reported and observed altitude reaches 300 feet, however, the controller will request clarification from the pilot, and if the difference is not rationalized or corrected, the controller will consider the altitude-reporting function of the transponder inoperative.

Automatic altitude reporting was conceived and implemented to serve the needs of air traffic control. The initial planning for an ATC system utilizing transponders and altitude encoding was accomplished by the President's Air Coordinating Committee in 1953 when it established the operational requirement for the Air Traffic Control Radar Beacon System. This requirement, known by the rather unseemly acronym Atcrbs, was modified over the next five years to accommodate both civil and military hardware constraints and, by 1958, was approved and published as an international standard by the International Civil Aviation Organization. Automatic altitude reporting was envisioned as a more accurate means for ATC to determine altitude than the traditional policy of having a pilot report his altitude. Another objective was to increase the efficiency of the ATC system by reducing voice communications.

The relative safety of automatic versus pilot reporting of altitude will be debated for some years to come, and we have yet to observe much reduction in voice communications between pilots and controllers, particularly since each controller asks for a verbal verification of the automatic altitude readout seen on the scope. The benefits of encoding altimeters probably are more obvious to controllers because of the automatic features they and the ARTS III radar system that interrogates encoders provide, such as easier handoffs between controllers, the ability to monitor particular altitudes, and computed information about each target. Automatic altitude reporting undoubtedly gives controllers additional peace of mind.

Encoding altimeters do nothing, however, to increase the precision with which altitude is measured. Errors in measuring altitude are due to many factors, such as the calibration of the instrument, inaccurate measurement of static pressure (particularly at high Mach numbers and during rapid changes in altitude), and limitations of the basic law of physics upon which altimeter design is predicated.

This last type of error occurs because all altimeters assume a variation of pressure with altitude that exists only in the rarely encountered standard atmosphere. This error has no consequence on ATC, since it affects all altimeters equally. Calibration errors, however, vary from instrument to instrument, and they tend to be a percentage of indicated altitude, so this type of inaccuracy increases with altitude. The FAA recognizes this fact by stipulating, in TSO C10b, that the allowable tolerance for altimeter accuracy can vary from the perfect reading by + 20 feet at sea level, + 40 feet at 6,000 feet, + 80 feet at 10,000 feet, and + 200 feet at 35,000 feet. Errors due to measuring static pressure only affect aircraft that fly high and fast, and they must be determined through flight tests and calibrations, but these errors typically are in the 75-to-150-foot range at a Mach number of 0.9 and an altitude of 30,000 feet. Because all the errors inherent in an altimeter result in large inaccuracies at high altitude, aircraft flying above Flight Level 290 are separated by 2,000 feet rather than by the 1,000-foot stratification used at lower altitudes. More precise altitude determination, with greater consistency among all aircraft operating at adjacent flight levels, probably will be required to achieve a more efficient utilization of altitude for aircraft separation.

In addition to serving an ATC role, encoding altimeters soon will play a part in monitoring obstacle-clearance altitude. Between January and August 1976, all 64 automated radar-terminal systems will be programmed to provide a minimum-safe-altitude warning. When an encoder-equipped aircraft descends below a predetermined safe altitude for its sector of the approach, or deviates from its course during an instrument approach, the radar computer automatically triggers a visual and audio warning seen by the controller. There are 69 new ARTS III sites planned for the near future, and presumably these also will have the minimum-safe-altitude warning feature, which is made possible by programming the ARTS III computer system to detect an abnormal altitude situation. One wonders how long it might be before ARTS III-type equipment could be programmed to detect potential midair collisions.

Even without computerized warning features, it is nice to know that another pair of eyes—the controller's—is watching your altitude and keeping you honest. The advisories he is able to give regarding other aircraft, and presumably terrain, take on added meaning when he has altitude as well as azimuth and distance information. An obvious word of caution, however: automatic altitude warning is great; so is that extra pair of eyes that can monitor your altitude. But remember, controller involvement can never be as great as pilot involvement. It's your rump, so monitor your own altitude and use the controller as a welcome backup.

Now that the technology to transmit flight data, such as altitude, to ground stations is well established, what will the FAA do next? Future plans call for a greater use of data links between the air and ground, possibly with greater monitoring of flights by ground controllers. The FAA has been relatively easygoing about accommodating non-encoder-equipped aircraft in airspace where automatic altitude reporting is required, but the handwriting is on the wall. We probably will see greater pressure to enforce and expand the use of encoding altimeters in the future.

⑥ RNAV: BY GOD, IT WORKS!
by Stephan Wilkinson

I RESISTED the gadgety blandishments of RNAV as long as I could, even though *Flying* has area-navigation units installed in several of its airplanes. I figured it to be some strange, semi-useless creation birthed by avionics marketing men with little else to do—a collection of circuitry that was probably more trouble than it was worth. After all, I thought, who needs to fly as the crow does when the Good Lord gave us hundreds of VOR stations, so populous an assortment that any two places in the country can be connected by straight lines? That the straight lines often resembled the last tango never really bothered me. Besides, how in the federally regulated world would ATC ever handle hundreds of independent-minded pilots flying RNAV-direct from Hither to Yon?

If that wasn't enough, I had already glimpsed some of the special RNAV charts that one apparently needed—flimsy sheets with out-of-scale "schematics" of routes, doubtless requiring updating every other Tuesday, that told me exactly how to get from LAX to JAX or BUF to HUF but did nothing for my quest for an airway from Provincetown to Poughkeepsie. Forget it; I'll right-oblique, left-oblique from beacon to beacon, and if I don't get there before you, I'll at least get there with a lot less trouble.

I have been wronger in my life, but not much.

RNAV crept up on me slowly, taking me unawares. I began to use it as an airport finder—a useful enough function, but hardly a necessary one. After all, I had never failed to find an airport in the days before RNAV, even if I sometimes temporarily misplaced one or two and milled about in the haze doing airwork until I got reoriented. With the new RNAV set installed in our Cherokee Six, however, I would faithfully snake my way from VOR to VOR en route, then set up an RNAV waypoint atop the destination airport and fly the last leg of the trip direct to the waypoint. When the DME read two or three miles to go and the VOR needle was centered, I'd look over the nose and there, like magic, would be the runway. It never failed, though it took me awhile to learn that I had to start looking for the airport a bit before I got there: the RNAV was so accurate that if I got everything centered and waited until the DME went to zero, I'd be more lost than ever because the only way I'd then be able to see the strip would be through a belly turret.

General-aviation RNAV setups have four basic components: an ordinary nav reciever, a standard VOR course-deviation indicator, a DME, and the actual course-line computer. With the computer disengaged from the rest of the nav system, the nav radio, CDI, and DME function normally, in the fashion we have unwittingly grown to know and love: they are most useful only when going directly to or from a VOR station or vortac. Let's engage the computer, however, and find an airport with it. (If you can do that, you have mastered the intricacies of the unit and you have almost all the RNAV instruction you'll ever need.)

Flying Associate Editor Thomas H. Block runs a 100-acre farm in western Pennsylvania when he isn't flying for Allegheny Airlines, and he had had the good grace to set up housekeeping a two-minute drive from a little semiprivate field called Grove City. Grove City has no airplanes, no obvious hangars, and seemingly camouflaged runways. The only easy way to find it without RNAV is to hope you come close enough to spot the drive-in movie just north of the field.

Grove City turns out, however, to be exactly 27.2 nautical miles west of the Clarion Vortac, on the 274-degree radial. The RNAV computer has two settings, marked "distance" and "bearing," and you set the appropriate numbers—27.2 and 274—in the respective little windows. Set the nav radio to 112.9—the Clarion Vortac frequency—and turn on the RNAV computer. The CDI flags and needle will give a tentative jump, then go dead as the computer digests the information being fed to it. After 10 or 20 seconds, convinced that the Clarion Vortac has suddenly been relocated to the Grove City Airport, the VOR indicator and DME will come alive and begin to give to/from, left/right information to this "phantom vortac." Wherever you are at the time—north or sourth, east or west of Grove City, as long as you're within reception distance of the *actual* Clarion Vortac—you just twist the omni-bearing selector until the VOR needle centers with a "to" indication and fly that course to the airport.

As you do so, the DME mileage readout will click off the miles to the imaginary beacon that the RNAV computer has created. (Actually, in the case of Clarion, we usually set the waypoint, as these RNAV creations are more properly called, directly over Block's old stone farmhouse, to give him enough of a rafter rattler to get him to drive over and pick us up, but we won't give you those coordinates.)

That, basically, is how RNAV is operated: measure the distance and bearing from the nearest convenient vortac to the point where you want to go, feed those figures into the RNAV computer, and your VOR and DME will operate just as though the vortac has physically been moved to the waypoint your RNAV computer has thus created. With two exceptions, one of them often useful: first, the DME *groundspeed* readout will continue to compute based on the location of the actual vortac, which means the speed indication will be inaccurate unless the waypoint is on a direct line between (or beyond) you and the vortac; second, the VOR indicator needle will indicate your deviations left or right of course in "linear" rather than "angular" fashion. What this means is that in the RNAV en-route mode, a two-dot deflection of the needle always means the same thing—that you're two miles left or right of course— no matter whether you're 8 or 80 miles from the waypoint. Operating in the normal VOR mode, of course, the indicator gets less sensitive the farther it is from the station, and a two-dot deflection means 1 mile off course when you're 15 miles out and 4 miles off when you're 60 miles from the VOR.

I had long felt that flight-planning using Captain Jeppesen's en-route charts was about as simple as anything deserved to get, since the airway mileages between VORs were premeasured and there always seemed to be enough VORs around to create a semblance of a direct course. But eventually, even the business of adding up all those little mileage hexagons plus maybe a few nonairway legs that needed measuring against the scale began to get to me. One fine day, sitting in Chatham, Massachusetts with a pressing need to get to Lancaster, Pennsylvania, I hesitantly picked up a long straightedge, butted the number 24 and 25 Jepp charts against each other, and drew a dead-straight pencil line between Chatham and Lancaster.

I had lost my RNAV virginity. I would never be the same again. I had suddenly rediscovered what pilots knew back in the good old days: that is it possible to scribe a line on a map and track along it, free of the leapfrog tyranny of VORs and victor airways. Instead of the long, looping course that would bend me well clear of Otis AFB, then to Providence and Hartford and Carmel and around New York City, then southwest into a welter of beacons and doglegging airways, none of which seemed to go quite where I wanted to go, I had reverted to the student pilot's pencil-path mobility, back to the days when airplanes flew straight and true and were flying the shortest distance between two points.

There was a practical benefit as well, even in an area as VOR-pimpled as the Eastern Megalopolis. My RNAV round trip turned out to be 580 nm; going to Lancaster from VOR to VOR in the most direct possible fashion would have added 30 miles to the round trip, and going by way of the most convenient precomputed airways would have made it 40 miles longer (an extra 20 minutes in the Cherokee). But the numbers don't tell the real story, even though one could probably postulate sample trips from Truth or Consequences, New Mexico to Trial and Error, Arkansas and show that in less populated areas of the country, RNAV-direct navigation saves enough fuel to pay for itself in two years, or whatever.

The real beauty of RNAV is that it puts the pilot back in control. Strange as it may sound, RNAV is in many ways a wonderfully *human* system of navigation: you go where you want to go, ATC allowing, and you're suddenly, in one pencil stroke, loosed from the bounds of those tracks in the sky—tracks that are invisible yet in many ways as solid as the 19th century's iron rails. Ever since that day in Chatham, the chicken tracks on the charts have seemed so limiting, and the ease of drawing that straight line so liberating.

The greatest pleasure, though, is using an RNAV-direct route on an IFR flight plan—both because it's easy and because it occasionally gives you a chance to exert just a little of what authority is left you as a pilot. It requires that your RNAV installation be specifically and individually flight-tested and FAA-approved for IFR en-route and approach operations, but once it has been, your only remaining problem will be convincing the people who run the ATC system that you're serious. (This is less of a problem in other parts of the country than it is in the Northeast, where the profusion of established airways makes RNAV neither as common nor as useful as it is elsewhere.)

The mechanics of the filing process may take a few more words— occasionally less—than filing a sequence of airways, but it's simple nonetheless: draw the straight line you want to fly and establish waypoints along it at least every 80 nautical miles; you can use as many waypoints as you wish while you're en route, but you're only required to file one every 80 miles. File the route by giving the names of the vortacs that you're using to form the necessary waypoints, and the bearing and distance of the waypoints from each of these stations: "That'll be RNAV direct Norwich 190, 12 miles, Hartford 211, 15 miles direct, destination Stewart . . ." or, in written form, "RNAV D ORW 190/12 HFD 211/15 D." Under "type of equipment," you'll no longer write the familiar slash alfa or uniform but foxtrot—the abbreviation that tells ATC you have RNAV equipment aboard.

(There are hundreds of established RNAV-direct jet routes, already charted with specified waypoints, and to fly these, you simply file the name of the route—J-843R, for example, which goes straight from New York to Dallas. There are still only a half dozen equivalent low-altitude routes, however, all of them in the Far West, so unless you're flying between the nine airports they serve, you're on your own—which is what RNAV is all about, after all.)

Sometimes, ATC doesn't take too kindly to this concept within the low-altitude system. You'll occasionally find the FSS annoyed by any attempt to step outside the system, for apparently, they haven't yet had much experience in some locations with low-altitude RNAV IFRs. When I recently tried to file an RNAV-direct clearance via waypoints off the Riverhead and Providence

Vortacs, the FSS crossed out my clearance and said, "Let's just make that direct Riverhead, direct Providence." Another filing, this time by telephone, involved lengthy explanations of the fact that no, I did not want to go straight to the Hartford Vortac and then to a spot 15 miles south but direct to the HFD 180/15 waypoint. On the same flight, when I was handed over from one Boston Center sector to another, near Hartford, the new controller came on with his best it's-been-a-long-Sunday inflection, hysteria hiding in the back of his voice, and said, "Fifteen Romeo, do you realize you're *15 miles south of Hartford???*" He'd had his problems with little airplanes that weren't able to stay on the centerline of the airway, but to miss the vortac by 15 nautical was a bit much even for him.

RNAV has a few minor disadvantages, and one of them is that because the computer requires a greater number of programming steps than does an ordinary VOR receiver, there's a greater possibility of procedural error. The first time I filed RNAV IFR—it was a clear day, luckily, and I was doing it for practice rather than out of necessity—I climbed in the general direction of my first waypoint, but the course that the VOR indicator was commanding was a good 80 degrees left of what I knew it should have been. It took me several minutes to realize that I'd made the most elementary of RNAV errors, which was to leave the course-line computer mode selector in the normal VOR position rather than switched to "RNAV en route." The indicator was consequently taking me straight to the station rather than its waypoint.

With "RNAV en route" finally selected, however, the indicator went dead, and I blundered on, thinking that perhaps the RNAV mode required more altitude or less distance for adequate reception. Five minutes later, it dawned on me that I had drawn my waypoint using an ordinary VOR instead of a vortac, and by the time I'd measured out a valid waypoint, I was a good 15 miles off course.

Some pilots eliminate the possibility of making the first of those mistakes by always leaving the computer switched to the RNAV mode, even if they're flying between vortacs rather than waypoints. (The "waypoint" that the computer tracks to or from will be the vortac itself, as long as the waypoint distance selected is zero.) Leaving the computer on the RNAV mode has an added advantage. By the time the signal fed to the VOR indicator has passed through the course-line computer (as it must with the mode selector in the "RNAV en route" position), it has been damped so heavily that all but the most extreme of the vortac's transmission vagaries have been removed from the signal, and the needle usually stays rock-steady—even steady enough to be used by nav-tracking autopilot with virtually no wander or course scalloping.

Also, it makes the left/right needle read in linear terms, so that you know your position in relation to the airway—one mile left or right per dot of needle deflection—no matter how far from the vortac you are. If you want to be even sharper than that—as you would if you were making an RNAV approach, establishing a holding fix, or finding a small airport under poor-visibility conditions—there's an "RNAV approach" setting on the computer that quadruples the sensitivity of the needle, so that each dot of deflection is equal to a quarter of a mile off course.

One more "disadvantage" of which you must be aware with RNAV is that you no longer have the freedom from prohibited and restricted areas that airways automatically provide, and, of course, that specified terrain clearance altitudes are not necessarily valid for RNAV-direct routes in the general area of the established airways. The latter need only be a consideration in truly mountainous areas or for RNAV flights at very low levels, but the freedom that RNAV provides does carry with it a bit of increased responsibility for one's own welfare.

What about all those pesky RNAV charts you have to carry? Actually, you don't have to carry any at all; the standard low-altitude Jepps or NOS charts work fine (and are, in fact, what you should use for IFR work). You could even use sectionals, though they'd get mighty cumbersome on serious cross-county trips. Still, Jeppesen does issue an astoundingly convenient set of VFR RNAV en-route charts, only 11 of which are needed to cover the entire conventional U.S. They come in their own plastic pouch, with a plotter, and cost $28.25 ($13 a year thereafter for revisions) from Jeppesen, 8025 East 40th Avenue, Denver, Colorado 80207.

The advantages of using them are two-fold: since one RNAV en-route covers three times the area of a standard low-altitude chart, it's the rare trip that entails unfolding more than one of them, and flight-planning can be done in your lap rather than on the kitchen table or most of the FBO's counter. Also, the charts are unencumbered by victor airways but show every vortac with its 90-, 180-, 270-, and 360-degree radials extended 40 miles out from the station, with a mark every 10 miles. Waypoint plotting is simply a matter of determining where your course crosses convenient vortac cardinals and counting off the 10-mile ticks.

For serious IFR RNAV, you'll also want Jeppesen's RNAV approach plates, SIDs, and STARs, and you may be surprised to discover that there are already over 200 approved RNAV approaches in the U.S.—some at fields you may have been flying out of for years without being aware of them. The RNAV airport service costs another $21, and $15 a year thereafter for updating.

RNAV has many little benefits and uses that you'll discover as you go along, but one of the biggest and most obvious is that you won't be where all the traffic is. When you're flying on airways—unless it's solid IFR—there's always the risk that somebody else who's just as good at staying on the centerline as you are will climb or descend through your altitude, and probably nowhere except in a traffic pattern is your degree of exposure as great as it is directly over a VOR, where all the traffic converges. On an RNAV flight, you have created your own private airway, and you may never go near a VOR anywhere on your flight. In fact, if you want to be compulsive about it, you can RNAV an indirect route that specifically avoids VORs.

There are other, more realistic ways in which you can plan as short a route as possible without compromising safety considerations. You can lay out overwater flights to stay within gliding distance of land, for example, without having to make gross deviations from VOR to VOR to accomplish the same thing. A supercautious night pilot could plan an RNAV route with waypoints atop lighted airports all along the way that might create an airwaylike zigzag, but in the event of trouble, all the pilot would have to do would be to look at the course indicator and DME for directions and distance to safety.

The extra-added-attraction uses of RNAV are limited only by one's ingenuity, but many of them involve arbitrarily setting up waypoints at fixes that may officially be determined by DMEs, outer markers, NDBs, or crossing radials, and using the RNAV waypoints as an easily interpreted backup for the raw data coming from the ADF or number-two VOR. I recently found myself trying to hold at a locator outer marker, for example, with an inoperative ADF but a functioning marker-beacon receiver. The area of marker-beacon reception at my altitude (8,000 feet) was broad enough that my holding pattern was more of an ostrich egg than a racetrack, and the affair was complicated by the presence of another outer marker, for a nearby airport, that kept giving me false indications of holding-fix passage. Setting up a waypoint at the outer marker, which was at a clearly specified distance and bearing from a nearby vortac, as the terminal route on the approach plate spelled out, solved all the problems. I was also able to use the needle displacement of the VOR indicator on my outbound leg to show my actual distance in miles from the inbound course, and this plus the relevant DME information made it the next best thing to a horizontal-situation indicator.

RNAV can also be used as an easily interpreted backup during an ADF approach; set up a waypoint at the NDB location and crank the final-approach course into the VOR indicator, which automatically makes an ADF approach into a kind of VOR/DME approach. (You must regard the ADF as the primary information source, however, for the approach to be legal. In other words, if the ADF says you're left of course and the RNAV says right on, you *must* believe the ADF.)

RNAV also provides an easy way to deal with radar vectors—that iniquitous ATC procedure that starts out seeming so simple but often ends up turning you every way but loose, leaving you wondering if the airport is ahead, behind, or over in the next county. When a controller turns you to a new heading "for radar vectors" to an intersection, outer marker, or initial approach fix, and all the VOR needles begin to swing lazily over to their pegs, you simply set up a waypoint there and tell him you'll go RNAV direct. Unless he's vectoring you for traffic separation, he'll be happy to have you do your own navigating—and so will you, since the VOR indicator and DME will give you a constant indication of your position rather than the vague I-hope-he-knows-what-he's-doing feeling one gets from vectors. ("Do you think he's forgotten us? Could the airport really be at three o'clock if the outer marker's at nine?")

RNAV advertisements often stress the cold, practical benefits of the device—the time and fuel savings, the ATC benefits, the part that area nav will play in the traffic system of the 1980s—without ever saying just how much fun it is to use the gear. Or they explain the working of the course-line computer in such detail that one gets lost amid a welter of waypoints and phantom vortacs and never learns just how supersimple the whole system is. This is a shame, for there's one overused little ad slogan that should have been applied to the still-neglected but very useful course-line computer: try it—you'll like it.

7 AUTOPILOTS
by John W. Olcott

AIRPLANES ARE too difficult to fly. They demand a high degree of pilot proficiency to control them precisely, particularly under IFR approach and terminal-area operations, where the work load is high, the need to maneuver is great, and nearly constant attention is required to maintain heading, altitude, and airspeed. Trim changes associated with various gear, flap, and power configurations further add to the problem of stability and control. The difficult aspects of flight are so commonplace that pilots accept the required high level of control compensation as an inherent characteristic of flight. They practice, they adapt, and eventually, they become proficient, but the fact remains that the task of managing an aircraft as a transportation device while controlling it as a dynamic system can saturate the less-proficient pilot, and occasionally can tax even the skilled professional. Aircraft are sometimes just difficult to fly—period.

In this era of deteriorating public transportation and necessary energy conservation, business aircraft also are the most effective way to use a gallon of fuel and an hour of time for travel. The inherent utility in a general-aviation aircraft is a basic reason why the industry did not fall apart during the fuel crisis and why it continues to be strong in these present times of economic uncertainty. When times are tough and market resistance is high, the needs to reach customers promptly and to analyze problem areas quickly are greater than during good times. Businessmen are thus relying more and more on business aircraft, articularly IFR-equipped, sophisticated singles and light twins, to satisfy their travel needs.

A potential conflict arises from the difficulty of flight and the inherent utility of general-aviation aircraft. In order to extract the value that an owner anticipates and pays for dearly, his aircraft must be used on demand, which essentially means it must be flown IFR. Customers and business problems do not understand weather delays; neither does the comptroller who signs the big fat check for a company aircraft. Even for the most conscientious pilot, the pressure is great to use the plane, although that may mean shooting a 200-and-half approach at the end of a long day, whether or not the pilot may be a bit rusty.

Businessmen pilots are a singular breed. They must be proficient pilots and sharp businessmen, and both these jobs are challenging. It may be heresy to say it, but occasionally the businessman pilot lapses into being more businessman and less pilot; decisions sometimes are based upon his need to be somewhere at a particular time rather than his proven ability to conduct the flight well within the limits of current experience and proficiency. Too frequently, not enough time exists in a busy executive's schedule to obtain proficiency checks or to maintain a sharp edge of piloting skill.

Hence the need for aircraft that are easier to fly. The work load associated with terminal-area IFR operations, particularly just after an instrument takeoff and during a low precision approach, points out the need for the "ideal" aircraft that a pilot can manage comfortably without exerting considerable concentration and super proficiency.

When pilots contemplate the perfect airplane, two items high on their list of essential characteristics are solid stability and responsive control. This is because stability and control are the principal ingredients in a pilot's ability to utilize an aircraft effectively. Constant attention to the mundane tasks of holding heading and altitude increases a pilot's workload and distracts him from more important management functions that are necessary for a safely and efficiently conducted flight. Poor stability characteristics and uncomfortable handling qualities are annoying, fatiguing, and potentially hazardous, particularly when the air is rough or ATC demands that the pilot absorb a complicated amended clearance. Being able to concentrate on several challenging tasks at the same time may be the mark of the hairy-chested ace, and all IFR pilots who can fly the gauges, tune the radios, and satisfy ATC while flying solo can also be proud of doing their dance through the ceremonial ring of fire. Still, the need to cope with the aircraft's lack of stability while handling these other tasks at best compromises the ease and convenience of flight and, under unfortunate circumstances, can lead to some bad experiences.

Like rubbing your stomach while patting your head, mastering an awkward procedure may offer some personal satisfaction, but is not necessarily the best use of one's talents. General aviation, hopefully, has outgrown the infantile stage where ego satisfaction is its own *raison d'être*. The industry has emerged into an era where the transportation utility and energy efficiency of general-aviation aircraft are responsible for its acceptance and growth. The trend will be towards aircraft that are easier to fly, because excellent stability and control are critical to the ease, safety, and efficiency of flight. The ultimate is a vehicle that will steadfastly maintain any desired flight attitude without requiring constant attention by the pilot, yet will obediently respond when he commands it to maneuver. The tendencies to slowly diverge into a descending spiral or to wander from the chosen airspeed, heading, and altitude are absent in the hypothetical aircraft of the future. Fortunately for those of us who must fly in the present, these disturbing flight characteristics also are absent in those of today's aircraft that are equipped with an autopilot.

The analogy between ideal future airplanes and airplanes with autopilots is far more direct that may be immediately apparent. It is easy to speculate how the ultimate aircraft should handle, but for several reasons it is far more difficult, if not impractical, to design all the desired stability and control characteristics into the basic airplane. To start, there is a fundamental conflict between the requirement for solid stability and the desire for response control. Stability is the natural tendency of a plane to return to and maintain a particular attitude or flight condition. Control is the pilot's ability to change the aircraft's attitude or flight condition. A plane that is very stable may be difficult or sluggish to control, and a plane with powerful controls may be too sensitive. Considerable attention must be paid to the proper harmony between stability and control to achieve a pleasant and easy-to-fly aircraft.

As the performance envelope expands and the difference between minimum and maximum airspeed increases, the task of achieving acceptable stability and control thoughout the entire speed range becomes challenging. For example, designing sufficient control power into STOL aircraft is difficult because of the aircraft's slow approach and landing speed. STOL aircraft also suffer because the large vertical and horizontal tail surfaces normally used to achieve low-speed stability cause this class of aircraft to be overly sensitive to gusts and turbulence in cruise. Similar problems exist for conventional general-aviation aircraft with a large speed and altitude envelope. An expanded center of gravity range also causes stability and control problems for the designer. The down springs and bob weights typically used to expand the rearward CG limit make the elevator forces rather large for the landing task. They also have an adverse effect on the aircraft's dynamic stability. The axiom that aircraft design is a compromise is particularly apparent in the areas of stability and control.

The solution to the problem lies in autopilot technology. All of today's high-performance fighter aircraft rely on sophisticated autopilots or electronic flight-control systems to achieve acceptable flying qualities. The desired pitch, roll, and yaw characteristics are obtained electronically by stability-augmentation systems, which consist of flight-condition sensors, computers to determine what extra control deflections are needed, and servo actuators to move the controls for the pilot. Conceptually, these components are the same elements used in the design of any autopilot. With their flight-control systems inoperative, most high-performance fighter aircraft are difficult to handle and their operational effectiveness is severely compromised. The basic airframe of the YF-16, General Dynamics' entry in the exciting lightweight fighter arena, actually is designed to be unstable in order to achieve certain performance advantages associated with greater maneuverability and lower trim drag. Naturally, pilots are not expected to fly an unstable aircraft; just the correct amount of stability in harmony with the desired control is achieved electronically by means of the YF-16's flight-control system, which, in essence, is a sophisticated autopilot.

The trend in most modern high-performance aircraft is toward less natural stability and more reliance on autopilots and electronic flight-control systems for desirable handling characteristics. The concept of less stability in order to achieve better control and the quest for an autopilot that can rationalize the conflicts such a concept creates, however, are as old as manned flight. Early aviation pioneers, such as Lilienthal, Chanute, and Langley, designed their aircraft to be inherently stable vehicles. Since their aircraft would right themselves in the presence of most turbulence, the pilot's task was solely to steer them. Because of their inherent stability, however, these early machines lacked maneuverability and were highly responsive to atmospheric turbulence. The Wright brothers took a different approach and made their aircraft unstable but controllable. Their vehicles, which succeeded where their predecessors had failed, were much more maneuverable and were less susceptible to atmospheric gusts. The pilot's job was also made much more difficult and tiring.

Autopilot development followed relatively quickly after the initial successes of the Wrights and other early pioneers. The first autopilot, designed to stabilize an aircraft and return it to the desired flight attitude after a disturbance, was installed in a Curtiss flying boat and tested in the fall of 1912. The device was designed and built by the Sperry Gyroscope Company. It consisted of gyros to sense deviations of the aircraft from the desired attitude and servo motors to activate the elevators and ailerons; in concept, it was similar to the methods of modern autopilots. In June 1914, the Sperry Aeroplane Stabilizer won for its inventor, Dr. E. A. Sperry, a safety prize of 50,000 francs offered by the Aero Club of France for the most stable airplane. For the winning demonstration, Dr. Sperry's son, Lawrence, flew the Sperry Aeroplane Stabilizer-equipped Curtiss flying boat close to the ground under automatic control while standing in the cockpit with both hands over his head. His mechanic also stood and walked back and forth on the wing to demonstrate the autopilot's ability to cope with the large yawing and rolling moments created by his presence. In spite of this impressive feat, autopilot development progressed relatively unnoticed until 1933, when Wiley Post installed the first prototype of the Sperry pneumatic-hydraulic Gyropilot in the *Winnie Mae* for his eight-day solo flight around the world. It is said that Post held a wrench tied to his finger by a piece of string; when he fell asleep, the wrench would drop out of his relaxed hand and the subsequent jerk of his finger would awaken him so that he could monitor the progress of the plane and autopilot before dozing again. Autopilot usage increased during World War II as bulky units weighing close to 200 pounds were used for bombers and some fighter planes. One model of the Republic P-47, for example, was outfitted with an autopilot to relieve fighter jockeys from the strain of long-range bomber-escort missions. Autopilot development and general acceptance of automatic flight continued after the war. In 1947, an Air Force C-47 made a completely automatic, hands-off, transatlantic flight, including takeoff and landing. With some noticeable exceptions (such as Eastern), airline acceptance of autopilots grew throughout the 1950s. After Pan Am proved that autopilot-equipped aircraft saved seven percent in fuel over nonautopilot aircraft, even Eastern Air Lines changed its tune and adopted autopilots as standard equipment.

The first autopilot small enough and light enough to be considered for general-aviation use was the L-2, developed by Bill Lear in 1949. It was an electrically operated, three-axis unit that weighed close to 50 pounds and consumed 20 amps of current. With altitude hold and approach coupler, the L-2 cost nearly as much as the few sophisticated single-engine planes that had the Lear autopilot installed. In the late 1950s, relatively low-cost autopilots in the $3,000-to-$4,000 range and weighing approximately 10 pounds became available from Brittain Industries, Tactair (no longer in production), and Mitchell (now part of Edo-Aire). When adjusted properly, these early units served their owners quite well, although they often lacked the kind of reliability that would instill confidence. This was particularly so for instrument approaches demanding use of their relatively unsophisticated omni and localizer couplers. Minneapolis Honeywell offered the more sophisticated H-14 autopilot for light-to-medium twins, and Collins, Sperry, and Bendix supplied autopilots for airline and larger general-aviation aircraft. The basic Lear L-2 autopilot design was acquired by Motorola and subsequently sold to Bendix, where derivatives of the L-2 were offered for many years.

Just as the basic piloting tasks have remained essentially the same over the years, so have the basic functions of the autopilot. But it is significant that the general-aviation pilot now can select from a broad variety of models with considerable sophistication in terms of computational capacity and flight modes. The principal advances are attributable to modern electronic design incorporating LSI chips and digital-computer techniques that allow today's units to perform a variety of electronic tasks even though the associated autopilot elements are considerably reduced in size and weight. Sophisticated designs, which 10 years ago consisted of large pedestal controllers, vacuum-tuble amplifiers occupying one of two ATR racks, and heavy servos, now are sufficiently compact to be easily fitted into a Seneca II, Baron, or Centurion without compromising space, load, or electrical supply. These new autopilots also have added capabilities that compensate for many of the shortcomings inherent in early general-aviation offerings. The more sophisticated models provide filtering and computational techniques that eliminate much of the characteristic needle chasing often found during omni tracking and approach coupling. Features such as all-angle intercept capability, automatic ILS arming and capture, and airspeed compensation are available on some models.

While there have been significant advances since the L-2 days of the late '40s and '50s, it will be some time before general-aviation autopilots are designed along with the basic airframe, much as the Collins autopilot system was developed for the Lockheed L-1011 or as advanced fighter flight-control systems evolved side by side with those aircraft. But the trend toward considering autopilot and aircraft as interwoven elements is apparent in general aviation, as indeed it must be if we are to extract all the capability inherent in this class of aircraft. The way in general aviation has been foreshadowed by the extensive use of stability-augmentation systems on helicopters, which have significantly less stability than typical fixed-wing aircraft. The addition of SAS has been both welcome and a commonplace contribution in recent years. Most autopilots applicable to medium twins, turboprops, and jets incorporate a yaw-damper that provides increased lateral directional dynamic stability for the aircraft, and most pilots use the yaw-damper portion of their autopilots on a full-time basis even though they may use the entire autopilot only for limited portions of the flight. For many years, the Mooney incorporated as standard equipment a simple, single-axis autopilot to augment its spiral stability; the Commander 112A does so now. Several autopilots offer methods of momentarily cutting out the autopilot function to allow the pilot manual control.

Autopilots are a great help to the pilot, particularly for IFR operations. For the non-IFR pilot, an autopilot should never be an excuse to continue into marginal or instrument conditions, but one can only surmise that many VFR pilots might have not lost control of their vehicles during emergency descents through cloud decks or during 180-degree turns of the solid IFR conditions had they been in aircraft equipped with autopilots and had they known how to use them. The principal benefits of an autopilot accrue to the qualified instrument pilot who can hold his aircraft on course, on altitude, and on airspeed without an autopilot but who chooses to use the device so that he or she can concentrate on navigation, communication, weather, fuel management, and planning.

An autopilot is particularly useful in helping a pilot execute an approach. In order to set up the approach features of an autopilot, it is necessary to know which facilities and procedures will be used. Once the autopilot is programmed for the approach, so is the pilot. Since many of the general-aviation autopilots offered today can be obtained with a flight director, the pilot has the option of flying a coupled approach while monitoring the autopilot's function with the flight director's command bars, or of choosing to hand-fly the airplane with the autopilot's computer providing guidance information through the flight director. Although some pilots may question the advisability of flying a coupled approach because of the possibility of autopilot failure, the risk of the autopilot/flight director fouling up may be significantly less than that of the overworked and slightly rusty pilot dropping the ball. The ability of the autopilot to reduce the pilot work load during critical missed-approach situations also is a significant safety feature.

An autopilot can make the difference between a tiring flight and an easy, comfortable one. Fatigue in itself is dangerous and can lead to high pilot work load. When high work load and a few tense experiences cause a pilot to doubt the value of his aircraft, then the overall utility of the aircraft has been severely compromised.

The case for autopilots as a means of obtaining the level of stability and control that approaches the ideal is challenged only by the cost of including these valuable items as an integral part of the aircraft. Fortunately, there are autopilots to fit most applications, from simple wing levelers for stability assistance to sophisticated three-axis flight-control systems with many features. Considering the application, one can usually find a unit that offers value in excess of its cost. Considering the sizable acquisition price of sophisticated singles and light twins, for example, the marginal cost of a full-capability autopilot is relatively low, particularly when it provides a means of extracting more usefulness and value from the initial aircraft investment.

The reliability available in autopilots today or in the near future does not approach what is needed to relieve a pilot from the basic requirement to be proficient in manual flight operations, nor is autopilot reliability sufficiently high to eliminate the need for manufacturers to design stable airplanes for civil aviation.

As businessmen pilots by necessity rely more on the stability augmentations provided by black boxes, the present need for increased autopilot reliability will be magnified. Manufacturers must rise to the challenge. For the foreseeable future, pilots will be required to maintain an emergency, autopilot-failed flight capability much as they should maintain an ability to handle an engine-out procedure in multiengine aircraft. Just because a powerplant can fail is no reason to use only one of two available engines; so why not take advantage of the needed stability and control provided by an autopilot?

Today's autopilots possess sufficient capability to expand significantly the operational sphere of general-aviation aircraft. Particularly for the single-pilot, owner-flown vehicle, they provide an aircraft with the outstanding stability and control characteristics that even the most gung-ho pilot would praise if only they were achieved by means of mechanical pushrods linked to unservoed control surfaces rather than made possible by sensors, electronic

computers, and actuators. Modern autopilots, in particular, make single-pilot
IFR a less fatiguing, more reasonable, and safer endeavor by relieving the pilot
from the mindless, boring tasks that a simple servo can do (with more
precision, most likely) and liberating him or her to do what only the human
can accomplish—the mentally stimulating management functions that enable the instrument pilot to extract a full measure of utility from his aircraft.

8 FLYING THE NEW RADIOS
by George C. Larson

THE NEW AVIONICS systems are here, after various production, delivery, and debugging delays. To evaluate the new-generation equipment on the job, we found owners who had bought complete packages of each of three top-of-the-line systems. This is radio gear that you can buy today—not factory demonstrators but actual for-sale hardware.

You know all about the new technology by now; that the new generation of avionics is designed for easier, cheaper installation with fewer antennas. But you still want to know how that affects performance. It's nice that the Bendix BX 2000 hops out of the box ready to plug into either 14- or 28-volt airplanes, but does it work? Narco designed its nav receivers to incorporate receiver multiplexing of the localizer and glideslope—a real moneysaver—but how is it on an approach? The avionics manufacturers are using microprocessors and large-scale integration and other stuff that's hard to understand, but do they get the job done?

Here's a pilot's-eye view of three admittedly happy customers flying three representative panels: the BX 2000, the Collins Micro Line, and the Narco Centerline. We chose these systems because they are brand-new, they are available, and they represent three distinct levels of complexity. There are other systems as well—from King, Edo-Aire, Cessna, Sperry, RCA, and Genave—and we'll be trying them as soon as the gyros wind down from this series of test flights. If you're going to pay your money, it's nice to know you can take your choice.

BX 2000:
Bendix's Traveling Light Show

Nobody has to persuade Dave Joyner that there's an electronics revolution afoot; he's part of it—he sells CBs, radar detectors, marine radios, and closed-circuit TVs. Joyner is very comfortable around advanced electronics, which makes him a natural prospect for the ultrasophisticated BX 2000 line. His 1967

Baron C55 puts in about 600 hours a year flying him from his base at Charlie Brown Airport, in Atlanta, to major customers in cities throughout the South. On the day we traveled with him, he was trucking a load of closed-circuit TV equipment to Memphis for a sales demonstration.

The BX 2000 avionics family is the wildest set of radios ever to hit the general-aviation avionics scene. It does things that no panel-mounted avionics gear has ever done before, and it introduces some radical designs and components that have never been seen in *anything,* panel-mounted or remote.

The most dazzling Bendix box is the NP-2041A navigation computer programmer, which incorporates a microprocessor (a small but very real computer) that can handle through its keyboard all the radio management involved in navigating and communicating. The really stunning innovation, though, is this box's ability to converse with a Texas Instruments programmable calculator.

This trip began in Dave Joyner's office, and that's where most of the business of navigating took place. Joyner first worked out a direct route from Charlie Brown to Memphis International on an area-nav planning chart; he set up five waypoints offset from vortacs on the RNAV chart. Using worksheets supplied by Bendix, he recorded the pertinent data for each waypoint: frequency, offset, and station elevation, which, combined with the altimeter encoder data, let the computer eliminate DME slant-range error. Once the worksheet was filled in, it could be filed in his Jepp book with the destination airport, but first, Joyner chose to load the data into the TI calculator. The flow diagram on the sheet told Joyner which buttons to push; he loaded the data, imprinted it on a blank magnetic card, and then filed the card in a wallet he carries that holds all his usual routes already programmed and stored. While running an errand, Joyner filed RNAV direct Memphis, using the telephone in his car. For all practical purposes, the navigation chores were accomplished before the flight even began, for the BX 2000 would take over all the navigation and nav-frequency management from there.

"I never fly the airways anymore," said Joyner, "and since I've gotten this system, I always set up the outer marker or the final approach fix as a waypoint." The BX 2000 has opened even more doors for him, you discover as he talks. The previous day, he had missed an approach during a violent storm in New Orleans and shot it again with the waypoint set up on the end of the runway. He recited with glee the airplanes that had missed and gone elsewhere while he had gotten in.

The reason he bought the RNAV in the first place was to find Boca Raton Airport. Of course he knows where it is, but the Florida airport near an escape-from-it-all house that he owns is usually shrouded in haze. The overwater approach takes him right through a group of beachfront hotels, so he uses the area nav teamed with a terrain-mapping Bendix WeatherVision to pick out the hotels, and he loves to call ATC and tell *them* where the airport is.

He has about 65 hours on the panel now with only one malfunction: an incorrectly located wire in the hookup to the autopilot caved in the whole works, but Bendix's Fort Lauderdale facility had received him on only a couple of hours' notice and the people there had labored until two A.M. to rectify the error. That impressed him.

Loading the luggage took far more time than loading the waypoints. Joyner connected the Texas Instruments calculator to the nav computer programmer with a short length of cable, placed the little magnetic card with the route into its proper slot, pushed a button, and in about a tenth of a second, his route was loaded into the airplane's system. There was literally nothing left to do between Atlanta and Memphis but operate the airplane and punch up frequency changes for handoffs.

In the bumpy rain clouds, we talked about whether keyboard-operated avionics are any good in turbulence; you can hang onto a knob, the argument goes, but a keyboard gives you nothing to cling to. Joyner has found that he can grip the side of the panel with his fingers and work the keyboard with his thumb. Once or twice he either missed a number or punched the wrong button, and he had to clear it and start over, but it was little worse than turning past a frequency with a knob-operated radio.

Joyner's routine when ATC assigns him a frequency change allows him always to store the frequency he is leaving so he can return to it if he has to. He uses the number-one com to talk air-to-ground because its antenna is belly-mounted. He punches up the frequency on the keyboard, checks it, punches "enter," and sends it down to the number-two-com. The CN-2011A dual-nav-com box has a little button that flipflops the frequencies in both radios; push it, and the frequencies in the number-one and number-two coms switch places. By using it, his new frequency comes up on number one and the old one is transferred down to number two, where it's effectively stored, since he can switch it back with another touch of the same button. With this routine, Joyner always transmits through the more effective antenna.

Joyner pointed out some of the computer's traits: it figures groundspeed with a series of digital elements called shift registers, so it takes about six minutes to settle down on a number. It also corrects for slant range, however, so there's no dropoff as you cross a station. As we climbed, he switched the autopilot to the heading mode to steer around some angry buildups, and the computer updated with revised headings to the waypoint very quickly—no lag there.

How does he like the new Bendix all-electronic course-deviation indicators, with their light bars instead of needles? Just fine, he says. His number-one nav drives an HSI, but number two runs one of the new CDIs, which gives him a redundant glideslope readout as well as a backup for navigation. Joyner says he finds the light bars more sensitive than the typical meter needle— much easier to read and to fly.

Joyner volunteered a demonstration of the vaunted pulling power of the new ADF-2070. He tuned the Brooks Locator Beacon, at Memphis, while we were still 150 miles out, and initially, the ADF arrow displayed a stable bearing. Later, it wandered around in both normal ADF and the extended-range mode, no matter what station he tuned. There were plenty of thunderstorms around, but Bendix claims its ADF is immune to lightning. The ADF got mixed reviews on this flight, particularly since it wouldn't settle down on a bearing even within visual distance of Memphis.

In all other respects, the flight went perfectly. As he neared each waypoint, Joyner used the keyboard to engage the next one. The computer does everything: tunes the proper frequency and displays the correct course and distance. Would Joyner have been in a pickle if this do-it-all computer had not worked as advertised? No, for there is backup manual tuning for both navcoms; he would merely have reverted to airway navigation and gone on his way using the remainder of the system.

Throughout the flight, his workload was so low that he had more than ample time to consider the weather and to enter a sixth waypoint at the Oakville Intersection, which is the initial approach fix for the localizer back course to Memphis Runway 27. As expected, he got vectors near Memphis for spacing, but the computer provided him with continuous updates on bearing and distance to Oakville. It's no fun being dumped into an outer marker or an approach fix while you're still clean and fast. Joyner always knows where he is, and that probably means more to him than all the other features combined. Some of them, such as the unit's ability to load whatever its present position is as waypoint zero, were never even touched. He figures the RNAV alone saves him 10 to 15 minutes per trip.

If there is any improving to be done in Joyner's panel, it's hard to find. It wouldn't hurt if they could devise some way to allow the pilot to program the computer without distracting it from its navigating chores. Joyner had to uncouple and fly heading while he was programming the new waypoint for Oakville, but he did it early enough that it didn't matter.

Joyner shut the BX 2000 down on the ramp at Memphis, ready for two days of business, and when he returned home, he just fired it up and ran the waypoints in reverse. No need to load the computer up again.

You can't help but wonder whether this sort of sophistication frightens some people away; there have been plenty of magic boxes in the past that got too complex for their own good. The difference is, this BX 2000 system *works*. It takes on the chores that contribute most heavily to the pilot's task during any flight, leaving him or her free to handle unexpected weather and the airplane systems that need monitoring. In both of those respects, it is profoundly different from any previous avionics offered for light airplanes. If there is any prejudice against such fanciness in favor of the good old days, don't look for it in Dave Joyner, because he's become a believer.

Micro Line:
Collins Quality Updated

Gillespie Field, in San Diego, got a localizer recently; and to the people who frequent the airport, that little gift was better than Christmas. They used to fly the back-course 27 approach to San Diego Lindbergh, then break off to land at Gillespie, but no more.

Larry Winn is one of the grateful. He runs an industrial-supply company based in San Diego with a branch office in El Centro, just over the mountains, and he flies his 1970 Baron 58 between the two airports frequently.

He sold a 210 in 1972 to get the Baron. A year ago, he flew it to Crown Air, at neighboring Montgomery Field in San Diego, and had them install as much of the Micro Line as was available at the time: two navcoms, ADF, transponder, and audio-panel/marker beacon. Collins had yet to deliver its promised area nav and DME for that family of avionics, but they will soon, and he'll go for the RNAV. He has a King KFC 200 flight director and KN 65A DME, which he plans to keep.

When Collins first introduced the Micro Line, the feature that drew the most attention was the com radio's ability to store a second frequency in addition to the one on display. Much was made of the idea of getting two radios for the price of one, but since that time, the stored frequency has appeared in other makes, and opinion seems to be mixed on just how terrific a feature it really is. In fact, the storage feature has been eclipsed by another that at first took a backseat: the continuous RMI readout available from the nav radios. Almost every user praises it.

Larry Winn says he uses frequency storage regularly but that adjusting to it was a matter of "remembering that it was there." En route, he now makes it a habit to store the frequency he's leaving before tuning the new one he's been assigned. He also tunes and stores San Diego Departure and Los Angeles Center on the number-two com before takeoff, since he knows the departure communications routine by heart.

Nothing is more routine to him, though, than the trip home from El Centro to San Diego and a straight-in to Gillespie on the localizer-D approach. Its initial approach fix is an intersection called Barrett Lake, defined as a DME fix off Mission Bay Vortac or as the 177 radial off Julian Vortac, to the north; that's where the RMI feature comes in handy. As he flies the approach, Winn can simply tune Julian and just sit back and read the numbers off the nav window instead of fiddling with an omni bearing-selector knob. (If you doubt the worth of that feature, think back to all the times you've misread the scale and set the OBS 10 degrees off. An RMI does not make that kind of mistake.)

Approaching the coast with us aboard, Winn intercepted the Gillespie localizer course from the north and tuned Julian on number two. From that point to the touchdown, all he touched was the HSI. At our urging, he also dialed in the radials for the Barrett Lake and Lardy intersections on the number-two nav indicator so we could watch it work.

On the localizer course, the RMI readout for the radials off Julian was not so well damped as the needle on the CDI. The numbers jumped around—add two, subtract three, up one, down two—so even though we could keep track of our progress toward Barrett Lake, it wasn't exactly a steady march. There is a lot of high terrain in that neighborhood that gives most VORs fits, which probably explains why it looked as if we were dancing a tango.

At Barrett Lake, Winn descended to 4,800 feet, next step down a long staircase to Gillespie. Our next fix was Lardy Intersection, defined only by the Julian Vortac's 188-degree radial. Still the RMI danced around. (The fluctuation, though distracting, really doesn't affect the execution of the approach. Wait until the numbers no longer bracket the radial, and you can consider yourself inside the intersection and safe to descend.) Larry Winn flies to several airports that use VOR intersections as approach fixes, and he has come to appreciate the ease of flying the RMI.

His panel was very nicely installed, and Winn proudly displayed the neat wiring harness visible from the nose baggage compartment. Although there is an ADF in the airplane, Winn doesn't use it much; but the absence of a long sense antenna means he pays little penalty for this silent passenger. In fact, you can install the full load of these Collins radios without that old antenna clutter that sometimes looked as dense as the TV horns on an apartment roof.

The electronic numerals on the nav frequency display proved more than amply bright, even in direct sunlight, and of course at night they're downright assertive. You can get the old-style painted-on numerals and save money in the process, but there is a hidden benefit to the numbers in lights: since the tuning knob doesn't have the added mechanical task of moving the little wheels with the numbers on them, they can design it so you get a lot of ground covered in a minimal twist. It's a little like power steering, and I noticed that Winn was able to tune very quickly and accurately. It almost obviates the need for frequency storage, but with a toy like that recall switch, who can resist using it?

Winn says his problems since installation have been trivial. One of the digit light elements had to be replaced; it was, very quickly, by his local service facility. A problem he'd always blamed on his previous transponder cropped up again with the Micro Line in place, thereby unmasking what had really been an antenna problem all along. He keeps the Baron hangared at Gillespie, and in the withering heat of El Centro, he puts heat shields in the windows. No doubt, the care pays off.

Winn has his ADF and transponder stacked together, as do most Micro Line users. Some people complain that the two units look too much alike, but when asked whether he had ever squawked his ADF or tuned his transponder to the compass locator, Winn looked a little amazed at the question. Like most people, he knows where to find things in his own airplane.

Throughout our flight with him, Winn looked right at home with the airplane and the avionics. The approach, in particular, was made with an economy of motion that tells you the pilot is comfortable. The system fits him, and after a year with it, he and his avionics have an easy relationship. When asked about his plans, he mentioned some airframes he'd like to try, but he never once hinted that he'd switch his brand of radios.

When Collins first brought out the Micro Line, the avionics seemed so new and different that everyone speculated as to whether they'd be accepted by pilots, who tend to be fairly conservative in the way they select avionics. Larry Winn took his time, asked around, and in the end, he could see only a device that does more work for him and fits the kind of flying he does most. If he ever harbored any prejudices about sticking something all-new in his panel, they've long since melted away in hours of steady reliability, and that answered his only question.

Centerline:
Narco's New Faithful

Gary Pesnell calls his company Power Graphics. What better name for a firm that publishes full-color wall posters of 18-wheelers for trucker freaks to plaster all over their bedroom walls? Pesnell ground some gears himself—upshifted, in fact, from driving the monsters for 12 years to printing posters of them, and from the proceeds purchased a fresh-picked 1977 Turbo Arrow III as the Power Graphics corporate flagship.

When Pesnell ordered the airplane, Piper told him they wouldn't install a Narco Centerline system until the 1978 model year, so Pesnell found somebody who would: Blairstown Avionics, in New Jersey. Pesnell insisted on Narco because his previous airplane had a Narco com and a competitive make side by side, and the non-Narco radio kept going away. So far, the pilot has 500 hours, the airplane 60, and the radios 40 hours of flying, and nobody's complaining. Pesnell attributes much of the credit to the installation. A nice, raw Arrow is the kind of thing small-airplane avionics installers dream about, and Blairstown did it up right. Pesnell pointed out such refinements as cooling air from *both* wing roots—double the normal supply—and he believes strongly in the worth of field installations: "Here's a shop that has to live with the radios they installed. If we had factory radios, we'd be just another airplane that happens to have Narcos."

Pesnell bought the full house: two com 120s, dual nav 122s (integral receiver indicators and glideslope plus marker beacons on both—"resale value," he says), a DME 190, AT150 transponder, ADF141, and audio panel.

The Centerline expresses in the look of the boxes the company's philosophy of simplicity. The Narco boxes are stark, simple, beefy. In fact, they look much like some of the new hi-fi equipment that strives so hard to imitate the "black box" look of traditional avionics. One of the nicest touches in the new system is the large knobs and controls, all sized so you can get a grip on them; no teensy thumbwheels or precious little hide-and-seek buttons on these radios. The transponder has an ident button the size of your thumbnail and comparatively huge knobs for code selection, and the knobs on the navs and coms are so big you could use them for tiedowns.

Pesnell also fell in love with the backlighting in the boxes. "You end up finding excuses to fly at night," he said. Even the Narco logo lights up from the back with a soft blue glow, and that big-airplane-style illumination is a definite ego booster. It's produced by applying silkscreen masks to a translucent plastic front cover so that with only a couple of bulbs behind the front of the box, you can light up all the functional words and numbers.

Each com radio also has a "transmit" light that glows when you key the mike, a feature fellow pilots will appreciate; too bad more radios don't have some way of alerting the pilot that he is putting out a carrier that's jamming everyone else. The transmit indicator does not tell you whether anything is leaving the antenna, but the light does flicker in proportion to carrier modulation, telling you you're accomplishing that much, at least.

We left Morristown Airport, New Jersey, climbing out along the localizer, and both marker audios sounded off like a horn duet. The only way to shut them down was to use the muting feature, which hushes them for what Pesnell later described as "the fastest 12 seconds in the world." It's on his list of corrections to be made; the dealer says it's a simple adjustment.

We flew a VOR approach to Sky Manor, New Jersey, for Pesnell is most impressed with the Centerline's absolute accuracy on low-level nonprecision approaches. He feels the system is well matched to his type of airplane—the high-performance single that will operate frequently out of airports with VOR and ADF approaches. The needle damping on the indicators was very pleasing, and later, on ILSes to both Allentown and Morristown, we were impressed with the prominence and smoothness of the glideslope needle's rectilinear path in the middle of the dial; it's exceptionally easy to read.

The economy and design simplicity of the new multiplex receiver for the localizer and glideslope functions—it divides its time between both, but very quickly—turn out to be mostly academic: you'll never notice anything different in the way the indicators operate.

It was on the VOR approach, too, that the DME 190 looked best. This DME—and its relative, the DME 195 (a remote-mounted box with a smaller indicator in the panel)—are a quantum leap up from the old-generation DMEs that took their sweet time settling down to a valid reading. The 190 locks up right *now,* and on an approach like this one, you appreciate its ability to switch rapidly to groundspeed, render an instant verdict, then switch back just as quickly to distance. (The 195 gives both readings simultaneously, plus time to station.) The 190 box takes a lot of panel space, and you have to tune it to use it—slaving off the nav tuners is optional—but it is one of the most competent units available for the money.

Narco retains the "independents" philosophy in the Centerline. There are no one-and-a-half radios in this system. It has its advantages when it comes time to pull out a box. Instead of yanking a whole navcom when only the nav needs service, you can pull only the ailing box and fly away with the other functioning.

However, by combining the nav receiver with the indicator itself, you have to design all the tuning equipment into the indicator. The Centerline indicator has an OBS knob, a pair of frequency-select knobs, a frequency window, localizer needle and scale, glidescope needle, to/from arrows, marker-beacon lights, LOC and GS flags, and OBS scale. By the time you've packed all that into one dial face, it's a busy place. Pesnell said it doesn't bother him, but he admitted he'd recently had a glideslope flag flying and never noticed it.

Another aspect of "independents": with all those on/off switches ranging across the panel, you might want to consider an avionics master-power switch. (Pesnell passed that option up, however, for he didn't like the thought of every radio in the airplane being dependent on a single switch.)

The frequency-select knobs on the com radios seem rather odd to us, though Pesnell never commented about them. There are three: one on the left for the big numbers to the left of the decimal point, then two concentric knobs on the right, the larger for the hundredths and thousandths of megaHertz and the smaller one for tenths—the opposite of what someone new to the system might expect. One could make the argument, though, that the smaller knob of the right-hand pair is farther forward, therefore the first one you touch.

Pesnell's Centerline system is functionally traditional. Narco is fond of labeling some of the new functions on the market right now as "bells and whistles." Pesnell's panel offers him sophistication, but of a different sort. If you believe that simplicity can be a form of sophistication, you understand the Centerline approach. These are traditional avionics but highly evolved, the product of many lessons; and with the advent of a multiplexed receiver, that ought not to be minimized. It cuts the cost of building a nav receiver so that somebody like Pesnell can shop for dual glideslopes.

Narco believes there are plenty of people like Gary Pesnell around—people who are not interested in new features but in solid performance from the radios they're used to seeing. In Pesnell's case, he's had few problems. Just after our flight with him, the DME signed off, but the dealer says it's a simple fix. Pesnell took it in stride; he expects a few early glitches. He's already nestling comfortably into the Turbo Arrow. Ask him, and he'll tell you he has no doubts that he's made the right choices.

⑨ HOW A WING WORKS
by John W. Olcott

DOES IT really matter how a wing produces lift? Knowing when to push or pull the proper controls so that the airplane stays upright—now *that* matters. But after memorizing just enough aerodynamics to pass an FAA written, why bother to know more?

Why? Because a better understanding of what makes an airplane fly makes us better fliers. The forces that keep us safely aloft may be invisible, but they are not magic. Understanding the science makes us better aviators and provides a depth to our appreciation of the beauty of flight.

A wing is a wondrous thing. Used properly, within the laws that nature has decreed, a wing will render faithfully just enough lift to support an airplane's weight. Abuse those laws and the results are predictably bad, sometimes even disastrous.

As a wing is propelled through the atmosphere, the molecules of air it encounters are moved aside and forced to flow past the wing's contoured shape. When such movement and redirection of the airflow occurs, forces are created that act on the surfaces of the wing. Air near the wingtips can move sideways to make room for the intruder, but most of the air encountered by the wing must flow under or over it. If the wing is deflected at some angle as it moves along so that the air is forced to parallel the bottom surface and thus is redirected downward, an upward or reactive force is produced. A hand extended out the window of a car and deflected against the air experiences a similar force, not because the air "strikes" it but solely because the air flowing past it has been made to change its direction.

That is only part of the story, however. The force created by a wing is much greater than what can be produced by a nonairfoil shape such as a deflected hand or a piece of board. Something else is happening to amplify the forces, and the wing's curved upper surface is the source of that something extra.

Like the air flowing under the wing, air that goes over it also tries to follow the surface contour. Furthermore, these molecules of air attempt to make the journey along the upper surface in roughly the same time that is required for the air to traverse the wing's lower surface. But the top surface is curved, like the back of a porpoise, so that air must change its direction and accelerate as it attempts to slide over the longer, gently arched path. The air increases in velocity as it flows toward the thick portion of the wing, just as water flowing in a brook speeds up as it bends around a rock lodged next to the bank. After passing the thickest portion of the wing, the air starts to slow down as it heads toward the trailing edge.

To illustrate what happens to the air, consider a vertical slice of the wing extending from the leading edge to the trailing edge and paralleling the path a stream of molecules takes as it flows past the wing. What we are examining is an airfoil—the cross-sectional shape of the wing. Also, picture the air rushing past the airfoil, rather than the wing moving through the air; the forces generated are the same, and the same principles of nature apply.

Since some of the air moves under the wing and some moves over it, there is a dividing point where a molecule of air would hit the airfoil if it could not decide which way to go. That point is known as the forward stagnation point, and it is important because its location on the airfoil has much to do with how the air flows around the wing and thereby generates forces. There also is a rearward stagnation point, essentially located at the trailing edge, where the molecules of air rejoin their neighbors after parting at the forward stagnation point.

If an airfoil is symmetrical—like a long, thin teardrop—it has no camber: its top half is exactly the same shape as its bottom half, and a line (known as the chord line) extending from its leading edge to its trailing edge divides the airfoil into two equal parts. When a symmetrical airfoil is pointed directly into the wind—when it has no inclination—the air flowing past it divides equally, and the forward stagnation point is located exactly at the center of the leading edge, where the chord line starts. The rearward stagnation point resides at the trailing edge, as it always does, provided the inclination of the airfoil is not too steep and the trailing edge tapers to a sharp point. Therefore, two molecules of air that were adjoining as they approached the airfoil would part company at the forward stagnation point, accelerate, decelerate as they traveled along their respective surfaces of the airfoil, and rejoin as they reached the trailing edge. Forces would be created as the air moved around the airfoil, but the contributions of the top and bottom surfaces would be equal in magnitude. Consequently, the forces perpendicular to the flow of air would cancel each other, and the lift (which by definition is the net force perpendicular to the flow of air) would be zero.

When a symmetrical airfoil is inclined slightly to the flow of air so that the chord line makes a small angle—10 degrees, say—with the direction of the airflow, the forward stagnation point no longer is located at the middle of the leading edge. It moves to a lower position, still foward on the airfoil, but closer to its lower surface. The rearward stagnation point, however, remains at the trailing edge. Therefore, the air that flows over the airfoil must curve around the leading edge and travel over the upper surface before it can join the neighbor it left when they parted company at the forward stagnation point, the neighbor having taken the shorter route along the lower surface between the forward and rearward stagnation points.

Since nature requires these neighboring molecules to reach the rear stagnation point at roughly the same time, the air flowing over the longer upper-surface route has to travel faster, thus creating the low-pressure area that is responsible for the majority of the airfoil's lift. This relocation of the forward stagnation point as the airfoil is inclined is why a symmetrical shape can produce lift.

Camber, which is the degree of upper-surface curvature compared with that of the lower surface (essentially, the amount of upper-surface bulge), causes the forward stagnation point to be situated lower on the leading edge than it would be if the airfoil had no camber. Thus, a cambered airfoil at zero inclination to the airflow behaves very much like a symmetrical airfoil that is inclined upward a few degrees; it produces lift.

The angle of inclination between the airfoil's chord line and the direction of the airflow is defined as its angle of attack. Therefore, angle of attack is the angle between the wing's average chord line and the direction in which the wing is moving through the air, since a wing consists of an airfoil shape repeated over the length of the wing, and the forces created are the same with the air moving toward a stationary wing or with a wing moving in still air.

An airfoil or wing will continue to generate more lift with more angle of attack, provided the air is able to make the trip from the forward to the rearward stagnation point. When the angle of attack reaches too high a value—about 16 to 18 degrees for typical general-aviation unflapped airfoils, but 3 to 4 degrees less for the same airfoils with flaps, and considerably greater angles for ones with special leading edge devices—the air no longer can follow the upper surface contour. At those larger angles of attack, the forward stagnation point moves down to the vicinity of the airfoil's bottom surface, far from its low angle-of-attack position near the lower portion of the leading edge. From the stagnation-point position, the air no longer can turn the corner around the leading edge and still have sufficient energy to negotiate the route along the airfoil's upper surface. Being unable to follow the shape of the airfoil, the air separates from its normal path and cannot produce the velocity distribution over the upper surface that is needed to generate lift. This inability of the air to follow the upper surface shape causes a stall.

The movement with increasing angle of attack of the forward stagnation point down the airfoil's leading edge and eventually to the forward portion of its lower surface can be seen in what is called a smoke tunnel. Small streams of smoke injected into the tunnel flow along with the air so that its path becomes visible. As the angle of attack approaches the stall value, the streams of smoke continue to split at the forward stagnation point (by now, situated nearly on the airfoil's lower surface) and flow over the leading edge and along the upper surface. But they start to depart from their smooth path near the trailing edge, and if the angle of attack is increased further, the smooth flow over the airfoil's upper surface will break down entirely and the smoke lines will spill off the airfoil in a disrupted and diffused manner.

General aviation's most popular stall-warning indicator senses the movement of the forward stagnation point to indicate an impending stall. A small tab on the wing's leading edge closes an electrical circuit to a stall-warning horn whenever the stagnation point has moved below the tab's location, indicating that the wing is approaching the stall angle of attack.

In addition to angle of attack, the amount of lift in pounds generated by a wing depends upon the density of the air, upon the wing area, and upon the velocity at which the wing is moving through the air, lift varying as the square of the velocity. But none of those factors has the influence of angle of attack, for if the air cannot complete the trip smoothly around the airfoil, it does not matter one bit how fast the wing is traveling, how large it is, or how dense the air might be.

Therefore, the factors that affect the air's ability to follow the contour of the airfoil or wing are very important. Obviously, angle of attack is one; the others are the shape of the airfoil's leading edge, the camber or curve of the airfoil, and the distribution of the airfoil's thickness along a line, called the mean camber line, that lies midway between the airfoil's upper and lower surfaces. The shape of the trailing edge and the roughness of the wing's surface, particularly those portions that lie ahead of its position of maximum thickness and near the trailing edge, also are important.

Pilots have more control over these physical characteristics than they may realize. Flaps, for example, behave as camber-changing devices; they significantly reduce the angle of attack where no lift is produced. Compared with an unflapped wing, therefore, a flapped wing can provide the same lift at a lower angle of attack or it can fly more slowly at the same angle of attack and still produce the same lift. The stalling angle of attack on a flapped wing is reduced three to four degrees because the air can tolerate just so much of the camberlike effects of flaps before it is unable to follow the upper surface contour, resulting in separation and a stall. Highly flapped aircraft, such as a Cessna 206, often are landed flat, even nosewheel first, because extended flaps so reduce the body attitude of the aircraft that pilots fail to rotate the nose sufficiently for a tail-low landing.

Leading-edge shape occasionally is under a pilot's control, but more often leading-edge devices work automatically, as they do on the Helio Courier. Recently, manufacturers have modified the leading-edge shapes of airfoils to influence the stall characteristics of existing aircraft designs. Familiar examples are the Grumman American AA-1, the Cessna 172 and 182, and the Century III models of Gates Learjet.

In a gross sense, wing-surface roughness is under the pilot's control, since he or she can keep the leading edge free from mud, frost, sharp dents and ice. The trailing edge also should be kept clean. Ice, in particular, will decrease the angle of attack where the stall occurs, because its roughness inhibits the ability of the air to flow smoothly toward the trailing edge. Even if the air does make it around the leading edge and over the thickest portion of the wing, the roughness caused by ice, frost, or mud can disrupt the flow sufficiently that small pockets of separation occur near the trailing edge, causing excessive drag even though the wing is not fully stalled.

Angle of attack is the only factor that is entirely under the pilot's control. It can be varied faster and more effectively than can anything else that affects lift, including velocity. An appreciation of angle of attack and an awareness of its value allow a pilot to extract lift and safety from a wing.

Explanations of how a wing produces lift can take any of a number of precise and complex forms, but the meat of the matter inevitably will deal with angle of attack; no other aspect of aerodynamic theory applied to wings is more important to a pilot.

10 THE TAKEOFF
by Richard L. Collins

WITH A TRICYCLE-GEAR aircraft, the takeoff seems one of the simplest phases of flight. Just steer down the runway until the airplane reaches a predetermined speed and then fly away. Simple, but a lot of pilots still have trouble with this basic initial part of flight. Crosswinds, high density altitudes, soft surfaces, mechanical problems—all can affect takeoff technique and performance.

Takeoff considerations should start in the preflight planning and last until the airplane is 500 feet above the ground. But few general-aviation pilots trouble to make performance calculations for every takeoff. That might be okay as long as you have guidelines that define the situations where performance calculations must be made.

We're talking about singles here, so let's consider an A36 Bonanza, an airplane with a broad range of loading options. Gross weight of an A36 is 3,600 pounds. According to the general performance specifications in the book, 2,040 feet are required to take off and climb to 50 feet, at gross weight, sea level, on a standard day. The conditions for a normal takeoff from the pilot's operating handbook are full power before brake release; a paved, level, dry runway; no wind; flaps up; and landing gear retracted after liftoff. The prescribed airspeeds at gross weight are 71 knots for liftoff and 78 knots at 50 feet. These figures are for a new airplane—clean as a whistle with brakes that don't drag and an engine that is at its best. The procedure of applying full power before brake release isn't normally followed by most pilots.

A 50-percent margin is good for takeoff: this gives generous allowance for flaws in technique, engine wear, slight uphill gradients that are not readily apparent, and other conditions that might affect performance. It's nice to be 50 feet high over the end of the runway on climb; so distance to 50 feet is a good number for required runway length. If I had an A36, I'd think of 3,000 feet as a minimum runway length at gross weight with moderate temperatures (50° to 70°).

Heat the air to 100°, and the desired runway would be 3,900 feet. Take off on an 86° day in Denver, and a 6,000-

foot runway would be desirable. Fly from grass or a soft surface, and even more distance should be added, as per the pilot's operating handbook or conservative rules of thumb.

These numbers are not absolute minimums: they are guidelines that hoist warning signals. In those conditions, a runway any shorter would send me to the books to see if I should lighten the airplane.

Eyeball the airport. Say the runway is 3,000 feet long and obstructions dictate that the departure be handled precisely. According to the manual, the takeoff roll will be about 1,250 feet. Add to that for contingencies and say 1,500 feet. Check the halfway point on the runway. If the airplane is not ready to fly away there, at the prescribed 71-knot liftoff speed, it is not performing as per the book. Abort the takeoff.

In light airplanes, we seldom use runways so short that our takeoff is a lot more than half their lengths, so there is usually room to stop if a takeoff is aborted at any reasonable time. When a takeoff run will require more than half the runway, it's extra important to calculate a go/no-go spot and abort if the airplane isn't up to speed at that point. In this case, there might not be room to stop before reaching the end, but the trip into the rough would be at a minimum speed. For each moment we persist past the point of no return on an impossible takeoff, the speed at which we go off the end increases—a key factor in who gets hurt how much.

Casing a marginal runway might also offer some alternatives. If there are obstructions only at one end, a downwind run toward the clear end might be your better bet. Unfortunately, not all pilot's operating handbooks give a means of calculating the effect of mild downwind components. However, a useful rule of thumb is to increase the runway length by 10 percent off each two knots of downwind component.

To summarize preflight takeoff planning, the key is *knowing* that the available runway is adequate and how to tell if things aren't going according to plan. Rough estimates are not permissible unless the available runway exceeds the estimated requirement by at least 50 percent.

What about flaps for takeoff? The A36 Bonanza handbook does show figures for an obstacle takeoff using 20 degrees of flaps. But this chart contains a warning: "Obstacle takeoff is not a recommended procedure, as it utilizes speeds at or below power-off stall speed. In the event of an engine failure or wind-speed fluctuations, a stall may occur which may cause uncontrolled contact with the ground." That's enough to make me forget about that procedure. It doesn't, however, completely rule out the use of flaps, which would be especially handy for getting the airplane off after the shortest possible ground roll on a rough or soft field.

In other airplanes, the best flaps policy is whatever the handbook specifies for a normal takeoff. My Cardinal RG book recommends 10 degrees of flaps and an initial climb speed of 65 knots. That speed is 10 knots above the power-off stall in that configuration, so it provides some margin. The desirability of flaps is apparent when you try the takeoff clean: the airplane runs much farther, the pitch-up attitude required to fly seems excessive, and the airplane lacks the buoyant feel it has at liftoff with 10-degree flaps.

Before every takeoff, conduct a little mental review of how it is to be done. What are the liftoff and initial-climb speeds? When should the gear be retracted? Power is not in question because it should remain at the takeoff setting until the airplane reaches 500 feet. The flaps should stay at the takeoff position until 500 feet is reached, and the airspeed should remain at that value shown for initial climb after a normal takeoff or for best angle of climb, whichever is higher, until the airplane is 500 feet above the ground.

Fret not about noise. Altitude is a good sound buffer, and this procedure will put the airplane as high as possible in the least possible distance. On many airports, you can reach 500 feet before crossing the boundary.

Fret not about the engine. If yours isn't up to 500 feet of climb wide open, best leave it at the shop. Even engines that have a time limit on takeoff power give a generous allowance for climb to 500 feet.

There's little to say about directional control except in a crosswind. With no wind, steering it down the strip is usually simple; in a crosswind, a challenge can appear, not from any lack of controllability but from overcontrolling by the pilot.

There's seldom a crosswind of constant direction and velocity, so as we trundle down the runway, the effects of the wind will vary. This makes tracking the stripe a bit more difficult and can lead to some slight swerving on the runway. This could encourage overcontrolling as the pilot's feet get out of phase with the swerves. Or it can tempt the pilot to lift off before the predetermined speed is reached, to get away from that evil old runway. That best liftoff speed would be higher than normal because in a crosswind we want enough speed to *know* that the airplane is going to remain airborne. A variable wind can cause airspeed fluctuations at liftoff as well—one more reason for a bit of extra speed.

Regardless of how squirrelly an airplane may feel on the takeoff roll, though, lifting off at a low speed is not the answer. It can only make matters worse. Once off the ground, the airplane is likely to drift; if it's too slow, it might touch back down—perhaps in an unfriendly place, like a snowbank beside the runway.

At the correct liftoff speed, rotate the airplane precisely to the initial-climb attitude, then after a moment or two of flight check the airspeed indicator for verification of your attitude. It takes very little experience with a given airplane to learn the appearance of a proper climb.

When should the gear be retracted? There are many opinions on this, and some airplanes require special procedures. On Cessna Skymasters, for instance, the gear should be left extended until obstacles are cleared. The drag created by the opening of the gear doors is greater than the drag of the extended gear, so you want to be up and over the first hurdle before accepting that momentarily higher drag.

If your airplane's manual has no specific recommendation on when the gear should be retracted, consider all the circumstances and do what seems most logical. Certainly you'd never want to retract the gear so quickly that you wouldn't have time to get it back down before reaching the ground on an aborted takeoff. Some people say that the gear shouldn't be stowed until a landing on the runway no longer would be possible, but that often seems wasteful. Most airplanes don't climb very well with the gear extended, and after takeoff on a long runway, leaving the gear down can mean an extended period of inefficient climb, which in turn means the airplane will make more noise over more people on the ground. It uses a few extra drops of gas, too. There are too many disadvantages to leaving the gear down for a long period after takeoff. Just don't pull it up too quickly; wait until the airplane has ascended a hundred feet or so.

What if some ailment overwhelms the engine during initial climb? The drill always has been to continue straight ahead, altering course only slightly to avoid obstacles. This is still the best way to go if something happens in the first 500 feet of climb. To turn 180 degrees at twice the standard rate would take 30 seconds. Power off, an airplane in the takeoff configuration at its best glide speed, wings level, would probably descend at nearly 1,000 fpm. That's 500 feet shot right there. But the wings wouldn't be level and the airplane wouldn't be at best glide speed. In fact, it would be perilously near the stall. In the A36, the initial climb was at 78 knots. If that speed were maintained and the airplane were rolled into a 30-degree bank, the margin above stall would be *one knot*.

In initial climb, good discipline is demanded just to transition from climb to glide attitude in the case of an engine failure with the wings level, and a turn that can be shown to be impossible on paper surely isn't worth trying.

The proper climb procedure gives you the best shot in case of trouble after achieving that initial 500 feet. The airplane is put on as high a perch as possible in as little distance as possible. After the transition to cruise-climb, a return to the airport after an engine failure might not be impossible and would be subject to the pilot's judgment. How far is it back to the airport? How large a turn would be required? What are the alternatives?

The pilot who lifts off, then below 500 feet pulls the power back and settles into a flat climb at a higher than necessary speed is not taking advantage of the airplane's potential performance. Normally, this causes no problem; given trouble, though, it can cut the options.

There are some miscellaneous items to remember on takeoffs: if you hear any unexplained noise on the takeoff roll, abort. It's probably an improperly secured baggage or cabin door, or a seat belt or coattail hanging out. Those problems are better fixed on the ground. Any abnormal engine sound or instrument indication should also dictate an abort.

Be wary of the effects of excessively nose-high attitudes on the takeoff roll or immediately after liftoff. At low speeds, drag increases rapidly as the nose is raised—this can only prolong the takeoff. An airplane will often stagger off the ground in a nose-high attitude but will not fly out of the ground cushion. In ground effect, which extends upward a distance equal to about half the span of the airplane, there is an interaction between the wind and the ground that helps the airplane fly. The effect is lost above that level. The proper takeoff procedure in a tricycle is to run in a *level* attitude until the computer liftoff speed is reached. Then transition to the liftoff attitude.

Good takeoff technique doesn't seem important until you need it, and beautifully executed takeoffs don't bring the ego satisfaction that accompanies a perfect landing. But starting out right is just as important as finishing with a squeaky flourish.

11 MANEUVERING SPEED

by John W. Olcott

IF YOU GATHERED a crowd of proficient pilots and asked them what was the best speed to use when flying through dangerous turbulence, you'd expect to find a consensus. But apparently there is none—at least not among operators of light aircraft.

Commenting on his own rough-air procedures, an experienced ex-military pilot mentioned that he recently penetrated a small thunderstorm in a sophisticated four-place single. Concentrating solely on the aircraft's altitude and ignoring its airspeed, which was at its normal cruise value, he pushed through the maelstrom while everything in the cockpit was tossed like croutons in a Caesar salad. Maintaining control was his only consideration; it never occurred to him that he should reduce his airspeed.

Another seasoned pilot, who spends many hours aloft delivering new singles and light twins for a major manufacturer, remarked that his technique for penetrating turbulence is to slow to 20 knots below the aircraft's maneuvering speed, concentrate on the artificial horizon, and try to keep the wings level with the ailerons. He thought the reduced airspeed made it impossible for gusts to overstress his aircraft, but he admitted that on three occasions he had been unable to prevent extreme bank angles—80 degrees or more—even with the ailerons fully against their stops.

These two techniques—maintaining cruise speed and significantly reducing airspeed—are reflections of the controversy that surrounds the question of how best to cope with extreme turbulence in light aircraft. The fact is that following these procedures would expose an aircraft to the two major risks of flying in turbulent conditions: fly too fast and gust-induced loads can exceed the aircraft's structural design limits; fly too slowly and the turbulence may upset the aircraft.

For a given gust intensity—expressed as so many feet per second—the gust loads imposed on an aircraft depend primarily upon the vehicle's forward velocity and to a lesser extent upon the aircraft's altitude, weight, wing area, chord, and sensitivity to lift changes with changes in angle of attack. All light aircraft certificated since 1969 are designed to withstand 50-foot-per-second gusts at the aircraft's cruising speed, but thunderstorms can produce far stronger gusts. Unless speed is reduced, it is therefore possible for intense turbulence to cause structural damage.

Gusts impose loads on an aircraft by creating momentary changes in angle of attack. These changes bring about increases or decreases in the aerodynamic lift produced by the wing. The greater the aircraft's speed, wing area, and sensitivity to angle of attack, the greater the lift generated by the gust. Offsetting the gust-induced aerodynamic force is the weight, or inertia, of the aircraft; the heavier the aircraft, the more its inertia reacts against the gusts.

The structural loads felt by the aircraft as it encounters turbulence depend upon the ratio of these aerodynamic inertia forces. Newton's second law of motion states that force equals mass times acceleration; therefore, the G-load felt by the aircraft will depend upon its weight: the heavier the aircraft, the lower the G-load for a given gust intensity, wing design, and airspeed. It is the ratio of aerodynamic forces to weight forces that would enable a 707, with its very high wing loading, large wing chord (which has the effect of reducing the abruptness of the gust), and lower lift-generating sensitivity to changes in angle of attack (because of the 707's swept wing), to penetrate the same level of turbulence yet experience less gust-induced loads than a Cessna 310 flying at the same speed. Likewise, a Cessna 310 experiences lower G-loads than a Cessna 172 when both encounter the same intensity gusts at the same speed; for a 50-foot-per-second gust encountered when flying at 120 knots, the 310 would experience a gust-induced load about 40 percent lower than would the 172.

Unless the manufacturer specifies some other technique for gust penetration, the best procedure is to fly at the aircraft's maneuvering speed—Va— since Va guarantees that the aerodynamic forces caused by full and abrupt movement of the controls will not impose G-loads that exceed the design limits of the aircraft. Although not specifically developed as a turbulence-penetration speed, Va also guarantees that gust-induced loads will not exceed the vehicle's design conditions.

Why? The reason relates to stalling speed, which increases as the square root of the load factor experienced by the aircraft. Maneuvering speed is equal to the design stalling speed times the square root of the load factor you do not want to exceed. Therefore, if flight controls are moved rapidly against their stops (as can be the case when a pilot is attempting to keep the aircraft upright in heavy turbulence) or the aircraft is experiencing large gust-induced loads, the wing will stall and relieve the aerodynamic forces imposing the G-loads.

No matter how large the gust, an aircraft being flown at its maneuvering speed will not experience G-loads in excess of its structural limits. Furthermore, aircraft are tested to withstand 1.5 times their normal design loads for a period of three seconds, which is more loading than an aircraft flying at Va can experience from any gust.

Maneuvering speed, however, is a function of aircraft weight, because stalling speed varies as the square root of weight. For a 10-percent reduction in gross weight, the stalling speed and, consequently, the maneuvering speed are reduced by about five percent; with a five-percent weight reduction, maneuvering speed is about 2.5 percent lower. Since manufacturers usually publish maneuvering speed only for gross weight, the pilot should calculate and be aware of the proper speed for the aircraft's actual weight when penetrating areas of heavy turbulence.

Even now, I can hear the howls. Some readers will argue that you can fly faster, not more slowly, at a reduced weight since total pounds of force fail wing spars, not G-loads, and total force is the product of G-loads times the aircraft's weight; although the gust loads are higher at a reduced weight, the force felt by the aircraft may be less. Others will agree that an aircraft should be flown more slowly when it's lighter because the reduced weight usually represents fuel that has been left out of wing tanks; without the weight of that fuel counteracting the aerodynamic forces acting on the wings, bending loads at the wing root will be higher than at full gross weight.

Both positions have their elements of truth, but neither explains nor refutes the position that a lighter weight requires a slower speed for protection against turbulence.

As fuel is consumed from wing tanks, the distribution of inertial loads changes according to which tanks are used. Burning 100 pounds of fuel from the tip tanks of a Cessna 310 has a different influence on bending loads at the wing root than does burning 100 pounds from a Travel Air's mains, which are located between the fuselage and the engine nacelle. To determine the precise influence of fuel consumption on relieving bending loads at the wing root, specific structural tests that simulate the exact internal weight distribution and the corresponding spanwise lift distribution would be necessary. They are not conducted, however, since flying at maneuvering speed assures that the aircraft will stall, thereby relieving the aerodynamic loads induced by gusts before structural damage can occur, regardless of the strength of the turbulence or the spanwise distribution of weight. The protection provided by flying at Va works because the stall acts as a safety valve; it is not based upon detailed evaluations or structural tests at midweight spanwise and chordwise loading conditions.

Since reducing an aircraft's weight lowers its stalling speed, reduced weight also lowers Va. Hence the need to fly more slowly, at the maneuvering speed for the aircraft's actual weight, if you encounter extreme turbulence.

If flying more slowly protects against structural damage from gusts, why not fly *much* more slowly—well below maneuvering speed? The slower the speed, the greater the risk of an upset due to reduced aileron and rudder effectiveness. A stall can also be caused by a less enthusiastic gust, although any gust-induced stall should come and go quickly as the aircraft responds to the turbulence. The maneuvering speed that corresponds to the aircraft's actual weight represents the safest compromise between flying too slowly for good controllability and too fast for structural safety.

How effectively can a light aircraft flying at Va penetrate extreme turbulence? You can be assured that the structure will stay together, but controllability still may be a problem. Airspeed indications will fluctuate wildly, so considerable attention to maintaining the proper pitch attitude and power will be necessary. Smooth yet prompt applications of aileron *and rudder* will be required to maintain a wings-level attitude. The ride will be hell, and the risks are high because you may lose control and be unable to remain upright at Va.

So don't be misled. Although Va is the speed to fly in extreme turbulence, do not consider maneuvering speed a guarantee of thunderstorm survival.

12 ATTITUDE + POWER = PERFORMANCE
by Thomas H. Block

It's said that Orville yelled across the sand dunes at Kitty Hawk, "Tell me, Wilbur, which do I use to control the airspeed, the throttle or the elevator?"

SITTING BEFORE me is an old piloting proposition: throttle controls altitude and elevator controls airspeed. Also before me is a letter emblazoned with the official emblem of the Federal Aviation Administration; upon it, I read the words: "Power controls *speed* and elevator controls *altitude*." I must now weigh these contradictory claims and decide which school of thought knows where it's really at. Make your decisions and place your bets, please. Stand up and be counted with either the old-time civilian airmen, on the side of elevator controls airspeed, or the FAA and military people, who say elevator controls altitude. Let us descend hand in hand into the pit of sticky verbiage and viscous obstinance.

The concept of using elevator to control an aircraft's speed has its origins in aviation's hazy past. Somehow, somewhere—perhaps under the wing of a biplane in some farmer's field, or in the back room of a World War I barracks—the idea of connecting the action of the elevator to the indications on the airspeed gauge first evolved into a recognizable form. Until well after World War II, most pilots controlled airspeed with the elevator; altitude with throttle; and used needle, ball, and airspeed to stay upright.

Slowly, pilots began to discover a new instrument: the artificial horizon. The military flight-training instructors were the first to make this discovery, although even they had allowed years to slide by before realizing the advantages of teaching pilots to control flight by attitude rather than performance gauges. The concept of attitude instrument flying was born, with its litany of "attitude plus power equals performance" as the crux of the new teaching experience. The which-controls-what relationship was soon to be reversed.

Many of the participants of the era say, "What we had been doing with our airplanes forced us to change the theory to match reality. If you watch closely, you soon discover that pilots control their altitude with the elevator, not with the throttle."

Sit in on an FAA flight-instructor revalidation course and you will hear the same thing. "The elevator controls not speed but altitude," they preach to scores of pilots who have yet to be baptized in this new faith. Some walk away as converts, while others absolutely reject what they've heard. Many cannot decide. Is the FAA attempting to legislate a change in the laws of physics, or have many airmen simply been following a procedure that's both a day late and a dollar short?

The truth is that neither elevator nor throttle independently controls airspeed or altitude. Moving the throttle simply creates more or less thrust; the elevator merely adjusts forces so that the angle with which the wing bites the air can be changed at the pilot's will.

When an aircraft's lift exceeds its weight, it gains altitude. Greater lift can be had by increasing either speed or the wing's angle of attack. An increase in airspeed can follow if one pushes in the throttle or lowers the nose. About the only completely accurate statement that can be made about the use of cockpit controls is simply that the elevator controls the attitude of the aircraft, and the throttle controls the power. If this sounds familiar, look back to the concept of attitude instrument flying for the clue: attitude (elevator control) plus power (throttle control) equals performance. Nowhere in the attitude-flying concept is there a mention of which aspect of performance is controlled by which handle. Obviously, the intent of the idea is to show that elevator and power combine to create performance; pilots from either of the two schools who argue otherwise are simply wrong. Even the much-touted FAA *Instrument Flying Handbook* is quite specific: "The pitch instruments are (among others) the altimeter and the airspeed indicator . . . The altimeter gives you an indirect indication of pitch in level flight . . . The airspeed indicator presents an indirect indication of pitch attitude . . . In this handbook, the altimeter is normally considered the primary pitch instrument during level flight."

If nowhere in the body of attitude-instrument-flying doctrine is it specifically alleged that specific aspects of performance are tied to specific individual controls, then why did the need arise to postulate such a connection in the first place? One reason was to provide a handy rule that could be used to expedite pilot training and guarantee a workable level of safety from pilot actions in various situations. Armed with their rule, say the advocates of each school, a new pilot can be trained more quickly and effectively and an old pilot made to fly more safely.

The FAA contention that we should all think of the elevator as the altitude controller has advantages in the sphere of training. One can present a logical case to show that 99 percent of the times when a pilot pulls or pushes the yoke, his intention is to capture a certain altitude or rate of change. Think about a typical flight and the truth in what the FAA says seems obvious. Cruise flight and instrument approaches are two examples of times when the elevator is apparently used to manipulate altitude while the throttle controls the airspeed. When ATC requests a flight to slow down, the pilot invariably comes back with the throttle, not on the wheel, thus further verifying the FAA observation that "this is how we really fly so it's the best method to teach."

The elevator-controls-airspeed people can be heard rising to the battle. "What about on takeoffs, or when the engine quits—or in a glider?" The FAA patches up their shot-through blanket with a qualification: "The concept works whenever power is both *variable* and *available* for the purpose of controlling speed. That covers most of our flying time, therefore making it the better concept to teach. Naturally, if the power is not variable and available, then the pilot must use the elevator to adjust his speed."

"How about spins?" ask the elevator-controls-airspeed group.

"Well, that's not a normal situation," counters the new school, "and our concept is used primarily as an aid to teaching." That sounds like qualification number two.

"See? Controlling airspeed with the elevator is safer because it always works, whenever there's any altitude left."

"But we don't really fly by controlling airspeed with elevator, so why teach it?"

"Because it's safer to use!"

"But it's harder to teach!"

Stop. Both groups are wrong. Their error lies in the premise that there was ever a direct connection between elevator and throttle on the one hand, airspeed and altitude on the other. We have erred because we have not thought about how we really fly.

Pilots control either airspeed *or* altitude with the elevator, their choice depending on which facet they're more concerned with at the moment.

Analyze a pilot at work and you'll see for yourself. During takeoff, he cares about gaining altitude, but he is more concerned about maintaining the proper airspeed—and the elevator assumes that important task. During cruise, the main concern becomes one of maintaining a constant altitude, and it becomes the role of the elevator to satisfy this new need. Normal descents usually are executed by reducing the power to an acceptable point, then flying a specific airspeed or rate of descent by playing elevator to achieve either one of these goals. While the major concern of the pilot may alternate between airspeed and altitude, one fact that remains constant is his manipulation with the elevator to accomplish these tasks. The FAA analysis that says we spend 99 percent of our time controlling altitude with the elevator is true simply because the real world of aviation is more precise about altitudes than airspeeds. We fly at *precisely* 8,000 feet but we accept whatever forward progress we can get.

Flying is actually accomplished by the concerted use of attitude (elevator action) plus power (throttle action). The best device to use to control the forces of flight is the elevator, because power is so difficult to adjust precisely. This hypothesis addresses another myth, this one surrounding flight control in turbojets. Since the throttle of a jet engine is less positive in its effect on flight (due to spool-up and lack of "propwash"), pilots find themselves adjusting the elevator more often and the power less.

The FAA is correct when it says people are best trained when they learn the methods that they'll really use. Provide a system that is both safe and effective, though, and everyone will be covered all the time. Rather than confuse the issue with synthetic techniques, both schools would be better off abandoning old lines and stressing the sole fact that stands out loud and clear: the throttle is an important adjunct, but the real business of flying is done with the yoke.

By teaching pilots to control their most pressing performance requirements by using the elevator, we have given them the most sensitive, accurate, and dependable device in the cockpit to take care of their greatest momentary need. Stall or spin recovery, cruise flight, and flying the glideslope to a bare-minimums landing are jobs that belong in the control department—which, quite literally, places those tasks directly in the hands of the pilot.

13 THE SECRETS OF SHORT-FIELD LANDINGS

by Thomas H. Block

THERE IS MORE to putting an airplane into a pea-patch runway than just guts and hope. Short-field landings usually mean flying the aircraft at the limit of its performance capabilities, and such landings often test a pilot's ability to analyze his situation and decide how close to his airplane's limits he wants to go. Knowing the length of the runway is obviously important, but it's just as important to size up the overall situation at the field. Approaches come before landings, so the pilot must first consider any obstructions that may protrude into his planned path. He must also note the runway-surface condition and accommodate his plan to the wind and weather conditions that pertain.

The airplane itself is the only factor that can be even remotely considered a constant in the short-field equation; however, landing-distance charts provided by the aircraft manufacturers actually describe near-perfect performance by new airplanes in the hands of experienced test pilots, and they should be treated with respect. These performances may be equaled or even bettered—but probably not with any pretensions toward superior airmanship. The secret, if it can be called that, is deliberate and judicious deviation from the two standards that are the principal variables in determining landing distance: touchdown point and touchdown speed.

The simplest form of short-field landing—one in which the size of the runway is the only consideration—is also the most straightforward to perform since reaching the intended point of touchdown and maintaining the intended airspeed are most easily managed. The manufacturers' charts are based on set criteria—for instance, 50 feet of altitude over the threshold with airspeed 30 percent above stall—but these standards need not be followed if the pilot is willing to cut into recommended and time-tested safety margins.

Any airspeed that can be flown safely 50 feet in the air can be flown with equal safety all the way down final. This is the main reason why experienced pilots advocate going to the selected approach speed while still some distance from touchdown. To fly an approach at erratic airspeeds, especially when the speeds being toyed with are nearer to stall than the customary 1.3 times Vso, is both unnecessary and stupid. There is no valid reason for not being well established at the selected attitude, power, and aircraft configuration while still a few miles from the point of confrontation with the ground.

The shorter the field, the more the pilot will extend flaps, slots, slats, and whatever other devices he can find to help him lower his aircraft's stalling speed and therefore fly it more slowly. How close to edge of stall the pilot can get is determined solely by his courage and his will to survive; he may, if he is so inclined, wallow down final to the serenade of the stall-warning horn.

It is vastly safer and easier to fly such an ultraslow approach through calm skies. Turbulence that disrupts the wing's smooth airflow commands additional airspeed to prevent gust-induced stalls. A surface headwind is always a help, but its effect can be diminished by accompanying turbulence, which forces the pilot to add a gust factor to his approach speed. Still, there is nothing quite so effective as a healthy flow of pushing wind, and some really short rollouts can be accomplished when 20 or more knots lie ahead of you.

While pursuing the slow touchdown, pilots have discovered one more device: pushing the throttle in to lower the stall speed. The slowest airspeed attainable comes just before the onset of a full-power stall, so any increase in engine power above the idle setting will yield corresponding decreases in the minimum speed at which the airplane will fly. This device can be used to magnify the margin above stall on an approach at a set airspeed or, if you're not hewing to a particular airspeed, to get closer to the airplane's absolute minimum flying speed.

The adage against flying low and slow should be forgotten when one is attempting to drop into a tiny airport. The dual advantages of lowered stall speed and increased ability to steer for a touchdown on a precise point of runway are reasons enough for mastering the power-on approach; a pilot who knows how to power the airplane on can aim at a runway and perform the heralded drag-it-in maneuver to maximum advantage.

The region of high-angle-of-attack, high-drag flight wherein excessive thrust is needed just to balance drag forces and sustain stable flight is called "behind the power curve." It can be demonstrated by flying in that stable region just before the break of a power-on stall where neither altitude nor airspeed is changing in spite of the combination of holding the nose high and maintaining full power. Airmen must tiptoe through this area, however, under threat of reprisal from the twin demons of stall and unchecked descent.

Theory says that a pilot who flies final approach with the sounds of maximum engine power and the stall-warning horn chorusing in his ears will be at absolute minimum speed during touchdown. Such an exaggerated condition is far too dangerous for life-loving people, but it represents the ultimate extent to which "power-on" can lower flyable speed. Experienced short-field pilots sometimes employ part of this area behind the power curve by using a measure of nose-up attitude and engine power, always leaving themselves enough reserve power to vary the rate of descent—or to call the whole thing off if need be.

Terrain and obstacles often prevent the drag-in technique from being a workable solution to short-field problems. A hybrid technique using a power-off approach with the airspeed just above stall provides both a faster rate and steeper angle of descent. To a highly courageous few, this type of final approach becomes little more than an incipient power-off stall held until very near the ground, at which time recovery is executed by firewalling the throttle—a move that simultaneously lowers the stall speed further and begins to accelerate the aircraft. Obviously, timing becomes a crucial factor in this exciting procedure, and any airman who attempts one but pushes power either too little or too late will find that he has nothing left to flare with but body English and nothing to stop his descent but the strength of his oleos. Both are likely to prove painfully inadequate. The flare-with-power gone sour becomes the ultimate example of a hard landing.

Pilots flying twins have it even tougher on short fields. The low airspeeds they must strive for when they attempt to get into tiny runways will often be less than the aircraft's Vmc speed. The engines will probably be grinding along at moderate power settings, and the pilot is really gambling that neither powerplant will sneeze. Recovery from below-Vmc engine failure in a conventional twin requires immediate power reduction on both engines and a forward, altitude-sacrificing push on the wheel to get speed back up to where power may again be safely added. Flying a twin below Vmc is a calculated risk, the adverse consequences of which can quickly turn all aboard into fond memories.

Once an airplane has been dropped onto the runway both as near the threshold and as slowly as possible, there remains the job of bringing this Cannonball Express to a halt before it reaches the far end. Runway surfaces like tall grass or mud help the airplane to slow down, but they may very well grab the wheels so quickly and firmly that the machine plunges over on its nose. Unless the pilot plans on trucking his aircraft out of there (or possibly mowing the lawn or waiting for a hot, drying sun), the takeoff that awaits him after a landing into soft stuff may be virtually impossible.

Many factors can influence braking or detract from the ability of the tire to grab the runway. Hydroplaning is not solely a turbine-era problem, and the effects of even scattered patches of ice and snow on a short runway will never be forgotten once they are experienced. Stomping hard and causing the wheels to skid is a much less effective braking method than pumping the brake pedals rapidly until the airplane has slowed.

The easiest short fields are those clean, hard, dry, and unobstructed runways where the pilot can select an appropriate airspeed and get his house in order a few miles out, spending the time on final with power up and airspeed down. The worst locale for a landing is a tiny runway surrounded by trees and wires on a day when the crosswind of the adjacent hills is kicking up a fuss and the windshield and runway are covered with rain. Between those two extremes lie situations in which the amount of runway needed to stop will depend on both the airplane and the pilot.

Experienced pilots fly short-field landings this way: come up on downwind with the landing checklist nearly complete and the airplane flying relatively slowly, so as to provide time to look over the arena about to be entered. By working your line of sight back from the touchdown point (the painted runway marks on the long practice runways provide the best targets), make note of obstacles that will affect the angle of descent and therefore the amount of engine power you can use during final. Put the landing gear and flaps down, set the prop control and do whatever other cockpit fiddling is required while it's still early, so your complete attention can be devoted toward runway, obstacles, and airspeed.

Hold off turning base until there is absolutely no doubt that the final approach will be long enough to allow precise control of both airspeed and rate of descent. More short-field landings are fouled up by turning a hot, high, and harried downwind-to-base-to-final than for any other reason. Extend the pattern; there is no rush. Airspeed in the pattern should be slow but somewhat higher than the selected final-approach speed to keep from stalling during turns and to minimize the ill effects of any bad aileron/rudder coordination. Ten knots faster is enough in most light planes if banks are no greater than 30 degrees and the ball stays somewhere near its allocated spot.

Turn final, reduce airspeed to its targeted value, and begin a descent. Airspeed control is maintained by elevator action; engine power controls rate of descent. Get low as quickly as obstacles will allow, then use the throttle to lessen the rate of the remaining portion of the descent. Quick descents can be made in whatever stages are necessary to match the irregularities of the obstacles, but don't get lower than obstructions will allow; that may seem an obvious enough observation, but it is a common trap, literally, to fall into. The urge to dive for the threshold after passing the last obstruction may be overwhelming, but it can only foul up the landing. Keep airspeed steady with the elevator and angle of descent where you want it with the throttle. Flare with a pull on the column (or a blast of the throttle if airspeed is very low), and touchdown should be on the spot and at the speed you aimed for.

By trying different techniques and by looking behind after each try to see how much runway he uses, a pilot can develop a realistic, practiced, and scientific method for landing on a short runway. It is so much better a response than flying wallowing, prayerful final approaches and hoping they'll end happily.

14 HOW TO STAY HEALTHY IN A TWIN

by Richard L. Collins

PARK A TWIN and a single on any ramp and pose the question, "What happens when an engine quits on either of the machines?" Simple on the single: it comes down—all the way to the ground. But what about the twin? A nonpilot might say that the twin will stay up. Some pilots might say the same. A light twin can deceive you, though: there are times when it needs both those engines. And when one of them fails, there is always a thin line between possible and impossible, between success and failure. That's why twins have a fatal-accident rate because of engine failure that is twice as high as the rate for singles.

My visit with the FAA light-twin experts at Oklahoma City was revealing. They are young and sharp people, a generation removed from the "license the survivors" group that used to require almost suicidal maneuvers during light-twin training. Gone is the virtual requirement to do Vmc demonstrations within 500 feet of the ground—a feat that caused many a tragedy. They have backed away from stalls, too, now requiring only the approach to a stall. The emphasis is on planning and knowledge of the airplane's performance capabilities, and it's a good thing. It's also good that they have an open mind on the subject—a far cry from the way it used to be.

Have they gone too far, though? Has light-twin training become too tame? Will we go from a situation where most of the accidents happened in training to one where most will happen out in the real world? That's something to consider, but at least there's a realistic attitude on the part of the FAA, and that's half the battle. The other half has to be fought by the pilot, in the airplane.

We'll try it in a Cessna 340A, a pressurized light twin that is close to the ultimate in capability for the owner-pilot. When they get much bigger, there's usually a hired pilot involved. The professional pilots break them too, but let's think in terms of the nonprofessional—the person who strives to fly well but has other things to think about besides.

To begin, let's load the airplane to near gross weight. You don't want to? Sorry, but if we're going to peer into its performance envelope and pull out its secrets, it has to be heavy. Let's do it with fuel and sandbags, though—no passengers.

I just happen to have a handy little card on which to list all the important speeds as well as the airplane's performance capabilities under these conditions. Important speeds: 91 knots, safe single-engine speed—a performance speed for Cessna twins used both for liftoff and to get up and over the first hurdle; 95 knots, best single-engine angle-of-climb speed, for the most altitude per distance traveled; 100 knots (blueline on the airspeed), best single-engine rate-of-climb speed, to gain the most altitude for the time involved; 108 knots, best multiengine rate-of-climb speed. Minimum engine-out control speed—Vmc—is 81 knots, but Vmc is a meaningless and grossly overrated number when you're considering the *performance* capability of an airplane, so we won't even list it. Vmc relates only to the amount of rudder available to offset the output of one engine.

Takeoff weight is 5,914 pounds—76 pounds below gross. Performance capability: a normal takeoff will require 2,580 feet. The distance required to accelerate to 91 knots and then stop will be 3,220 feet. The distance required to accelerate to 91 knots, lose an engine at liftoff and then climb to 50 feet is 5,370 feet. Available single-engine rate of climb is 250 feet per minute. Be aware that those figures are from the pilot's operating handbook and are dead accurate only if you use maximum effort in a new airplane.

The available runway is 3,825 feet. If one fails before 91 knots, there's no question: abort. It might go off the end of the runway because the run wasn't made from a full-power standing start with brand-new brakes, but at least it would be a soft ride through the fence and into the mud.

If one fails after 91 knots and before the gear starts up, we'll also abort. That'll mean a more vivid ride through the fence and deeper into the mud.

If one fails after 91 knots and after the gear starts up, a crucial decision is at hand. Flat country around here, but if an engine failure occurs between the time the airplane passes through 91 knots and the time it reaches 500 feet and at least 108 knots, a return to the airport is open to question. It might be necessary to deposit the airplane more or less straight ahead. Willing to accept that risk? Okay, let's go. Let's do a briefing like that before every takeoff, too. If I'm solo, I'll just talk to myself.

On the first takeoff, we'll depart normally, no engine-out-exercise. The FAA Academy procedure for a normal takeoff is to lift off at not less than Vmc plus five—we'll use 91 knots—then accelerate to multiengine best-rate-of-climb speed and climb to 500 feet above the ground using that speed and takeoff power. On reaching 500 feet, we'll accelerate to cruise-climb and reduce to climb power. That's a good procedure for every takeoff, in a single or a twin; it is purely dumb to reduce power early. Over 80 percent of all engine failures occur when power is changed, so it's silly not to climb high enough to have some options before reducing power.

Up at 6,500 feet now, we'll look at Vmc demonstrations and stalls. This is a turbocharged airplane; full power is available to 20,000 feet, so Vmc will be substantially the same indicated airspeed at this altitude as at sea level.

One engine throttled, reduce the airspeed gradually and use every bit of that rudder I need to keep it straight. It takes a lot of push. Some pilots get into trouble in twins because they don't use full rudder; instead, they try to make it go straight with the ailerons. Use of aileron against the yaw created by a dead engine actually tends to increase the yaw, and yaw at the stall makes an airplane want to spin.

Look at this nice airplane: it stalled before directional control became a problem. The rudder held it straight into and through the stall. Ailerons were held almost neutral. On the next one, I'll hold some aileron against the roll and won't use all the rudder—well, it still does okay, but my instructor-pilot friend in the right seat says he mishandled it in that manner at lighter weight and the airplane acted as though it wanted to spin. This would be more likely at lighter weight because the stalling speed is lower and the stall would come with less rudder effectiveness.

Angle-of-attack reduction—lowering the nose—has handled every loss-of-directional-control situation so far. Be ready to reduce power *if necessary* to maintain directional control, but remember that elevator is primary, power is secondary. Don't think in terms of chopping the power, either. That would be a bad thing to have stored in mind if an engine were to quit with the airplane a couple hundred feet high on initial climb. If this 340A were at 200 feet climbing at 91 knots, and if one engine quit and the pilot abruptly chopped the other, the airplane would hit the ground very flat and very hard. In that situation you'd want to get the nose down (primary) and reduce power as and if necessary (secondary).

There is one situation in which all power should be chopped, promptly. If the airplane is mishandled at the stall to the point that it cracks off into the beginnings of a spin, salvation lies in getting rid of all the power and using normal spin-recovery techniques. Power tends to flatten a spin; power on just one side increases the yaw that would make a spin develop fully.

Looking into the dark corners of a twin's characteristics need not be overstressed in training. They should be approached with care and only with a true expert in the airplane. Some twins are friskier when they stall than are others. Discipline in speed control is what really counts. If ever I get this bird that close to a stall down near the ground, the curtains would be all but closed, so the best lesson is to stay out of the dark corners.

Head for the airport now. Landings will be *simulated* single-engine. On some, we'll pull one engine back to simulate zero thrust, and during others, we'll just limit power on both engines to 45 percent—about 20 inches of manifold pressure and 2,400 rpm, which equals the power we'd get with one engine running wide open. Doing it that way keeps them both warm.

Every multiengine approach you ever fly should be planned so that less than half the airplane's total power will get it to the airport. None of that business of dragging it in with a lot of power on both engines. Fly normal approaches with, say, 35- or 40-percent power on final, and an engine-out approach will feel "normal." Well, almost normal.

Gear down at the usual time and place. Start down the imagined approach slope. The runway is long, but we'll consider it 4,000 feet in length and shoot for a landing on the first third. There, now, going though 500 feet above the ground, consider the airplane committed to a landing. No go-around past this point; if a donkey eating peanuts walks out on the runway, we'll just use the grass to one side. Also consider the airplane committed to a landing the minute full flaps are selected.

A few of these approaches worked well. Now let's look at how people load themselves into the boat on a single engine approach. We're always worrying about the engine failure on takeoff, yet that's actually but a part of both the challenge and the problem. Only a small percentage of the engine failures occur on takeoff, but *every* engine failure has to be followed by an angine-out approach.

We'll do it incorrectly this time. Gear and *full* flaps down too far from the airport. Look at the sight picture of the approach slope. At a constant airspeed and pitch attitude, the position of the runway in the windshield is moving higher, indicating that the airplane will reach the ground before it reaches the runway. The urge to pull the nose up—to control altitude with the elevators—is very strong. Try to stretch that glide. Slow it 10 knots below the blueline just to see. That gave a momentary indication of stretching the glide, but now the picture suggests that the airplane will reach the ground even farther from the runway than before. Blueline is an absolute minimum on approach. With gear and flaps down, the airplane has no available rate of climb at this speed, only a minimum rate of descent. When no climb is available, the speed for minimum angle of descent is above the speed for minimum rate of descent—the opposite of what is true when climb is considered. Stay a knot or two above blueline.

Getting rid of some drag is the only remedy left. The flaps have the most drag. Retract them. Oh my, what an awful sinking spell. The airspeed wandered a bit above the blueline there, and the sinking spell was reflected in a 400-foot loss of altitude. Maybe it would have worked better if the flaps had been only partially retracted.

You can get rid of the landing-gear drag without any sinking spell. You might wind up landing that way, but wouldn't a gear-up landing on the runway be preferable to a gear-down landing in the woods?

The big message, however, is that it had best be done right the first time. Second chances don't come easy when you're flying an airplane this large with only 310 total horsepower available.

My check-pilot friend is pulling an engine on every takeoff. Windmill it, go through the shutdown procedure, and then take it back to zero thrust to simulate a feathered prop. The engine cut comes as the gear starts up. On this one, we have the failed engine set a little below zero thrust, and it is apparent we won't make it; the terrain is outclimbing the airplane. Determining the numbers in advance is important, but if you end up in a hopeless situation, it should be obvious from what you see. Look at the airspeed and look out the windshield: the airspeed is on the proper value, but the airplane is headed for the ground. Accept that, maintain the airspeed, and put it in the best-looking field ahead. Engine-out twin accidents are serious because most are of the stall-spin variety. The reason they are stall-spin is because the pilot fails to maintain airspeed while trying to extract performance that simply is not there.

It is also plain that the business about controlling altitude with the elevators plants a seed that could kill you in an engine-out situation. Airspeed is so important, and I surely don't see any way to keep it on a healthy value with anything other than the elevators when you have only half your engine power available.

Let it get a little slow, a little below the blueline, and your head is in the alligator's mouth. Get the nose just a tad high and watch what happens to the airspeed. If it's low or decaying, the nose *must* be lowered, the angle of attack *must* be reduced. This relationship is critical, and if the pilot can't accept that discipline, there's likely to be trouble whether he's flying a single or a twin. It is extra important in the twin because of the strong temptations associated with having some power but not enough to fly in the manner to which you have become accustomed. In the twin the probable choice is full power or half power; in the single, it is full power or no power. The record shows that pilots perform better with the latter option. Shouldn't be true, but it is.

Would it be possible to do some of this demanding, low-power multiengine training in a single? Feathering procedures and the results of asymmetrical thrust can be experienced only in a twin, but, really, the most important thing is the ability to fly the airplane precisely, plus an understanding of the effects of the drag of gear and flaps on the approach slope. We could have done about half of this today in my Cardinal RG, using a maximum of 45-percent power on the approaches.

During these engine cuts on takeoff, don't be in too big a hurry to feather.

Sure, cleaning up is important, but first establish control of the airplane at a pitch attitude that will result in a safe airspeed on one engine. Then go through the dead-foot, dead-engine routine: bang—there it goes; it's the right one. Left foot is pushing on that rudder pedal, right foot isn't working. Now, without looking at the levers (that can be distracting), put your hand on both throttles and pull the one on the right side—the same side as the dead foot—back. That didn't change anything, so that engine is verified as dead. Now, hand on the prop controls. It was the one on *this* side, verified by foot and throttle. Call it the right one or just think of it as "the one on *this* side." It's hard to remember left and right during the bad moments, as my old drill sergeant at Fort Chaffee can testify.

If everything is done properly up to the point of failure, feathering is the only mechanical chore left. The checklist was run carefully before takeoff; I know that all switches, valves, bells, and whistles are as they should be. The airplane was lifted off at the proper speed, and if the engine fails before I start the gear up, it is an automatic abort. Once the gear starts up, its drag is no longer a factor; all that's left is to feather the propeller. I get only one chance at feathering the correct one, so it's important to be methodical about the whole thing.

All that zero-thrust and reduced-power flying was good exercise, but what about the real thing? How about an ILS to minimums under the hood with one actually feathered? If one engine ever quit while I was tooling along IFR at FL 210, things would seem easy at first, but that critical approach might become the task of the day. Should I turn to my passengers and say: "Now, be calm, I have never shot an approach to minimums with a prop actually feathered, but I have a lot of *simulated* engine-out landings in my log?" No way.

Okay, the right one is feathered. You say you might use the autopilot for at least the beginning of the approach, to unload the pilot? I'll turn it on, just to make you happy. Look at that, it's goosey in pitch. There's no way to use just the heading/nav portion of this autopilot, so we'll dispense with it entirely.

Cleared for the ILS approach. No radar here, so I'll use the ADF to ensure localizer interception before reaching the outer compass locator and monitor DME for approximate distance.

Running level with 75 percent power on one, gear and flaps up, indicating well above the blueline (best single-engine rate-of-climb speed, which will be a reference speed all the way in). I won't go below that speed. In fact, I'll stay above it. You cannot store altitude on an ILS; you have to fly the glideslope. If I get it on that glideslope at minimum safe single-engine speed and then sag below the descent path, I'll be flying headlong into a black and aceless hole.

By the marker now, start down. Cleared to land. Extend the gear. Note how much power it takes to track the glideslope with the gear down while flying about 15 knots above blueline—a bit over half throttle. That isn't bad; there's some power left for dealing with the wind shears and sinkholes of the world. That extra 15 knots is also money in the bank to trade for altitude. I have drag to add for more rate of descent but I have little to cash in for more up.

One thousand feet above the ground. Go to approach flaps but no more. That extra 15 knots looks even better now. Add a little power to keep it there. Five hundred feet above the ground now, committed to the landing. Still have 15 knots in the bank. One hundred above DH. A touch above the glideslope, on the localizer. The feathered prop to my right is very real. What would I do if the clouds were also real instead of simulated, and I didn't see a runway at DH? I sure as hell wouldn't go around. I would keep those needles centered and put this airplane on that airport. No go-around below 500 feet on one engine. No exception.

Decision height. Look up, runway, go to full flaps now, powering back, the float will be longer and landing will be a little different with one feathered and the other engine idling. Don't think about that. Just fly, using controls as necessary to land.

Good firm arrival. No doubt that we are on the ground. Brakes now, try for the first turnoff. Made it. There was no wind, and I turned off 2,850 feet down the runway. That extra 15 knots of airspeed sure didn't hurt anything, and it might have saved my tail in a pinch. And it was the only extra factor in my favor during the approach.

Soda-pop stop's over now; let's go back out to fly again. The realism of that approach with one feathered was worth a hundred with zero thrust. There's simply nothing like the real thing.

You say you want to do an actual engine-out on takeoff? Okay by me.

There's 91 knots. Lift off. Failure. Lower the nose. It's the right engine. Identify, verify, feather. The airplane seemed to keep climbing a little after the failure, then to sag. Its upward inertia was probably worth 50 or 100 feet. Now the right prop is feathered, we are 100 feet high, the airspeed is on blueline. And yes, tower, that right prop certainly is feathered. We told you we were a training flight, and we'd like to come around and land.

Couple of hundred feet a minute rate of climb. No obstructions. Clearly there would be no call for an off-airport landing. I'll climb straight ahead to 500 feet, then come back around and land on Runway 13, the runway of departure. You wouldn't? You would bring it around, starting now, and land on Runway 36? Okay, we'll do that.

This is the real thing again. Even though the airplane has unfeathering accumulators, it might not be possible to get that engine back on the line instantly if we got into trouble. In a right turn, blueline plus five, maintaining altitude in the turn, maybe 300 feet high. There's the runway. The gear is coming down. I am almost to the point where I can glide to the runway. Come on with the flaps, on to full flaps, power back a little, blueline plus a couple, hold that, the sight picture suggests we'll land about 500 feet down the runway.

Another firm arrival. Turn off 3,000 feet down the 4,250-foot runway.

That worked well. Project the situation to a day 20° warmer, an airplane 300 pounds heavier and a runway with rising terrain off the end, though, and I'd have had to land the airplane somewhere out in those fields. It is as important to know the forced-landing spots in a twin as in a single; more important, perhaps, because the lower stalling speed of the single makes it easier to fit into a smaller space.

The 340A has good engine-out manners, but regardless of the airplane, mental discipline is the key to long life in a light twin. A good small twin is so serene and comfortable as you fly along on one engine that it's difficult to accept the possibility of an off-airport landing. But if the time were ever to come, it would have to be accepted in advance. If not, the airplane would make the acceptance for you. Justice is swift when the laws of performance and aerodynamics are flouted.

What about the business of actually feathering propellers at tender moments? We were operating that twin Cessna in situations where the allowable mistakes could be counted on my thirteenth toe. The check pilot had to be supersharp and had to know the airplane—an absolute requirement for this sort of thing. Also, the flying was done at a big airport where there were no congested areas beneath the flight path.

We've been flying today on the dark side of what is usually done in multiengine training and proficiency operations, but I'd rather do it this way than place great faith in simulated training and the hope that neither engine is likely to fail; I opt to take risks in training to minimize risk elsewhere. The record shows that the average twin pilot is twice as likely to die because of an engine failure as is the average single pilot, and I just don't want to be an average twin pilot.

It is easy to see why that average twin pilot does so poorly after an engine failure. True, every current light twin has the performance to get to an airport except in the rare instances when an engine fails over high terrain or before the airplane reaches an altitude of 500 feet, but they don't allow room for mistakes. The twin demands perfection in the planning and execution of the engine-out approach and landing. Let the airspeed get 10 knots slow, put the flaps or gear down prematurely or give up altitude before the magic moment, and the airplane might well be in a position where an accident is inevitable.

So I'll be back for the same exercises three times a year. With them, I'll feel there is at least as little risk to life and limb from engine failure in a twin as there is in my single. There might be even less.

15 FIVE WAYS TO SAVE FUEL

by John W. Olcott

A MODERN AIRPLANE is an efficient transportation device—much more efficient than an automobile, for example. When operated at the proper power setting for maximum miles per gallon, a typical single-engine, four-place retractable can go about twice as far on a gallon of gas as a car. Even a fast twin such as the Cessna 310Q, with a total of 520 hungry horses to feed, is capable of about 10 mpg, while lighter twins such as the Beech Travel Air and the Piper Seneca can squeeze nearly 14 miles out of a precious gallon of fuel. In terms of passenger miles per gallon, light aircraft with reciprocating engines rival even a 747 for efficient use of fuel. Recent CAB operating figures indicate that the Boeing 747-100 quaffs about 3,350 gallons per hour. Assuming a 520-knot cruise speed and a 350-passenger load, the big bird requires 1.6 gallons to transport one passenger 100 statute miles. A fully loaded 310Q cruising at maximum-range power needs only 1.69 gallons to carry one passenger 100 miles, and a four-place single needs about 1.4 gallons to do the same task. The lightplanes we fly do an excellent job of moving people and goods efficiently, a fact that should not be overlooked during this time of fuel shortages.

The secret to obtaining maximum fuel economy is knowing at what airspeed to fly. Naturally, proper leaning is essential, and an EGT probably is worth its weight in gold these days. However, even the best leaning techniques will not allow you to cope with the fuel shortage if you insist upon cruising at top speed.

Typical general-aviation aircraft have recommended cruising speeds and cruise-power settings far in excess of what is required for maximum fuel economy. Maximum economy occurs at maximum-range airspeed, which is achieved at a relatively low power setting, usually less than 50 percent; for a fully loaded Cessna 310Q, for example, maximum range and therefore maximum fuel efficiency occurs at about 45 percent of rated power. Pulling 75 percent in a 310Q at 2,500 feet produces about 7.5 mpg, yet at 44 percent power at 7,500, you can improve that by nearly 33 percent. Your cruising speed will only be about 20 percent slower, a very small inconvenience for the ability to survive in a time of fuel cuts. Cruising the 310Q at 45-percent power instead of 65 percent increases miles per gallon nearly 17 percent at 7,500 feet, and that includes the climb. On a 500-mile trip, the slower, more efficient speed only adds about 30 minutes to the trip, and the slower speed is quieter and generally more enjoyable.

In theory, maximum range—and thus maximum fuel efficiency for propeller aircraft—is achieved in still air at the speed where the ratio of lift forces to drag forces is maximum. In practice, that exact speed is rarely identified as such in the flight handbooks, but it is not absolutely necessary to know it precisely to enjoy the benefits of better fuel consumption. It wouldn't be a bad idea, nonetheless, for manufacturers to begin issuing their individual recommendations for most fuel-efficient operation, starting with specific power settings and airspeeds. Flight handbooks usually present range figures at power settings varying from around 40 to 75 percent. If you study these charts, you will be able to identify what power setting produces the best range. Use that setting, but select the lowest rpm that is approved by the manufacturers. Propellers are more efficient at low rpms because the blades are taking a bigger bite out of the air and the tip speeds are lower.

There are several concepts to keep in mind when attempting to save fuel on a cross-country trip. First, the speed for maximum fuel economy is not the minimum power required to keep the plane aloft. Minimum-power setting produces maximum endurance, which is fine for holding patterns and flights where your aim is to prolong the time you can spend aloft, but the relationship between how much fuel you are burning compared with how fast you are proceeding toward the destination is what determines best economy. Second, the best fuel-efficiency speed can be influenced by headwinds and tailwinds. You can afford to fly slower and save fuel with a tailwind, but you must fly faster into a headwind. To illustrate this point, consider flying into a headwind the velocity of which is equal to the zero-wind maximum-range speed. No progress would be made toward the destination; thus, miles per gallon would equal zero. But if you flew faster, at least you would be getting somewhere, and the miles per gallon, of course, would be greater than zero. As a rough rule of thumb when flying into a headwind, add enough extra power to increase the recommended maximum-range speed by an amount equal to the headwind component, and when flying with a tailwind, reduce power sufficiently to fly at the maximum-range speed less half the tailwind component.

The indicated speed for maximum range does not change much with altitude, but it is altered slightly by weight. For a 5,000-pound light twin, the indicated airspeed for maximum range should be reduced about one percent for each 100 pounds the plane is under gross; for a 2,500-pound single, reduce the recommended power so that the maximum range IAS is slowed by about two percent for each 100 pounds under gross. As the plane burns off fuel, the indicated airspeed should be reduced accordingly for maximum mpg.

Fuel efficiency will be better when the plane is light, so don't carry around any unnecessary weight. From the viewpoint of getting more passenger-miles per gallon, however, make the most of each trip by using all available seats. Also, with the uncertainty of being able to refuel completely at the destination airport, it is a good idea to keep the tanks full.

In practical terms, fuel efficiency is not affected by altitude. What slight benefit there is in flying higher in an unsupercharged aircraft generally is negated by the fuel and time required for the climb. Cruise altitudes should be selected on the basis of winds aloft and the anticipated duration at the cruise altitude. If there is a headwind, stay as low as possible to minimize the time spent at high fuel-flow rates needed for the climb and to maximize the amount of the flight at lower altitudes, where the headwinds should be less. Remember to increase your power setting just enough to add the velocity of the headwind component to the best-range indicated airspeed.

If there is a tailwind at altitude, go after it, provided you plan to ride it long enough to overcome the extra time and fuel needed for the climb. Climb time should be kept to a minimum; thus, in theory, best-rate-of-climb airspeed should be used. In practice, not much is lost by cruise-climbing to altitude, and forward visibility is considerably better at cruise-climb angle of attack. In the interest of safety, use factory-recommended cruise-climb numbers; it costs little in fuel efficiency. Select your cruising altitude by considering: 1/the difference between indicated cruise airspeed and climb airspeed, 2/the duration of the climb, and 3/the ratio of the climb fuel flow to the cruise fuel flow. Multiply all three factors and divide by the amount of tailwind available at the selected altitude. (Remember to use knots for airspeed if you use knots for the wind.) The result is the number of minutes you'll need to remain at altitude to gain back extra fuel used to climb there in the first place. If it's to be a short trip, there is no point in wasting a lot of time and fuel to climb up after it.

For example, consider a Cessna 310Q climbing at 120 knots IAS and burning 28 gph. Taking off at gross weight, the plane could be at 10,000 feet in about 12 minutes, and it would have consumed about 6 gallons of fuel for the climb. Once established at a 45-percent power cruise, the 310 would be indicating 128 knots and would be sipping 17.8 gph. The product of airspeed difference (8 knots), duration of climb (12 minutes), and ratio of climb fuel flow to cruise fuel flow (27/17.8) is 8 x 12 x 1.57, or 151. Therefore, with a 10-knot greater tailwind at 10,000 feet than at sea level, the pilot needs to spend about 15 minutes at altitude to make up for the extra fuel he used for the climb.

For this example, even a slight tailwind aloft is worth pursuing, but for another aircraft with a larger difference between climb and cruise airspeed and a lower rate of climb (the fuel-flow ratio usually will be about 1.5 and thus doesn't differ much between planes), the numbers might indicate that it is better to accept a 10-knot nudge at 3,500 feet than to struggle up to 10,000 for a 25-knot push.

For the letdown, it is better for fuel efficiency to remain at the selected cruise airspeed and reduce power for the descent. The most efficient speed is determined mainly by the aircraft's aerodynamics; attempting to use more power than is needed to hold that speed is technically inefficient. Plan your letdown so that rpm and manifold pressure stay within safe operating limits, thus assuring proper cylinder-head temperatures, and start the descent so that you are at pattern altitude close to the destination airport. The power you don't have to add at the end of the ride adds to the total fuel efficiency. Reducing manifold pressure five inches results in a 500-fpm rate of descent at cruise airspeed, and figuring two minutes for each 1,000 feet of descent makes it easy to calculate when to start down, provided you have some notion of your ground speed.

Whenever possible, fly the most direct route VFR. IFR flights require 10 to 15 percent more air time, which takes a big bite out of fuel efficiency. When IFR is necessary, telephone the tower—or radio them before startup—to confirm that they have your clearance and that there will be no delays on the ramp. Remember, anything you do to minimize unproductive engine time saves fuel and adds to the efficiency of the lightplane.

16 STRETCHING IT
by Peter Garrison

FLYING AN airplane efficiently must be an end in itself; if you try it in spite of yourself, you will end up not doing it at all. Efficient flying can produce large savings of money—enough to pay half the cost of an engine overhaul. But saving money is not reason enough for such a quest; for one thing much of the savings must be plowed back into the equipment that is necessary for careful flying in the first place, such as an EGT gauge. Operating an airplane is a costly pastime, and skipping one trip will save more money than flying efficiently on ten others; so to be honest, the best reason for seeking efficiency in flight is the love of it. You have to feel that to use less gas, produce less smoke, make fewer waves is beautiful; then, all the rest will follow.

The first element of efficient flying is the airplane.

Some planes are easier to get the most out of than are others. Instrumentation is a big variable. Proper mixture control, for instance, is most easily accomplished using EGT and CHT (exhaust-gas temperature and cylinder-head temperature) gauges and a vernier mixture control. The EGT should have an expanded scale, with 25-degree ticks not less than a sixteenth of an inch apart. The type of gauge that shows a range of 1,000 degrees is not sensitive enough for fine leaning.

A constant-speed prop is also a useful tool for increasing engine efficiency. Fuel injection usually provides better fuel distribution and a more uniformly lean mixture than does carburetion.

Fuel-flow gauges, on the other hand, are mostly an invention of the devil. Though they are marked off in percentages of power and other glyphics that would lead you to believe they can be used for setting the mixture, these gauges are actually quite useless for this job. I am not really sure what they are for; their utility and entertainment values are equally slight. The only kind of fuel-flow gauge that is both interesting and useful is the digital type, which measures actual flow, not pressure; it is still no good for leaning, but for flight-planning purposes, it gives precise insight (especially while en route) into the rate at which fuel is being used and the amount remaining aboard.

There is a right instrument for each task; it is the one that provides the most direct window to the information we need. For setting the mixture, for instance, the EGT is the rational choice; it gives a sensitive, precise real-time picture of the combustion condition inside the cylinders. The CHT gives a direct reading of the suitability of the chosen mixture for the flight conditions. On the other hand, the unholy conspiracy of mp, rpm, fuel flow, and a power chart is merely a contrivance for multiplying errors in the pursuit of an approximation.

Apart from having the best, most informative instrumentation in his airplane, the pilot can do a few things to enhance the airplane itself. A well-maintained engine with good compression and proper sparkplug gaps and timing will be slightly more efficient than a tired, leaky one; a clean wing, free of annihilated bugs and bird droppings, will be slightly more efficient than a dirty one (though mere dust or an unwaxed surface makes no difference, the claims by makers of "speed waxes" notwithstanding). A propeller that has been dressed and sanded to a velvety smoothness will help; all the power that drives the airplane is conducted through the propeller, and a nicked prop is as bad as a dented wing. Good door seals can also be important.

All these slight benefits add up, I don't know how much, but say to five percent, give or take four. The difference isn't dramatic—if it were, we would see more people sanding their props and shampooing their wings every weekend—but it's bound to be there.

Loading also has some effect on fuel consumption. Because conventional airplanes are designed to fly with a downward load on the horizontal tail most of the time, and because that load produces some drag (just as the life of the wing must also take its toll of drag), it is desirable to keep the center of gravity fairly far aft to reduce the amount of work the tail has to do. Disposable loads that can go anywhere should therefore go as far aft as possible—into the baggage compartment rather than onto the backseat, for instance. Beware of loading too far aft, however, since stability is affected adversely. Usually, loading is of little practical value; most people are not such ruthless pursuers of efficiency that they would load their passengers into the backseat and ride alone in the front just to get rid of some trim drag. Some pilots slide their seats as far back as possible once they are cruising, with their feet on the floor; they claim it does worlds of good, but I doubt it. I expect it would do a little good—another one of those "slightlies."

Another "slightly" can come from carrying only as much fuel as you need for a trip and refueling more frequently while taking on less fuel each time. Many of us fill up whenever we get low and then ferry hundreds of pounds of fuel around with us on a lot of short trips. At six pounds per gallon, 40 or 50 gallons are the equivalent of two passengers, and extra weight takes its toll in fuel consumption. Performance suffers too—and a lightly loaded airplane is more fun to fly than an overburdened one.

So much for the equipment: a well-maintained, clean, lightly loaded airplane will be more efficient than a rough, dirty, heavy one; a well-instrumented airplane will be capable of flying more efficiently, and its engine will get more considerate treatment; but the real differences are among pilots and their strategies. Between a pilot who does everything wrong and one who does everything right there may be a difference of 25 percent in overall fuel consumption. At current fuel prices that would amount to more than $3,000 over the TBO of a 285-hp engine—whence comes my figure of half the cost of an overhaul.

Good strategy begins with trip planning. On an extended business itinerary, avoid situations where it is necessary to get from one place to another at top speed. A more relaxed schedule, with travel taking place at the end of days spent on business, will avoid a heart attack and also encourage lower, more economical cruising speeds. Planning also can avoid doubling back or making excessive doglegs. While en route, flying direct legs between omnis spaced far apart—by using a little dead reckoning in the middle—will help smooth out the jigs and jogs of the airways, though the distance penalties of their indirections are smaller than you'd think. It takes a real fanatic to justify an RNAV on the basis of fuel savings, except perhaps in the Northeast, where the shortest IFR routing between two points seems to be a convergent spiral festooned with holding patterns.

With range and routing determining the two-dimensional view of a trip, altitude lends the third dimension. The choice of an altitude at which to cruise is one of the thorniest and perhaps most controversial problems for efficient flying, because it requires us to weigh a number of rather obscure variables against one another.

On the one hand, most pilots are aware that the true airspeed available from a given percentage of power increases with altitude. It follows that the mileage is greater at altitude, and it can be seen that the best mileage is obtained, for each percent of power, at the altitude at which that percent of power is the maximum available. For a normally aspirated engine, this means that altitudes between 10,000 and 16,000 feet will render the best combination of low fuel consumption and high speed, at a power setting consisting of full throttle, a lean mixture, and the highest possible continuous rpm.

High rpm is not classically associated with good economy; the faster the engine turns, the more horsepower it wastes in overcoming its own internal friction, and so "correct" procedure calls for selecting a low rpm setting and the highest manifold pressure that the engine manufacturer allows at that rpm—such information can be found in the engine operator's handbook. The old saw that the engine should never be run oversquare—that is, with an mp numerically higher in inches than the number of hundreds of rpm—is extremely conservative; it avoids the potential problems of excessive cylinder pressure, all right, but it is as though in order to avoid wolves, one were to stay locked up in a fortieth-floor apartment in Manhattan.

With a turbocharged engine, it would be possible to fly at high altitude, where the speed is, and still keep the rpm low and the mp high, where the economy is. With a normally aspirated engine, a compromise must be made; and since the speed gained from a higher altitude is greater than the efficiency loss suffered from a higher rpm, we have cast our lot with altitude, except under certain limiting conditions.

Sometimes we may not want to go as high as we would to get the maximum speed from our selected cruise percentage of power. An assigned IFR altitude, cloud layers, an icing level, a passenger with a cold, head winds aloft, or a trip too short to justify the long climb—any of these might cause us to fly at a much lower altitude than the optimum. In this case, we can turn to our second line of tricks for saving fuel: control of manifold pressure and rpm.

The engine runs most efficiently when the throttle is wide open; to start with, then, we always would like to cruise at full throttle. Next, we would like to cruise at the lowest rpm that fits within engine limits and produces the speed we want. In practice, this means that at altitudes between 5,500 and 8,000 feet, you can get about 55- to 65-percent power at between 1,800 and 2,100 rpm and full throttle. At still lower altitudes, full-throttle operation has to be abandoned in favor of maintaining low rpm and a moderate percentage of power.

The absolute maximum range of a pistol-engine airplane is achieved at the altitude at which the maximum power available produces an indicated airspeed equal to or slightly greater than the sea-level best-rate-of-climb speed—since this speed is very close to the best L/D speed, which is, theoretically, the ideal for maximum range. In principle, the range at this indicated speed would be the same at all altitudes, though the distance would be covered more rapidly and with a proportionately higher hourly fuel consumption at high altitudes; but in fact, the engine runs slightly more efficiently at high altitude because the back pressure on the exhaust system is lower, and so the range at altitude is perhaps five percent greater than it would be at sea level. This most economical altitude happens to be quite high in most cases (though the lower the performance category of the airplane, the lower the altitude) and for a typical 285-hp retractable single, it might be up around 15,000 feet or higher. It is not often practical to climb to this high an altitude; the fuel used in the climb may not justify the mileage gain over a more moderate compromise altitude, such as 10,000 feet. If there is a tailwind aloft, however, a climb to 15,000 or higher may be quite justifiable.

The common notion that fuel used in the climb is lost forever is only partly true. Much of, but not all, the fuel can be recovered during the descent, provided that: 1/the descent is stretched out over a longer period of time than was spent climbing, preferably twice as long; 2/the descent is made by reducing power, not by increasing speed; and 3/the climb is made in the most economical way. To climb economically, one should pick a speed equal to or slightly greater than the best-rate-of-climb speed and continually adjust the mixture to the "best power" range (100° to 150° F. on the rich side of peak), enriching further only if engine cooling requires it. There is a delicate balance among engine temperature, mixture, and forward speed. Usually, by the time you have passed 6,500 feet, you are using less than 75-percent power and can lean to a cruising mixture setting. A flat "cruise climb" gives the impression of hastening you on your way, but in fact, you end up at the same place in the same time as if you had climbed at best rate.

At altitudes above 10,000 feet, the rate of climb becomes fairly low in most normally aspirated airplanes, and the climbing seems to go on forever; but by this point, your true airspeed is fairly high and you can run lean without overheating, so a high-altitude climb is not as wasteful as it might seem. It is more difficult, however, to balance a very long climb with an even longer descent; there is a limit to how gradually you can descend.

With a constant-speed prop, the descent should be initiated by reducing the mp. With a fixed-pitch prop, you have to throttle back. At first, it seems as though, despite the power reduction, the airplane does not want to descend; then we usually make the mistake of reducing the power further, whereupon the descent becomes too rapid. A reduction of 200 rpm or two or three inches of mp will eventually settle into a good, gradual descent; be patient.

From the fact that the absolute maximum theoretical range of the airplane is essentially unaffected by altitude, you might infer that there is no point in climbing for economy at all. Indeed, most of the legendary long-distance flyers, such as Max Conrad, select low altitudes for their flights. Conrad flies right on the deck. This is a perfectly correct procedure, especially with a heavily laden airplane that will not climb well. Assuming no increase in engine efficiency with altitude, the only sacrifice is speed. It is always possible to get more or less the same number of miles per gallon at sea level as at altitude, but at sea level, the true airspeed for a given mileage is much lower than at high altitude.

Tailwinds obviously contribute to fuel savings; they contribute most if, rather than cruise at our habitual speed and save time, we reduce power and make good at the same groundspeed as we would in still air. On the other hand, we have to speed up for a headwind; the only rule of thumb I have heard on this subject is to increase speed by a quarter of the headwind component. I do not know how that figure was derived, but it is probably better than no figure at all.

By the way, the Creator has deigned to make the majority of winds disadvantageous to aviators, perhaps in reprisal against their presumption. Even a direct crosswind results in some effective headwind, and if the wind is strong, the crosswind can be slightly from the rear and still slow us down. Slow airplanes are more vulnerable than fast ones. Hence arises, no doubt, the widespread feeling that there is always a headwind.

Winds are the great complicating factor in altitude selection. The question is always asked, "How strong does a tailwind have to be to justify climbing a few thousand feet higher to get it?" The answer depends on the trip length. Also, the relationships between wind speed, trip length, and fuel consumption while climbing are so involved and so variable from airplane to airplane that it is useless for me to try to evolve a rule of thumb. I simply proceed by instinct; I will climb fairly high for any trip, because I know that higher up I will get good speed, smooth, cool air, and low fuel consumption. I will go to 10,000 feet for a trip of an hour or more and to 15,000, if I am alone and have a good oxygen supply aboard, for a trip of two hours or more. (I find high-altitude flying—though my plane is normally aspirated—very comfortable and aesthetically pleasing, and so I tend to favor it even when it is perhaps not defensible on grounds of fuel consumption.) On the other hand, for a flight of half an hour or 40 minutes, I generally climb to somewhere around 6,000 feet, set full throttle, 2,000 rpm, and the leanest mixture I can and try not to think about the comparatively low true airspeeds I'm making. A really short trip might call for a storybook power setting like 21/1,800, just for the sake of feeling virtuous. The gas gets ridiculously low—120 knots on an airplane capable of cruising 175—but the mileage goes up as high as 24nm per gallon.

I use a strong headwind as an excuse to see high true airspeeds; with a headwind of 30 knots or more, I just fly at between 70- and 75-percent power. You have to sometimes just to get your juices going a little.

Whether these procedures are correct I don't know; they are probably typical of what pilots work out for themselves in the altitude-selection puzzle.

The aim of flying efficiently—a modest enough one, you would think— seems to lead us into a tropical proliferation of internecine variables. Back pressure would send us up, manifold pressure down; our pocketbooks urge us to slow down, our buttocks to speed up. Amid the wind, the cloud, the angle of attack, and the brake mean effective pressure, we hardly know where to turn. Fortunately, a simplification is at hand; it resembles Nirvana in that, once we arrive at it, we discover that we have been there all along.

A study of parameters for long-range flying made at Lockheed during the development of the P2V Neptune patrol airplane brought out the fact that the ideal maximum range of the airplane, the so-called "Breguet range," after the "Breguet equation" by which it is calculated, demands a protocol that pilots dislike: namely, that as the plane gets lighter, it must fly slower; and so, as the destination draws near after hours of flying, pilots are obliged to go slower and slower, and defer, it must seem asymptotically, the termination of their ordeal and the surcease of their sorrow, not to mention a visit to the john. Furthermore, even if a mean speed were selected and flown continuously, it was too slow; nobody could stand it. The Lockheed investigators concluded that by flying at 60-percent power, rather than the 35 to 40 percent required for the best-range speed, a range not much inferior to the theoretical was obtained, and it was done in much less time than in the idealization of M. Breguet.

Oddly enough, this is pretty much what we have been doing all along. Perhaps not quite 60 percent; perhaps 65. But it is not a difficult thing to go from 65 to 60; not half so difficult as from 75 to 45.

About 10,000 feet turns out to be a good compromise altitude with a constant-speed prop, and usually somewhat higher for a fixed-pitch—it depends how close to redline the maximum continuous rpm is. They've been getting higher lately. If a high altitude is undesirable for one reason or another—freezing levels are a typical good reason—then a good setting might be something like full throttle and 2,050 rpm at 6,000 feet, with peak EGT; it is quiet and comfortable, producing a moderate but respectable speed and a low fuel flow; try it—you'll like it. Conversation is easy at low rpms, and engine life is probably lengthened by this kind of operation. Measured by tach time, it certainly is; most tach-hour meters assume a much higher cruising rpm than this.

It turns out, after all, that the most important elements in efficient operation are, as you would expect, simply power setting and mixture. So far as power setting is concerned, the more you can stand to slow down, the better your fuel efficiency will be, all the way down to the best-range speed, which is only a little more than the best-rate-of-climb speed. As for mixture, once again, I rely on my practical approach toward efficiency. I think it is essential to stop flying at the "best-power" mixture setting, as is typically done in fixed-pitch airplanes by setting the mixture for maximum rpm. Instead, the mixture should be as lean as it can be without losing smooth operation, provided the manufacturer's recommended procedures are not violated. Use best-power mixture only during the climb, and then only above 3,000 feet. During the descent, reduce power rather than increase speed; and let the engine run as lean as allowable. It will get leaner and leaner as you lose altitude, but so long as it runs smoothly, there is no harm. (Continental recommends against mixtures leaner than 25° F. on the *rich* side of peak, without explaining how this rather fine margin is to be maintained without an EGT probe at each cylinder. I myself ignore this recommendation entirely; but then again, it is my ass and my warranty that are in jeopardy.)

Like economical driving, economical flying is just a matter of good habits; it can be practiced by anyone willing to shed his addiction to jackrabbit starts and breakneck speeds. It is unrealistic to expect pilots to vie with one another in splitting hairs over optimum this and optimum that; there are too many other factors to be considered in flight planning, many of them more compelling than the desire to save a pint of gasoline. A small airplane will, however, yield gas mileage competitive with, and in some cases greater than, those of automobiles, while carrying similar loads at much higher speeds; and with the cost of gas high and undoubtedly going higher, there is increasing reason—even apart from the beauty of it—to try to wrest a few extra miles from each gallon of gas.

17 SOARING TECHNIQUES FOR POWER PILOTS

by William Langewiesche

TURBULENCE IS a nuisance. It increases a pilot's workload and reduces a power plane's performance. Yet to the glider pilot, turbulence is the breath of life. Turbulence means free energy, waiting to be converted into altitude, speed, or distance. The trick the glider pilot uses to cash in this energy bonus is an easy formula: spend more time in the updrafts and less time in the downdrafts. The way the glider pilot goes about doing this depends on the kind of turbulence he's flying in.

Most turbulence is thermal—air rising because it has been heated by the warm ground. Some glider pilots visualize thermals as columns of ascending air much like smoke from a fire; others think of them as rising bubbles. It doesn't matter. The important point is that the area of rising air is narrow, maybe only a couple of hundred feet across. Glider pilots search out thermals by holding a constant airspeed and watching their virtually lag-free rate-of-climb indicators. The slight increase of Gs as the lift shoves the glider up is another clue.

When the glider hits the thermal, the pilot rolls into a steep bank. If the lift is sufficiently wide and strong, and if the pilot can turn tight enough to stay in it, he or she goes up. That's where the glider's slow speed is an advantage; it can spiral at 40 knots or less—just off the stall, in fact. This keeps its turning radius small and allows it to stay within the thermal.

Unless the thermal is particularly wide, a power plane attempting to thermal-soar is carried wide by its speed and flies out of the lift. If you can stay in the rising air, you'll gain as much as an extra 1,500 feet per minute of climb. It's fun. But the circling costs time that you can't make up even if, back on course, you dive away from the free-gotten altitude to gain extra airspeed. (I've tried it.) If you want to get somewhere in a businesslike way, thermal circling is of no use.

There is another thermal technique that is fairly new even in the soaring world: dolphin flight. It has been discovered that on many days, you can use updrafts while flying a straight line. This is obviously the kind of thermal soaring that most directly relates to our everyday flying. How does it work? The glider pilot beelines it for the destination, flying as fast as possible through the downdraft. When he enters a thermal, he hauls back on the stick, pulling as much as two Gs. He zooms upward, of course, but more important, his airspeed falls off sharply. He stays slow as long as the thermal lasts—the slower you go, the longer it lasts. Then out the other side and into some sink. Forward comes the stick and up shoots the airspeed. It often has to be done so violently, in fact, that the pilot floats weightless in zero Gs. The result of this violent porpoising is that the pilot gets the maximum benefit from the updrafts and loses as little altitude as possible to the downdrafts.

Flying a power plane is not like flying a glider. At most altitudes, we have power to spare and don't desperately need the free energy in the atmosphere, and we often have to maintain a constant altitude for traffic separation. A well-held altitude will also give the passengers the smoothest ride. Aside from that, we move too fast through the thermals, so our dolphining would be out of phase: by the time we got the airplane slowed, we'd be into the sink and then would charge right through the next thermal with the speed gained in the downdraft.

However, there are times when dolphining may be justified. Picture yourself in a low-powered single in the high desert country of New Mexico. It's a rough, hot summer afternoon, and you're carrying a heavy load. The poor little airplane doesn't have enough power. Your takeoff was agonizingly long, your climb has been poor, and the engine is running uncomfortably hot. On top of all that, there is a mountain ridge ahead; you're going to have to coax the airplane up to 12,000 feet to clear it. Here's where the glider pilot can help you tap the energy in the sky. Of course, you could stop in a good updraft and circle up to altitude, but you'd waste a lot of time. So you begin to porpoise through the thermals. When you hit the updrafts, pull back to the best-rate-of-climb airspeed. In the downdrafts, don't try to climb; lower the nose and accept a sink. Your engine will appreciate it, too. And you end up at 12,000 feet. No sweat. Nature has been working for you.

When you think about it, holding a constant altitude in rough air is inefficient. In the name of "holding an altitude," we push the stick forward in the updrafts and pull it back in the downdrafts—just the opposite of what the glider pilot does. We prolong the downdrafts and shorten the updrafts. Yet if we want to get the most out of our airplanes, we shouldn't maintain altitude. We should ride the ups and downs, even exaggerate them. There is a wide range of dolphin flight available to us. The degree of porpoising we may choose on any day will depend on the limiting factors of passengers, airspeed, traffic, regulations, and weather.

Updrafts are sometimes organized into "streets"—long series of thermals extending with the wind in a straight line. Ideal for the glider pilot, they are another source of cheap energy for those of us with engines. The streets are marked by strings of regularly spaced cumulus clouds.

You might also decide that you don't want to put up with all that bouncing around and porpoising. You've had a hard day, and you don't feel like a hard ride home. Well, just as you can choose areas of lift, you can pick areas that glider pilots avoid. "Blue holes" are characterized by a noticeable lack of cumuli—organized swaths of blue cutting through the haphazard landscapes of clouds. In hot, dry country, they will give you the smoothest ride in the neighborhood.

There is another kind of lift that often goes unnoticed. Standing waves are a relatively new phenomenon to the soaring world, and they are even less familiar to power pilots. When fast-flowing water passes over a boulder in a river, a series of waves is formed on the downstream side of the underwater obstruction. The crests of the waves are parallel, long, evenly spaced, and stationary with respect to the riverbed. Substitute mountain for boulder and air for water, and you have a good picture of standing waves in the sky.

There are several tipoffs to the presence of a wave. Waves recur in the same locations, invariably the lee side of certain mountains. You need strong winds aloft, blowing roughly 90 degrees to the mountain and increasing substantially with altitude. The waves are often marked by lenticulars—long, narrow, "lens-shaped" clouds. The size and definition of the clouds are not a function of the intensity of the wave, for you can see beautifully developed lenticulars in ridiculously weak waves. Just as the wave is stationary over the ground, so is the cloud. If you watch carefully, you can see the cloud forming along its leading edge and dissipating on its trailing edge.

The wave system from a mountain can extend amazingly far downwind. The waves coming off the Great Smokey Mountains, for example, can affect your performance hundreds of miles to the east, where there isn't a mountain in sight.

The lift and sink associated with waves are so smooth that many pilots have been flying through them for years without knowing it. But waves also produce some of the worst turbulence you'll ever hit. Just as there are whirlpools and eddies below and to the sides of the waves in a river, so there are "rotors" in the atmosphere. The rotors are narrow rollers of furious turbulence, spinning rapidly at the level of the peaks, or below. A typical rotor may be a mile long and only a few hundred feet across. The most violent ones are associated with the primary wave, but you'll also find them farther downstream. They are sometimes marked by clouds that appear to be dark, wispy, dissipating cumuli. They look innocent, but they mean trouble. One glider pilot flew into a particularly violent rotor while doing wave research off the Sierra Nevada—wearing a parachute, fortunately. He regained consciousness in a free fall, the glider having disintegrated around him.

Glider pilots use waves to soar to heady altitudes. They regularly go to 31,000 feet off Pikes Peak, in Colorado. Paul Bikle set the world sailplane altitude record (46,267 feet) in a wave off the Tehachapi Mountains in California—which are not even very high.

If a glider pilot feels conditions are right for wave lift, he will approach the mountain into the wind. He knows he's found the wave when he finds himself in velvet-smooth lift or sink. Then it's just a matter of feeling out the area of maximum lift. Since the wave is stationary over the ground, so is the best lift position; once the pilot has found it, his job is to keep the glider exactly over that spot. Because the wind is invariably strong, this doesn't present much of a problem—nose into the wind and do some S-turns. Up he goes—a smooth ride all the way.

If the spirit moves you, wave soaring in a power plane is quite possible. In the West, it helps to have oxygen because of the high altitudes. One pilot I know was in a Cessna 182, northwest of Laramie, Wyoming, when he saw a lenticular downwind of a nearby 11,000-foot mountain. He turned toward the mountain, flying into the wind, looking for a wave. He ran into the rotor at 10,000 feet. Pencils, coins, and charts floated around the cockpit. Then he was through the bumps and into an area of smooth lift. He started to S-turn grabbed the oxygen, and throttled back, carrying just enough power to keep the engine warm. Fifteen minutes later, he had to break off because of the positive control area at 18,000 feet.

If you're flying the high mountains of the West in a low-powered airplane, you can find yourself with some pretty marginal climb performance, perhaps because you have unwittingly flown into a smooth, wave-induced downdraft. Even if this is not the case, you may want to take advantage of wave lift.

Waves become really useful to power pilots flying downwind and more or less parallel to a mountain range. If you're going in the right direction, you can get free energy by flying among the wave-induced updrafts. I was flying a 172 down the western side of the Sacramento Valley last year, en route to San Francisco from the north. To my right lay the mountains of the Coast Range, which sit between the valley and the Pacific. The winds aloft were from the west at 30 knots. Well-defined lenticulars stretched out above and ahead of me. I moved closer to the mountains to see what I could find, and the waves were indeed there. I let them carry the airplane to 11,500 and leveled off with a good 20 knots' extra airspeed. Smooth going all the way down the valley. People do the same thing motoring up the eastern side of the Appalachian Mountains.

Because the wave is so smooth, the only way to suspect you're in one when you aren't expecting it is by a discrepancy between your power setting and airspeed. If you find yourself losing 15 knots in smooth air with a standard power setting, it might pay to move to the left or right. If you're in a wave downdraft, you can be sure that the lift is nearby.

Waves are friendly. They treat you gently, and once you understand them, they can give you some spectacular rides. But people do occasionally get into trouble with them. Let's say that we're flying east with a big tailwind. Ahead lie some peaks, forcing us up to 13,000 feet. We have no oxygen aboard. As soon as we're over the mountains, we're tempted to descend immediately, at a high airspeed, to a more comfortable altitude. The wave itself won't be a problem, but if we run into a rotor, we may be in trouble. We should delay the descent a few minutes longer, then start down at a moderate airspeed with the seat belts tight.

Pilots stick together, depending on what kind of flying they're doing. Airline pilots, military pilots, glider pilots, lightplane pilots—we're all guilty of grouping into tight little fraternal bands. This kind of kinship is understandable, of course. But every now and then we should peek over the fence to see what's happening in our neighbor's yard.

18 THE ACCIDENT-PRONE PILOT

by Thomas H. Block

HOURS AGO, all the aircraft had been hangared or tied down for the night, doors locked, and line men long gone. I drank the remainder of my beer, motioned for the bartender, and ordered another round.

"Why did I do it?" my drinking partner asked, dejected and bewildered at the same time. "I just don't know what I was thinking of," he continued, barely raising his voice and his eyes. Almost three hours had gone by since he had passed within three seconds of killing himself in a light twin.

"Dick Miller, the guy owns Miller Construction, was the only passenger. I've taken him places before, and during our last trip, he told me he was considering buying an airplane for his company. We talked about airplane types and two-engine safety requirements over lunch, so during climbout today, I flipped the right-engine fuel valve off to demonstrate how this airplane could be flown after engine failure. Anyway, when the right engine quit, I reached up and feathered the left engine! I finally got the right one going again just before we reached the trees."

The unusual ingredient in this confession was that the pilot had many times refused to actually feather engines during practice at low altitude, a condition he considered unnecessarily risky. Something, however, had caused him to abandon his previous standards and literally fly in the face of what he considered dangerous. That lapse was quickly followed by an error so gross it seemed almost as if he feathered the left engine purposely. The basic change in personal standards and habits was most alarming. The man himself seemed the most surprised by the incident; he just couldn't believe that he had committed such a ridiculous aeronautical sin. "My brains must've been out to lunch," he said, and at first, that appeared to be the only possible explanation. Halfway through his next beer, however, my friend began to remember other occasional acts that, in retrospect, also seemed exercises in mindless stupidity. What was it that crept in and out of his flying that caused incidents like this? Could it happen again? Why would a normally intelligent pilot suddenly do something so grossly dangerous and stupid?

"The pilot's basic error of inadequate preparation was magnified when the flight reached its destination airport area and entered a holding pattern. . . . When the flight's holding time was extended, the fuel remaining on board was then enough for approximately one hour of flight. The pilot, however, elected to remain silent rather than tell approach control about his situation. Considering the superior instrument-approach facilities at the destination airport, and the pilot's lack of knowledge of the traffic and weather conditions at the alternate, his decision to shortly thereafter ask for and accept a clearance to that alternate when the aircraft had 47 minutes of fuel remaining was questionable. The pilot should have known that the proper decision was to disclose his fuel situation and obtain assistance by any means necessary. . . .

"The pilot made his crucial error when the aircraft had 43 minutes of fuel remaining. He again elected not to disclose his fuel problem, although he was now faced with a new clearance requiring 25 minutes of holding prior to beginning the approach to the alternate airport. This decision intensified the problem to a critical state and narrowed the available solutions to only one—complete reliance on a successful approach to the alternate. This reliance is shown clearly by the low level of concern in the pilot's communications before the approach—as contrasted to the great concern reflected when the approach was missed. . . .

"The Safety Board determines that the probable cause of this accident was fuel exhaustion resulting from inadequate preparation and erroneous decisions by the pilot in command. . . ."

"Accident-proneness" is a favorite expression when describing areas of questionable pilot judgment. The description seems to apply only to those airmen who have had several crashes. The fact is, though, that average pilots pass through this condition periodically, and the only factors that stop actual accidents are luck and good fortune. A man reminded by the control tower that his gear is still up considers the affair only an incident, yet the accident was avoided only by the controller's convenient glance. The expression "incident-proneness" is a better evaluator of good pilot judgment, since circumstances outside the pilot's control may intercede and prevent 9 out of 10 horrendous incidents from culminating in accidents. A few bad breaks, however, and incident number 10 may prove fatal.

Very few pilots have no incidents. These few and widely scattered zero-incident individuals can certainly be labeled "excellent"; they are the airmen who have their human factors—that personal combination of emotional ingredients that colors the way reality looks—thoroughly under control. The art of piloting is, for the most part, the art of personal control. Almost any individual with average reflexes and reasonable hand/eye coordination can learn to physically manipulate an airplane after a proper amount of instruction. Success from then on depends on how well a pilot can keep it all together, use what he's learned at the proper moment, and maintain a calm and realistic attitude toward the world outside his windshield.

Accidents caused completely by structural failure or other inevitable catastrophes are no reflection on a pilot's judgment. Regardless of who was in the left seat, the flight's outcome would have been identical. Yet although it is both tempting and understandable for airmen involved in an incident to say, "It wasn't my fault," fully 95 percent of all accidents fall in the category of behavioral or judgment areas. Only 5 percent are basically unavoidable accidents—an encouraging evaluation if you regard it as an indication that better pilots will have drastically better safety statistics.

Those strange periods of accident/incident-proneness pass through a person's character like waves on a beach. The less frequent the incidents, the better the pilot. Somewhere along the spectrum is that nebulous average pilot—the man who has neither more nor less problems than his peers. As the occurrence of incidents per man goes up, the gray boundary between average pilots and those who are unmistakably accident-prone is crossed. (At the extreme of accident-proneness is the suicidal individual. FAA estimates, which are admittedly almost a guess, indicate that around two percent of fatal accidents are voluntary.)

The statistical advantage of leading average pilots toward excellence is overwhelming. Incidents, accidents, and fatalities would plummet if more pilots were sharper more of the time. But why do average pilots pass through periods of incident-proneness with such inevitable frequency? There must be a reason.

Countless pilots have blamed countless accidents on pilot fatigue, since it is obvious that performance quality goes down as the fatigue level goes up. The connection between tiredness and accident susceptibility is equally obvious: the man who flies while fatigued has a greater chance of making silly mistakes. The mechanics of sleep are still a mystery, yet it appears that the more tired you are, the easier your mind can literally be tricked by either internal or external events.

A pilot's working environment and his cockpit conditions are the stage upon which many of these events are played, and that environment is therefore a factor in accidents; it works with fatigue to form a vicious combination. The more "unreasonable" the cockpit engineering, the sooner a fatigued pilot will retract gear rather than flaps, or turn the cross-feed on when he wants it off. Flying an airplane you basically dislike or distrust, the additional mental energy used while worrying about your machine, will cause the fatigue factor to snowball. Anxieties brought about by a poor cockpit layout do a lot to derogate proper piloting performance. In one World War II aircraft, an outrageous number of after-takeoff crashes led to the discovery that reaching for the airplane's radio controls caused a powerful vertigo sensation that often produced distrust, disorientation, and unnecessary corrective action. Also, repeated failure of this plane's primary instruments put the pilots in a double bind: whether instruments worked or not, the pilot still felt uncomfortable.

The potential for visual illusions and vertigo is always present, though its causes can vary in different airplanes. The disparity between what the airspeed and attitude indicators say and what your body feels can often be more than just annoying. A pilot's experience (plus other factors) will override his internal sensations—but the strength or weakness of this overriding judgment will determine how much anxiety, distrust, and poor performance result from these illusions.

The eye is a camera that photographs everything; the brain decides what we "see," however, leaving out that which judgment considers superfluous and emphasizing items that seem important. Since every perception— whether something is seen, heard, smelled, tasted or felt—is filtered by judgment, everything affecting that judgment also affects the sensation. Illusions are incorrect evaluations, and they vary widely. In some instances, information is refused because it "doesn't make sense"; in other cases, what one actually sees is modified by illusions of what one was expecting to see.

The more anger, fear, or hate in a pilot, the greater the chance that his perceptions will be colored by those emotions. If you desperately want to see a runway appear at minimums or a fuel gauge stop its inevitable slide toward "E," your mind might be accommodating enough to grasp the slightest clues and imagine the rest, thus providing an illusion or hallucination of the desired object. Perception is greatly influenced by emotional state: it provides the ears with the sensitivity to hear automatic rough when a single-engine airplane heads out to sea; it provides the eyes with blinders to shade out lighting and towering CBs when the press of get-home-itis borders on overwhelming.

"Human factors"—the catchall phrase that says a person's total character is stamped on everything he touches—is a universally understood concept. When a man growls over a stack of bills, his wife knows that now is not the time to ask about vacation plans. When there's a ballgame on TV, the grass might not seem quite as high as it did yesterday.

Providing you have an understanding wife and neighbors, the consequences of these emotional lapses are negligible; after all, we're "only human." Penalties can quickly become more severe, however, if you find yourself taking your bad day at the office out on the car's gas pedal—though it would be difficult to locate a man who has not swerved, sped, or tailgated under the stress of some personal problem. Also rare would be the pilot who has not ''gotten it off his chest'' through his airplane.

Everyone is subject to personal problems and emotional distress. Some people let off the resulting tension by bowling, fishing, or going to the movies. Others find a satisfying outlet in soaring, aerobatics, or instrument ground schools. To accuse the average pilot of recklessness for taking a proper airplane to a proper altitude and doing a series of approved aerobatics just because he first had a yellout at home would be grossly unfair and inaccurate. If a pilot had a domestic scuffle and then borrowed a light twin to do hammerheads on an airway, however, we'd certainly raise an eyebrow. The difference between the two cases is judgment: the first man has his intact; the second's judgment is overriden by emotional distress.

The more a person loves, fears, or hates the object of his attentions, the more apt he is to lose sight of reality and thereby force his judgment to make decisions based more on what he *wants* to be true than what is *really* true. The more a particular emotion becomes consuming, the less a pilot will care about anything other than satisfying that consuming desire. If a pilot can convince himself that everything he holds precious requires a particular action, he might then find himself running back into a burning building, flying the Atlantic alone, buzzing a railroad train, or descending below minimums.

A young but experienced pilot was loading his helicopter in an open field near a girl's college. Several of the girls approached to watch. The pilot looked toward them, then walked casually away. He was struck by the moving rotor blade and fatally injured.

A totally accident-free and extremely conscientious military pilot with several thousand hours had a minor accident that was not entirely his fault. Nevertheless, he received a stern reprimand from his superior. During the next several months, he had three additional aircraft accidents, a skiing accident, and an automobile crash.

A professional pilot with an above-average record witnessed the after-takeoff crash and explosion of an aircraft piloted by a close friend. Shortly thereafter, he was involved in several incidents and one minor accident, all resulting from apparent wanton recklessness on his part.

A commercial pilot was told by his wife's doctor that her rate of recovery from an illness would be considerably increased if he continued to be away from home several days at a time. While landing at a distant but familiar airport on his next flight, he touched down substantially short of a well-lit runway, causing considerable structural damage to the aircraft.

The element that all of these pilots had in common was an outside emotional factor that overrode their judgment, colored their impressions, and decreased their skills. From then on, however, each of the pilots had an individual ''story,'' a personal blend of factors that led to an identical result:

accident-proneness.

Every human has a different and varying ability to cope with anxiety, stress, and mental discomforts. When that limit is reached, the brain will protect itself in ways that are often inconspicuous, insidious, and surprisingly detached from the central problem. For this reason, there can be as many categories of accident-proneness as there are pilots; each brings with him a complicated set of circumstances, and each can react to the problems of flight in an infinitely variable way. Like the labels "good person" and "bad person," the labels "good pilot" and "bad pilot" overlap, intermingle, and finally become meaningless as science discovers more about what people are like, and why.

The helicopter tragedy was analyzed as an "ego accident." The pilot was, at that moment, concerned more with his appearance to the girls than any other consideration.

The military pilot, who was admittedly a perfectionist, was reacting with guilt after (in his own eyes) having lost face.

Fear generated in the professional pilot by witnessing his friend's death was turned into a deliberate disregard for safety and rules. By exposing himself to danger, he alleviated his own considerable anxieties.

The commercial pilot was torn between powerful yet conflicting desires. His level of mental tension over his wife's needs and the requirements of his job made rational decision making nearly impossible.

When exploring human factors, each general statement leads to a more specific question. If we say a man is "concerned more with his appearance to the girls than any other consideration," immediately we can ask why—and if such a response is possible for us. The startling thing about most analyzed accidents is how terrifyingly human some of the causal errors are—how even amid the pilot's complex and extreme reactions, most people can reluctantly recognize a little piece of themselves. For whom does the word "perfectionist" not strike a personal bell? How many of us have mouthed the aphorism, "The best defense is a good offense?" Who among us has not gone blindly into a situation, letting his actions make the difficult decisions for him?

If indeed it's true that 95 percent of accidents fall in the behavioral or judgmental category, and emotions are the prime movers in displacing good judgment, then obviously a major design error in the construction of the typical pilot was the inclusion of emotions. Having emotionless pilots might work if we're ready to sacrifice one ingredient: motivation. The link between emotion and motivation is recognized if not quite thoroughly understood. An IBM 360 doesn't care if someone unplugs it; a pilot will react if you take away his food, his shelter, his airplane, or other components of his life.

Without motivation, there is no learning—a painful discovery for students and teachers. And since life is a dynamic process, without continued motivation there is *unlearning*. The skill of the sharpest stickman in the sky will recede if he loses his basic motivation: "He just doesn't care anymore" is usually followed rather quickly by "He's not the pilot he once was."

Fear of the unknown is a powerful motivation, and it's the sort of thing that often fades. With a few hundred hours logged, the unknowns of flight become familiar enough that the pilot isn't compelled by fear to spend time and effort acquiring knowledge. If the undermotivated pilot scares himself sufficiently (by stumbling into horrible weather, say), that fear is rekindled, and he sets out to again combat it by acquiring more knowledge—studying for an instrument rating, in this case.

The man who is not remotivated after committing a horrible mistake is a man who was scared either too much or too little. Too little means that in spite of the dangers, he got there anyway, and the emotion generated by the apparent success has stopped any fear. Too much fear and the occurrence becomes traumatic; like the defense of fainting, a person can just refuse to recognize an incident in hopes that it will somehow go away.

Most people—pilots included—are motivated by success. So many of the slides into accident-proneness, however, are caused by a misapplication of—and an overemphasis on—the word. When the safe pilot says, "I am going to Chicago," his emphasis is on "going." The accident-prone is wrapped up in the concepts involved with one of the other four words.

Many highly regarded professions require certain emotional indulgences (a quick temper, a domineering character, arrogance) if a person is going to match his job perfectly. Piloting, however, is among a select group of jobs that are very different. There may be isolated instances in aviation when indulging in certain emotional outbursts gives a momentary advantage, but that advantage is far outweighed by the restrictions any emotional outburst can put on a pilot's ability to see the world as it really is.

In several studies done on pilots rated "superior" and "outstanding," the words used most often to describe them were: careful, rational, curious, restless, distant, punctual, orderly, examining, and intelligent. In addition, two recurring phrases were: "a good balance of personality factors" and "enough flexibility when needed." Flying is one area in which even the slightest personality disorder in the pilot in command can place him and his passengers in a very dangerous situation very quickly. Piloting draws on the greatest capacities of the total human being; the farther a pilot is from an integrated, balanced personality, the more apt he is to place himself in unnecessary and harmful situations. Leaving the ground when your emotional outlook is off-center is like taking off with your cargo out of CG range. You might get away with it—but the consequences might well be life or death.

What are the rules by which a pilot can live in order to assure himself that he isn't taking off with his emotional needle ball askew? Can we list 5—or 10 or

20—surefire methods for flying safely and sanely and never busting minimums? Like whistling in the dark to ward off fear, people can construct "applied psychology" techniques that will often effectively deal with an emotional overabundance. "When in doubt, do a 180"; "Whenever you're rushed, sit quietly for five minutes"; and "Do the preflight twice when you feel upset" are training aids that encompass one basic recommendation: stop and think. Exactly *what* are you trying to accomplish with this flight? Exactly *why* are you trying to accomplish it? *Can* you honestly afford to indulge yourself to this extent?

Self-analysis is the method; putting judgment back on top is another way of expressing it. Happily, many times the answer to "What, why, and can" will be affirmative, and you should—in perfect safety and good conscience—go aloft and slap the airplane around as though it were the president of the phone company. If, however, your mind is so wound up that you can't even approach the "What, why, and can" questions, sit down and have another cup of coffee; you're not quite ready for flight yet.

Emotions themselves are not the enemy. Emotions, when properly directed, are the source of the energy that changes life from routine to sublime and allows a man to put out like he's never put out before. Just imagine for a minute, though, the consequences of the childish yet compelling act you may be about to perform, as you walk out to the airplane. After honestly asking what you're really doing, why you want to do it, and whether the results are worth the risk and effort, the desires that bordered on mindless will usually have lost much of their urgency. This brief exercise also weeds out the real pilots in command from their impostor colleagues. A real pilot says, "Why should I?" instead of the more pitiful, "Why did I?"

19 LOST
by Peter Garrison

IT WAS an afterdinner conversation about flying—one of those conversations that nonflying friends dread. A woman doctor was talking about her experiences in her first 50 hours of flying—50 hours that she had logged in a great burst of enthusiasm, in a couple of months' time, after having feared and detested airplanes for her whole life. It was very interesting for me to listen to her, because it brought back the forgotten experiences of the beginner. Instructors must be in touch with them all the time and see with some perspective the depth of their own accomplishment; but for me—I've never instructed, just flown year after year—there seemed to be nothing remarkable about flying at all; I suspected that beginners' difficulties all arose from mental blocks and that if I were an instructor, I would be able to make a skillful pilot out of a novice in five hours of careful psychologizing. And so, listening to this woman, I saw a little way into the distance that separated me from her; there were so many things that I had learned and forgotten I had learned that my perceptions bore no resemblance to hers.

At first, she said, her main anxiety had been that the airplane would keel over, lose its balance, perhaps end up on its back, and that she would be unable to right it. I had encountered this fantasy in other passengers; some of them felt that they had to hold themselves carefully erect and motionless, as though in a small canoe, in order not to disturb the equilibrium of the airplane. One of them would not, at first, even turn her head. All this makes sense when you think about it; the air appears like nothing, and if you are in a canoe balanced on nothing, what is to prevent it from turning over?

After she gained confidence in the airplane's willingness to stay right side up, the doctor's principal fear was of getting lost—"getting lost in the sky," as she put it, using an old phrase that somehow for me clarified her fear, since it evoked in me an image of finding oneself somewhere where no ground whatever was to be seen, only vertiginous blue sky all around. You would feel as though, having fallen asleep alone in a sailing boat, you awoke at dusk to find yourself out of sight of land. All this gave me a good chuckle; she told of being afraid of getting lost between Whiteman Airport, in Los Angeles, and Barstow, a desert town to the northeast, and I reflected that as far as I was concerned, you could practically see Barstow from L.A., so it was hard to understand how you could get lost between them.

A week later, I brought this up to an instructor friend, who said that his main fear on his students' behalf was, indeed, that they would get lost. Getting lost, he said, was the first in a series of problems that would lead, unless interrupted by good fortune or good sense, to a forced landing, or an overshoot, or fuel exhaustion, with fright and then panic bridging the steps.

He said that in his experience, students had two favorite places for getting lost: between the takeoff and the first checkpoint and between the last checkpoint and the destination. In both cases, they were looking out, trying to find their way, rather than concentrating on holding a heading. A heading, he said, was the most valuable thing a student navigator could have; it was everything; it was miraculous. He was always annoyed, he said, that his students could be so blasé about what always seemed to him a great revelation and wonder: that simply by holding a heading for a certain time, you could reliably arrive at a predicted spot. To the students, that this was so was not surprising, since they had been told it was so when they started; but to him, a pilot of years and thousands of hours of experience, who had invested time and again in exotic navigational radios and gnashed his teeth over their unreliability, the fact that you could get somewhere so accurately, with no equipment at all, was a cause for delight and astonishment.

He returned so insistently to the magical curative properties of a good heading that I later looked back into my own memories for verifying instances. I did not have far to look.

I was in Guatemala with a friend, Nancy. We had crossed the Caribbean—armed only with a heading, of course—from Key West to Yucatán. We had come down through Yucatán, always on headings and times over the endless and almost featureless carpet of jungle, to a place called Chetumal, and thence on a short flight to a well-disguised port of entry called Melchor de Mencos, on the border of Guatemala and Belize. Having deviated several times from my heading to avoid rainshowers, I almost missed Melchor, though it was only half an hour from Chetumal. Thence we flew to Tikal, a site of eerie Mayan ruins half digested by the jungle, and thence two days later to Flores, in central Guatemala, to refuel before continuing to Guatemala City, which is in the southern part of the country. We arrived in Flores at midday; but it transpired that no fuel was available, and it took several hours to negotiate to buy some from a private firm—an oil or forestry operation or something like that—at the airport. My companion was impatient, and so was I; so when the fuel was finally in the airplane, we took off without getting a current weather briefing.

I held a heading. The sky was overcast to the northeast, blue to the west; there were ragged layers of clouds and occasional showers; the air was smooth. After 30 or 40 minutes, we were close enough to the range of 13,000-foot-high volcanoes that lies just north of Guat City to see that there were solid buildups on them, 20-, maybe 30,000 feet high. I changed heading slightly to the west, looking for a low spot. As we came closer, I turned more and more to the west. After 45 minutes, we were heading west by northwest—back toward Mexico—and it was apparent that we could not reach Guat City, unless by flying all the way to the Pacific, following the coast southeast, and then flying northward across the lower mountains to the city; and we did not have the fuel, nor the time before nightfall, to do that.

Omni was useless; there was nothing within a hundred miles, except Guat City, which was silent to us behind the mountains. We had no ADF—which is essential if you plan to get lost south of the Rio Grande. Our only map was a 1:1,000,000 ONC and a Guatemalan road map on the same scale, and we had already gone off the bottom of the ONC. There was a bit more than an hour of daylight left. We had three hours' fuel at the rate we were using it. The western sky was bright with streaks of powdery blue among the snowy cu-nim; to the north and east it was gray and somber; below us enigmatic and multifarious rivers meandered, in no way resembling the map's sketchy indications; everything else was jungle. I turned back to a roughly northeasterly heading, slowly absorbing the fact that we were lost.

Time normally spreads, sways, and flows like a big anemone, reaching out and touching the past and the future along with the present; but when fear touches it, time shrivels like an anemone; it contracts itself into a lump of intense present. Being lost over that jungle meant that if we did not find Flores again—and if one of the patchy stratus decks had moved in over it, we would not—we would have a number of disagreeable alternatives. One was to head east to the Atlantic coast, or west to the Pacific, and arriving in darkness, hope to reach a major airport, or pick up an omni station before running out of fuel. One was to land on any bush strip or stretch of road, regardless of whether the airplane could take off from it again. Another was to ditch the airplane in a river if no dry alternative presented itself before nightfall; we had survival equipment and even a raft, and I had a naturalist's curiosity about the jungle to boot. And then there was the possibility of jumping, since we had parachutes; but my friend was not very eager to jump, and her principal fear, she later said, was that I would go to such lengths to save the plane that I would pass up a dozen perfectly hospitable clearings into which I could safely crash it, and finally oblige us to grab our toothbrushes and hit the silk as the engine sputtered away its last drops of gasoline, in darkness, over snake-infested foilage.

Here in the States, the possibilities are rarely so exotic. There are roads everywhere, and so it is usually possible to find some sort of road to land on in a pinch; maybe even a house to taxi up to. The worst penalty of getting lost here, usually, is embarrassment, unless, in your eagerness to avoid embarrassment, you work yourself into a more painful corner still. Yet in the mountains, or over the Georgia swamps, in a slow airplane, you would experience what you experience in other, less crowded and developed countries when you have flown yourself into a bind.

I sat up straighter in my seat; my feet pressed the rudder pedals more firmly; the engine droned on unperturbed. My only clue to our position was an improvised trace of our track that I had put on one chart, a line first straight, then hooking more and more toward the west, with some times written alongside it. If the trace were accurate, I knew our position; the only trouble was that the rivers made no sense at all. But what was there to do but ignore the rivers and trust the scribbled trace? So I took a heading from the end of the trace, estimated a time, and held that heading for dear life.

In such a situation, 30 minutes is a long time; it is wonderful how, in other circumstances, such a vast eternity can slip away unnoticed. At the end of 30 minutes, I thought, we would be able to see Lake Peten, on whose south shore the airport was located. There were other strips on the map, but I had not even seen the ones I had crossed right over on the way out. None of them amounted to much; even Flores, an airline stop, was a narrow line of semipulverized limestone mixed with razor blades—the standard Guatemalan paving material. The International Airport of Melchor de Mencos had resembled a rain-rutted driveway from which it was necessary to frighten horses and pigs and children before landing and taking off. The road—there was only one in this area—was a slice through the tall jungle; you could not see it at all if you were not above it. It was a strange place to fly, but exciting and beautiful in a way that the carefully structured United States would never again be; airplanes—an MU-2 landed at Melchor while we were there—with their white paint and gleaming metal, stared out of the mists of the dark green trees and the brown faces like marble statues fresh-sunken to the bottom of the sea. And so now we were alone, two of us, in this speeding, white pellet above the open jaws of the rainforest, with the clock ticking the minutes away and the sun sinking slowly.

When Lake Peten appeared, it was squarely in front of us.

Nancy had never fully understood all the ramifications of our situation—it was too difficult to shout my uncertainties over the noise of the engine—and now she was relieved, but like the novice pilot she was blasé, because things had merely happened as they should. I had said that a heading would take us back, and it had; so that was that.

But to me, it was, as my friend later said, like a miracle. Was not this a wonderful thing, that a number could take us where we wanted to go when there was no radio beam to follow, no recognition, no landmark, no spoor or smell, nothing at all but that number? We were saved by a Cartesian coordinate system: I head, therefore, I arrive. In the tail cone, a ring of iron had swung and settled, electrons had tirelessly swept through wires that I had strung through frames and grommets and connectors, and on the instrument panel, magnets had pulsed and trembled, and a needle had pointed to a painted cipher. What did all this have to do with Guatemala, I ask; how does this buy us our dinner at the Casablanca in Flores, rather than a handful of Hershey bars by a riverside in a nest of capybaras, coatis, and shrieking monkeys? Truly, this heading held a wonder like the beating of a heart, a friendly, fostering, motherly mystery.

Now I have one with me always. I used to scoff at the wind triangles on the written exams, because one never used them; but now they seem to me like trustworthy old friends, modest and quiet but perfectly loyal and reliable. I dead-reckon to San Francisco from Los Angeles, because it is the easiest way; all you have to do is hold a heading. And it's good practice, holding that heading, holding it so tightly that the whole operation becomes unconscious, and you awaken to your perfect heading from time to time with a little start of surprise. It's good VFR and IFR; it is the basis of all other navigation. If you have a rock-steady heading, then omni deviations mean a correction for the heading rather than a cut to bring the needle back. On the ILS, if you know your heading, you can ignore the scallops: the heading is like a straight western road that goes unbending to the horizon. Heading and time, heading and time: if you have them always with you, you will not get lost again.

20 THE FORGOTTEN JOYS OF PILOTAGE

by Norbert Slepyan

"WHAT ARE you doing? I thought I'd cured you of that awful habit!"

Caught real good this time, I defiantly respond with a shrug.

"You're hopeless. You're stunting your growth!"

I nod, resigned to the possibility.

"How are you going to grow into a real pilot if you keep doing things like this? *Why do* you keep doing things like this?"

I give him my inscrutable grin. "Because it's fun."

He shakes his head sadly, just as he did the first time, when he discovered how unworthy a pilot I am. Perhaps even more sadly, for he has genuinely tried to save me from myself. It's not easy for a man of his character to accept failure.

The first time happened during some casual hangar talk, when I happened to remark, ". . . and after I made the turn, I saw on the sectional . . ."

"What? You still use sectionals?"

His surprise and incredulity wounded me. He is, after all, a pilot I respect, a man who knows whereto he flies and whereof he speaks. He has ratings that, if framed, would make a lawyer's office look bare. He's flown aircraft such as I will never even see, much less fly, over hours of time I can never hope to match. He knows most things aviational and all things navigational. He's the High Professor of my aeronautical tutelage, and when his venerable brows briefly arched and then solemnly lowered, and his sharp blue eyes fixed me, I withered.

"Sectionals?" the HP said, his voice rising a dominant seventh. "Whatever would you use a sectional for?"

The tone of his voice was echoed in the sniggers of the other low-timers hunkered around him. *They* would never be so crude as to use a sectional!

"For not getting lost," I replied.

"Haven't you heard of low-altitude en-route charts? Jepp charts?" the HP pressed on.

"Sure, but . . ."

"You fly with navaids, don't you? Then why use sectionals? It's like navigating with comic books."

He went on and on about sectionals being so cluttered with colors and useless information that they are more in the way of a pilot than a help to him. And about how all the information a pilot with navcom really needs is in the Jepps and the NOSes. And about how most hip pilots with professionalism in their bones graduate to the low-altitude charts as soon as they've been liberated from all that misleading FAA-inspired training, with its foolish emphasis on following *sectionals*. He seemed to spit out the word.

It was grisly.

I spent that evening in reflection, staring alternately at the Jepp chart on one side of me and at my line-crossed New York Sectional on the other. The sectional was only a few months old, but I had already worn a hole in it with my constant unfolding and folding. There was the coffee stain, an instant island I had dropped into Long Island Sound one day, just beside Belle Terre Intersection. My many checkpoints—the familiar lakes, bridges, roads, air-fields, and drive-ins—were there, like so many faithful sentries along my most-traveled routes. Some bad moments were dourly reflected, too, part of the flotsam and jetsam of my infinitely long life as a low-time VFR pilot.

That pilots do not fly by pilotage alone, I knew well. I often flew the omnis but kept my gaily colored sectional always before me, my thumb tracking along it like the fleshy shadow of a blimp. Now it seemed that I had been doing it all wrong.

"Well," said I to myself, "say good-bye to all that."

Gently, and with a little less wrestling than usual, I folded the old sectional and tucked it away. I then turned to my new low-altitude en-route chart, the symbol of my coming of age as a space-era aeronaut. When I told the HP of my resolve the next day, his moustache broadened above his smile.

For a long time, I managed to be good. Like a piece of baby blanket that just can't be tossed away, my sectionals were kept in my flight bag, to be used sparingly on approaches to unfamiliar airports. But on my lap, where the action was, there usually perched my Jepp. Quickly, MOCA came to mean something other than a kind of coffee, and MEA something other than a reference to Mama. I kept to those straight and narrow victors, twiddled my knobs, and centered my needles. It was still all VFR, of course, but at first it seemed to take on enough of an IFR aura to make me feel veddy, veddy sophisticated. Nevertheless, as I concentrated on those colorless displays of lines and numbers before me in my Jeppesened airplane, I began to feel a sense of loss.

One day, as I was flying into Danbury, Connecticut, having dutifully homed in on the Carmel Vortac, I noticed something I hadn't been aware of before: on the fairgrounds that adjoin the field, a huge orange tent had been put up. In the sunlight, it was like a great beacon beckoning Danbury's pilots home. I began to wonder: how many other surprises and delights had I missed on all those flights when I had noticed the ground going by without really seeing it?

That led to another evening of reflection, or rather of recapitulation, of going back to the most basic basics, to before the omni training, before the first solo, before the first hour, back all the way to what made me want to learn to fly. The reasons lined up in their familiar array. Some looked a little vague and threadbare now, no doubt out of neglect, but two were particularly sharp.

One was my desire to capture even a little bit of the excitement and adventure of the old days of aviation, thrills that had filled my reading and fantasies about that unbeknobbed, ante-dialed time when the Fregs, Posts, Doolittles, Matuchas, Lindberghs, and Bennetts had huddled in their open cockpits, having only roads, bridges, towns, lakes, and rivers to fly by during the day and city lights, beacons, and what help the moon could provide by night. The other reason, still compelling by its presence, was my longing to lift myself into the air so that I might better perceive the ground.

With trembling hand, I opened my flight bag, knowing, fearing what was about to happen but unable to stop. My navigational backsliding was inevitable, the awful habit not yet kicked. I drew out the Detroit Sectional. Not New York. No, that would have been too much too soon. Detroit was rich enough. I turned the chart around and around and around—as in days of yore—and finally focused on the southern half. The greens, blues, reds, and yellows blazed up at me. The world again. Doctor! I can see!

From out of nowhere, my familiar course line abruptly appears at the eastern edge of this chart, aimed at Slippery Rock, Pennsylvania, where I sometimes visit a flying farmer who practices steep 720s on his tractor. The line begins over Montoursville, paralleling the writhing Susquehanna River until the river turns northward. South of the line, looking like primitively sketched range rings, are the great parallel arcs of Allegheny Mountain ridges and valleys; they are dappled with the yellows of mountain cities and the magentas of little airports. Seeming to waver all about are eerie, multicolored boundaries signifying controlled airspace. I have made several flights along that line, my thumb moving up hill and down dale, and I remember one particular flight when my curiosity, ineptitude, and stubbornness gave me a glimmering of the old days.

I am alone in a Beech Sierra at 4,500 feet, steering a magnetic heading of 281 degrees against a slight wind from the right. My course happens to pass over the Keating and Clarion Omnis, but I am determined to navigate solely by pilotage and dead reckoning. It is a clear day, except for some scattered clouds at 8,000 feet. All the way from White Plains, I have been averaging 155 miles per hour. Delicate springtime greenery has stretched below me on all sides as my pencil has ticked off the passing landmarks of New Jersey and Pennsylvania. I have put away the New York Sectional and have unfolded Detroit. The flight having settled into routine, my thoughts have turned to my destination, to my happy expectations of rapping with my friend, the flying farmer. Maybe, it occurs to me, just maybe this will be the day he lets me drive his tractor.

A buffet brings me back. Let's take another fix.

We should be coming over Montoursville and just after that, Williamsport, with the Susquehanna dropping off to the left. Hold on. Where's the river? My eyes dart back to the chart. That's what it says—that's what it's always said: there should be a river and some cities down there. And they've always been there before. Now, to right and to left, greenery, valleys, ridges and hills, scattered houses, all of it blending into a strange sameness, as if I were flying over an ocean. There should be a river. Do you see a river? No. Anything that looks like a sizable city? *No.* A highway? Where's the Interstate? To my right, I think I see something that might be a highway, but there are hills there, and whatever I'm seeing is hiding behind them, revealing itself only in short spurts.

I notice the DG: 262. Check the compass; it's steady on 262. Again I look at the ground, and see nothing. A wave of wooliness rolls through my head as I realize that I am lost.

A squeaky little voice begins to babble at me about being all alone up here, about never going to get down safe, about my being so stupid. My head keeps turning, my eyes wander over the terrain, and then all over the chart, seeking—what? There! What's that off to the right—that sliver of blue—the river? The Susquie! Is that this curve in the river drawn on the chart, then? Yeah! Yeah. That's a railroad north of it—is it? There's supposed to be a road alongside of it and—they've disappeared behind those . . . and I can't—no, I can't be to the north of the river if I'm flying in the 260s. Now calm down. Start again. Circle until you find something. No, go back and find Williamsport. Call flight service! Call your mother!

These thoughts rattle among the empty chambers of my mind, stirred and heated by the fuel of adrenaline. The frantic ricocheting dies down as the controller-in-residence within my head begins to assume command.

"All right, son," he says in a deep-voiced drawl, "let's settle down. Just tune in the Keating Vortac and see what radial you're on, and do the same with Philipsburg. This airplane is omni-scient, transponderized, and DMEtrified. Drop this pilotage business and find out where you are. You are not cleared to panic, y'hear?"

I reach for the tuning knobs but hold back. Check the fuel—good shape. Since I have these electronic backups and ample gas, why not try to work things out as if I didn't have them?

I realize now—and will kick myself very hard for it later—that I have made some classic blunders that I have been warned can get one lost and keep him there. First, I've been sitting fat, dumb, and happy, immersed in irrelevance instead of minding my business. When did I make my last check? Nearly 20 minutes ago—17, to be exact—well before reaching that big bend in the river. Where did the time go?

Second, I flipped. Didn't even grab a breath so as to take stock of things; just let my mind flop into the unknown. I'd forgotten that you've got to work from what you know to get what you don't know.

So I started looking all over the chart for landmarks—*immediately*, without really thinking where to look. My scanning must have covered 1,100 square miles. Working in such eager ignorance, I succumbed to wishful thinking. I wanted—needed—the flash of blue I saw to be the prominent branch of the Susquehanna someone had drawn on the chart, so I could at least say to myself, "Your are *there*." No matter how illogical that would be, how unlikely it was that I *could* be there. I wanted to match them up, so I matched them up. One reads the textbook cases of such errors and says, "No way that it will happen to me." Then, one fine, clear day, the dulls and deliriums settle into one's skull and suddenly lakes are being turned upside down; airports are uprooted from their foundations and Frisbeed to new landscapes where they are swiveled about to make their runways point the right way; highways are curved and straightened and intersected to a new order like charmed earthworms so they'll lie on the chart as others lie. Many such exchanges can be made, just for a little security.

Aware of the mistakes I have been making and that I am still flying the wrong heading, I force myself to be relatively methodical so as to be more analytical. I must first relieve the pressure of disorientation so that my speculations don't fly completely out of bounds. If I'm not where I ought to be, I still can't be too far away. My heading is 262. I visualize a line on the chart, angling downward from the one I am supposed to be following and rooted to the point of my last fix. The angle seems rather shallow, as if I were just a little off course, but that can't be right, for I would probably have been able to see the river as it proceeded southwestward. I've forgotten something. *Magnetic variation* and *wind correction*. On the chart, the variation is nine degrees west. Let's see, how does it go? East is least and—slow but sure. Right. Okay, my true course is much nearer to 253. If I still assume a four-degree wind correction to the right, I would get a truer course of 249—say 250. My second imaginary line puts me farther into the boonies, yet it's better than my first wandering efforts had done for me. I get out my small plotter (I knew it would come in handy some day) and draw the 250-degree line. I can't be at all sure that I am on that line, not knowing how long I've been flying this heading or what my actual track has been—probably some sort of shallow curve to the left—nor what the current wind is, but at least I'm inside a reasonably sized ballpark now and am not tempted to drag out all my charts, from Lake Huron to New Orleans, to find myself. In fact, the little cone taking shape looks reassuringly small.

How far have I gone in 23 minutes? A rule of thumb comes to mind: for every 60 miles you fly in an off-course direction, you will be one mile off course for every degree you are flying off course. That's assuming you've been flying in a straight line, but even if you haven't you can still make a reasonable estimate. Fly 60 miles off course on a 10-degree off-course heading and you will be 10 miles from where you want to be. Fly 30 miles on such a heading and you will be 5 miles off course. Okay, assume the 155-mph ground speed is still good, for the present—you've got to start somewhere, from the known. My computer says that would be 59 miles, which would put me *here* on my 250-degree line. The god of happy coincidences is with me. Making the proper adjustments to my original true heading of 281 degrees brings me to a true course of 268, so I should be 18 miles from where I had intended to be at the moment.

Very tidy, very textbookish, and probably wrong. What about all those assumptions that haven't been checked out—track and current wind and therefore true ground speed and time to this point are still unsure quantities. Yet, if the rule holds and my numbers are reasonably correct, I can't be too far off. I mentally draw a circle of 10 miles in diameter around the theoretical new fix, remembering that the circle will have to move as the airplane progresses over the ground.

From dead reckoning back to pilotage. I look for large landmarks on the chart, especially trying to locate them in groups. A single lake can fool you, or a single town or a single peak. One landmark makes a hint, two make a probability, and three can mean certainty. I would seem to be heading toward Mid-State Airport, which is flanked on its northeastern side by a lake shaped like the boot of Italy. It has a 5,700-foot runway and a 5,000-footer crossing it, so it should be quite visible. (Actually, I'm also approaching the Philipsburg Vortac, but we're still going to pretend that omnis haven't been invented yet.) This airport is in a valley, which may be a reason why I haven't spotted it yet—or am I becoming impatient again? Give it another five minutes. Meanwhile, if my figuring has been correct, there is a valley to my left along which runs a road flanked by a railroad track and a stream. Beyond that there may well be another low ridge and two small airports—Bellefonte Skypark and University Park. South-southwest of University Park, says the chart, is the large town of State College. I look outside the airplane to my left and *voila!* Nothing. Nothing out there looks like that. Just lots of green and rolling terrain. Well, back to the old. . . . "Now just hang on, son," the controller-in-residence says, "don't be getting fidgety again. Remember that the sectional shows things as they look from straight up and clearer than they'll be in hazy air. And the farther you look, the flatter things get. Let your eyes adjust."

The stream appears, and the road, but no tracks. Maybe a hint. Off farther to the left is what certainly looks like a community of some sort, and do I see an airport near it? Can't be sure. A hint-and-a-half. Wait; there are the tracks, seeming to emerge from underground and curving toward that town. As they do on the chart! No fooling. Probability. I glance to my right, and here comes certainty; the lake looks very flat as it lies beyond the Sierra's nose, with the protuberances that mark it as bootlike. And unmistakable are the crossed runways of Mid-State, a big beautiful X marking my spot.

Elatedly, I draw a line from Mid-State to where my original course crosses Du Bois-Jefferson County Airport, 47 miles away, a highly visible landmark to aim for. I work out my new heading and checkpoints and turn on course.

That I have passed Bellefonte Skypark in my meanderings is interesting, for Bellefonte is not new to the annals of pilotage. Long ago, eager to burst the bounds of their local airspace, pilots drew crude maps for themselves and their friends or wrote notes setting out what to look for in getting from place to place. Elrey B. Jeppesen wrote this about the approach to Bellefonte, circa 1920:

"After crossing another mountain range without a pass, Bellefonte will be seen against the Bald Eagle Mountain Range. On top of a mountain, just south of a gap in the Bald Eagle Range at Bellefonte, may be seen a clearing with a few trees scattered in it. This identifies this gap from others in the same range. The mail field lies just east of town and is marked by a large white circle. A white line marks the eastern edge of the field where there is a drop of nearly 100 feet."

That note and many others that Jeppesen wrote were published as a book, *Pilots' Directions, New York-San Francisco Route,* in 1921. Such notes were part of the beginning of the universe of navigation aids that are fondly called Jepps. Jeppesen & Co. has thrived. I see by my chart that the little mail field Jeppesen wrote about more than half a century ago is gone.

As the memory of that flight to Slippery Rock faded, I wondered why, of all the trips I have made, I recalled that bummer. The Detroit Sectional could have brought to mind many other perfectly jolly flights, with no wrinkles in them whatever. On the other hand, was it such a bummer? Perhaps it was just the turn-on I needed. It had been good to have my radio aids available to pull me out of my predicament had I really needed them, but without them, I had proved to myself that I still had resources of my own.

I repacked my flight bag for the trip I was to make the next day and then called flight service. At that late hour, the forecast was firming up as very good for the morrow, in fact, a pilotage pilot's day. Perhaps this would be as beautiful a day as the day of my first flight with my family to Cape Cod.

Recollections of it still flash on and off unbidden. The day CAVU and calm. Our Piper Arrow II, well equipped but its pilot using only the land itself as his omni system, his own eyes as his radar, his thumb on the chart as his DME. The destination is Hyannis, Massachusetts, by way of the North Shore of Long Island, two leisurely circlings of Shelter Island, and continued passage via the shorelines of Connecticut, Rhode Island, and Massachusetts. It's the easiest navigation of all, nothing but pleasure and almost blinding beauty—as on the leg from Shelter Island across the Sound to New London, Connecticut, which stands out sharply between the two river mouths that flank it.

Ahead of us is a two-masted ship also steering for New London. There is full sail atop her and a comet's tail of wake behind. "You are cleared to New London," says the controller-in-residence. "Your traffic is a two-master. You are cleared to overtake and pass at your discretion." Overtake her I do, passing off the white-crested sliver's starboard beam. The elements of the moment are harmonious: blue sea and blue sky; white ship trimmed with lines; white plane trimmed with lines; and the welcoming land ahead. Hoping the skipper will look up, I waggle the wings.

Crossing Narragansett Bay, we come upon a covey of small boats, flattened white triangles of sail moving like a flock of birds, in full regatta. Through the sparkling air, sharply lined fingers of rock and sand claw at Buzzards Bay. Above the Cape Cod Canal, I turn right for Barnstable Municipal Airport, at Hyannis, trying to avoid Otis Air Force Base, which is nearby. I'm tempted to steer to the airport under the direction of the Hyannis Vortac, but the sectional and the land show me the way instead: a highway stretching from beneath the spinner directly to Barnstable. Again and again it seems to happen that the land will help do your navigating if you give it a chance.

I remember another flight, this in the heartland.

We are low over the westward-pointing road as the sun approaches the horizon. The land undulates softly, its broadening shadows making forms suggesting primal rhythms of creation. All else seems to be motionless. The cows over which we pass must hear us, but they don't look up. To our right, a small lake eases by, its surface reflecting the dusky light. It will be good to land, to stretch, to address ourselves to supper, but I slow the airplane until it, too, seems hardly to move through the air. The engine drones more quietly. A familiar flash of green and then of white appears over the hillocks ahead, and the airport comes into view. Its runway lights are on by prior arrangement, just for us—the best of welcomes. As I swing onto downwind and then onto base, we catch a last glimpse of the road along which we have come and of the Midwestern hills over which we have just appeared. We land in the gathering darkness, and the land seems to embrace us.

Did it feel this good to the early ones, the pilots who followed the roads and tracks and their instincts beyond where the eyes of those at home could see them? Did it feel this good to them when they returned, with the shape and the feel of the land they had overflown embossed in their minds?

It must have felt better.

In instrument training now, I am beginning to see how vital those bland Jepps are, how they seethe with essential news and guidance. The Jepps speak of suspension in a world of abstractions, a universe of procedural and mathematical order. And the sectionals, those maelstroms of lines and colors and ragged shapes—as riotous as the earth itself—proclaim that flight, however high, is rooted to the earth, that flight is wonderful because its beauty is constant, from the ground up.

INDEX

A

Accidents, 305, 308–309, 330–337; human factors, 196, 197, 201, 330–337; IFR-related, 108–110, 175–177, 188–189, 215

ADF, 64, 165, 182, 225, 235–240, 241–245, 262, 275, 277, 278, 310, 340; approach, 184–187, 239, 244

Air traffic control, 7, 61, 73–76, 96, 111–113, 139, 150–152, 157–158, 169–173, 213, 253–255, 259–260, 263

Airman's Information Manual, 145, 173, 198

Airspeed: on approach, 301–303, 311; control of, 296–299; efficient operation and, 313–315, 319–320, 323; maneuvering, 292–295; multi-engine, 306

Alternate airport, 137, 138, 149, 156, 190, 193–195, 217

Altimeter, 197–198, 206, 297; check, 145, 252; encoding, 250–255; settings, 18

Altitude: awareness, 214, 250, 253–254; selection, 138–139, 315–316, 319, 321–323

Angle of attack, 284–286, 293, 307

Approach, instrument, 19–20, 61–62, 152, 165, 171, 174–195, 214–215; ILS, 113, 125, 171, 177, 179, 181–184, 189, 199–200, 205, 207, 238–239, 242, 310; minimums, 137, 175–177, 178–179, 182–183, 188, 189, 191, 194, 208; nonprecision, 66, 177, 178–181, 189, 205, 206–207, also see specific categories, such as "ADF, approach"; planning, 149–150, 152, 173, 181–183, 190, 209, 213–214; practice,

N

O

P

U

V

W